Progress in Colloid and Polymer Science · Volume 126 · 2004

Progress in Colloid and Polymer Science

Recently Published and Forthcoming Volumes

Surface and Colloid Science
Volume Editor: Galembeck, F.
Vol. 128, 2004

Analytical Ultracentrifugation VII
Volume Editors: Lechner, M. D., Börger, L.
Vol. 127, 2004

Trends in Colloid and Interface Science XVII
Volume Editors: Cabuil, V., Levitz, P., Treiner, C.
Vol. 126, 2004

From Colloids to Nanotechnology
Volume Editors: Zrinyi, M., Hórvölgyi, Z. D.
Vol. 125, 2004

Aqueous Polymer Dispersions
Volume Editor: Tauer, K.
Vol. 124, 2004

Trends in Colloid and Interface Science XVI
Volume Editors: Miguel, M., Burrows, H. D.
Vol. 123, 2004

Aqueous Polymer - Cosolute Systems
Volume Editor: Anghel, D. F.
Vol. 122, 2002

Molecular Organisation on Interfaces
Volume Editor: Lagaly, G.
Vol. 121, 2002

Lipid and Polymer-Lipid Systems
Volume Editor: Nylander, T.
Vol. 120, 2002

Progress in Colloid and Polymer Science
Editors: F. Kremer, Leipzig and W. Richtering, Aachen

Volume 126 · 2004

Trends in Colloid and Interface Science XVII

Volume Editors:
Valérie Cabuil
Pierre Levitz
Clande Treiner

Springer

The series Progress in Colloid and Polymer Science is also available electronically (ISSN 1437-8027)

- Access to tables of contents and abstracts is *free* for everybody.
- Scientists affiliated with departments/institutes subscribing to Progress in Colloid and Polymer Science as a whole also have full access to all papers in PDF form. Point your librarian to the LINK access registration form at http://www.springerlink.com

ISSN 0340-255X
ISBN 3-540-20073-8
DOI 10.1007/b10948
Springer Berlin, Heidelberg, New York

Springer Science + Business Media

http://www.springeronline.com

Typesetting: SPS, Chennai, India

Cover: Estudio Calamar,
F. Steinen-Broo, Pau/Girona, Spain

Printed on acid-free paper

Progr Colloid Polym Sci (2004) 126: V

PREFACE

The 16th Annual Meeting of the European Colloid and Interface Society was held in Paris in September 2002. Since the initial meeting, the motivation was always the same: to gather scientists, either physicists, chemists or biologists working in colloid and interfaces sciences. This proceedings including 41 written contributions is an overview of this mature but very open scientific field. As clearly demonstrated during the meeting, actual researches in colloid sciences impulse original developments in microfluidics and nanotechnologies. In the same track, physical chemistry of biological objects is a field where "colloidal scales" and confinement are essential. A huge "Terra Incognita" is still to discover and colloid and interfaces sciences should play an important role during this exploration.

Acknowledgement Editors want to thanks CNRS for it's financial support.

V. Cabuil, P.E. Levitz and C. Treiner

Progr Colloid Polym Sci (2004) 126: VI–VIII

CONTENTS

Molecular Self-Assemblies

Colloids in Biology

Progr Colloid Polym Sci (2004) 126: 1–5
DOI 10.1007/b93974

Md.H. Uddin
H. Kunieda

Effect of poly(dimethylsiloxane) chain length on the phase behavior of *A–B*-type silicone copolymer in water

Abstract The phase behavior and microstructure of *A–B*-type silicone copolymer, $Me_3SiO–(Me_2SiO)_{m-2}–Me_2SiCH_2CH_2CH_2–O–(CH_2CH_2O)_{51.6}H$, (abbreviated as POE–PDMS or $Si_mC_3EO_{52}$) in water was investigated for a constant POE chain length (EO_{52}) at 25 °C to ascertain the effect of PDMS chain length (*m* ranging from 5.8 to 52) on the phase behavior. Different microstructures, such as discontinuous micellar cubic (I_1), normal hexagonal (H_1), lamellar (L_α), and reverse hexagonal (H_2) liquid crystals, are formed depending on the volume ratio of POE to the copolymer and/or the copolymer concentration. In the short-PDMS-chain copolymer system, the H_1–I_1–W_m phase transitions take place with the addition of water, whereas, long-PDMS-chain copolymers form the L_α or the H_2 phase, which coexist with excess water. The effect of the PDMS chain length on the structural parameters and hence the phase transitions was investigated by using small-angle X-ray scattering. The effective cross-sectional area per copolymer molecule is almost the same (around 0.8 nm^2) for $m = 5.8$–12, whereas it increases from 0.82 to 1.13 nm^2 with an increase in *m* from 14 to 52. The new phase diagrams of water–$Si_{14}C_3EO_{52}$ and water–$Si_{52}C_3EO_{52}$ systems are also presented.

Keywords Silicone copolymer · Phase behavior · Liquid crystal

M.H. Uddin · H. Kunieda (✉)
Graduate School of Environment and Information Sciences,
Yokohama National University,
Tokiwadai 79-7, Hodogaya-Ku,
Yokohama 240-8501, Japan
e-mail: kunieda@ynu.ac.jp
Tel.: +81-45-3394190
Fax: +81-45-3394190

Introduction

Block copolymers form numerous morphologies in solution [1] and in the melt [2] depending on the volume fraction of each block and the product of the interaction parameter, χ, and the number of repeat units, *N*. This polymorphic behavior of block copolymers resembles that of aqueous solutions of surfactants in the liquid-crystalline phase [3] and the resulting structures have been found to have wide-spread applications [4]. The aqueous-based morphology control study is also a great topic of environment concern. Amphiphilic block copolymers are also called surfactants when their molecular weights are low. In the case of poly(oxyethylene) (POE)-type nonionic surfactants (C_mEO_n) [5] or copolymers [6], the layer curvature of the amphiphiles is successively changed from negative to positive with increasing EO chain length and the types of morphologies also change accordingly. The effect of lipophilic chain length on the phase behavior and self-organization is essentially ignored since the lipophilic parts of conventional surfactants are usually limited (C_{12}–C_{18}).

We have studied the phase behavior of *A–B* type POE–poly(dimethylsiloxane) (PDMS) copolymers (molecular weight ranging from low molecular weight surfactant to medium molecular weight copolymer) for different *A–B* chain lengths in the melt [7] and in solution [8, 9]; however, a detailed and systematic study of the effect of the PDMS chain length on the phase behavior and microstructure has yet to be done. In this context, we studied this effect of POE–PDMS copolymers in water.

Experimental

Materials

The series of copolymers with the general formula $(CH_3)_3SiO$–$[(CH_3)_2SiO]_{m-2}$–$(CH_3)_2SiCH_2CH_2CH_2$–O–$(CH_2CH_2O)_{51.6}H$ ($Si_mC_3EO_{52}$) were obtained from Dow Corning Toray Silicone Co, Japan. m is the total number of silicon atoms. Five different PDMS chain copolymers, $m=5.8$, 14, 25, 38 and 52, were used to construct the phase diagram at constant temperature. The samples of intermediate PDMS chains were prepared by blending two neighboring PDMS chain copolymers. The polydispersity index of the PDMS part, M_w/M_n, is 1.02, 1.19, 1.20, 1.06 and 1.07 for $m=5.8$, 14, 25, 38 and 52, respectively, where M_w is the weight-average molecular weight and M_n is the number-average molecular weight. The polydispersity index of the POE chain (EO_{52}) is 1.13. Millipore filtered water was used.

Phase diagrams

The copolymers were first melted and mixed and then water was added. Homogeneity was attained using a vortex mixer and repeated centrifugation through a narrow constriction in the sealed sample tube. The phase equilibrium was detected by visual inspection of the samples and with crossed polarizers for birefringence. The types of liquid crystals were identified by a video-enhanced microscope (Nikon X2F-NTF-21) and small-angle X-ray scattering (SAXS).

Small-angle X-ray scattering

The interlayer spacing of liquid crystals was measured by SAXS, performed on a small-angle scattering goniometer with an 18-kW Rigaku Denki rotating-anode generator (Rint-2500) at about 25 °C. The samples were covered with plastic films (Mylar seal method) for the measurement.

Results and discussion

Phase diagrams of water–$Si_mC_3EO_{52}$ systems as a function of temperature

The phase diagrams of $Si_{14}C_3EO_{52}$ ($f=0.64$, where f is the volume fraction of EO_{52} in the copolymer) and $Si_{52}C_3EO_{52}$ ($f=0.33$) in water constructed as a function of temperature are represented in Fig. 1a and b, respectively. $Si_{14}C_3EO_{52}$ and $Si_{52}C_3EO_{52}$ form lamellar (L_α) and reverse hexagonal (H_2) (POE chain is embedded in PDMS corona) ordered structures, respectively, above their solid melting temperature (around 50 °C). Since the PDMS chains are in a liquid or amorphous state up to a high molecular weight, the melting temperature of the copolymer corresponds to the melting temperature of the POE chain. In both systems, the solid melting temperature is sharply decreased with the addition of water

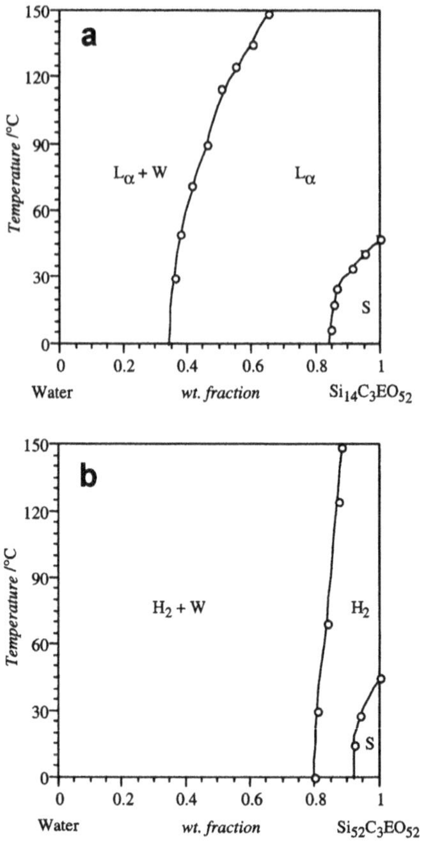

Fig. 1 Phase diagrams of binary water–$Si_mC_3EO_{52}$ systems as a function of weight fraction of copolymer: **a** $Si_{14}C_3EO_{52}$ and **b** $Si_{52}C_3EO_{52}$. L_α and H_2 indicate lamellar and reverse hexagonal phases, respectively. W is an excess water phase and S is a solid-present region

because of the hydration of POE chains of the copolymer. Both L_α and H_2 phases are thermally stable and do not melt up to the temperature examined (150 °C). The L_α phase can solubilize a considerable amount of water (65 wt% of water), whereas the solubilization is small (19 wt% of water) in the H_2 phase. Both phases coexist with excess water (W) phase beyond their solubilization limits. The aqueous phase behavior of the $Si_{5.8}C_3EO_{52}$ ($f=0.8$) and $Si_{25}C_3EO_{52}$ ($f=0.5$) systems as a function of temperature was reported earlier [10, 11].

The interlayer spacing, d, in the L_α phase was measured by SAXS at 25 °C. The half thicknesses of the polar $[d_H=d(1-\phi_L)/2]$ and apolar $(d_L=d\phi_L/2)$ domains and the effective cross-sectional area per copolymer molecule ($a_P=v_L/d_L$) were estimated, where ϕ_L is the volume fraction of the lipophilic part of the copolymer in the system and v_L is the volume of the lipophilic part per copolymer molecule [10]. The results are represented as a function of $1/\phi_L$ in Fig. 2. d and d_H increase monotonically, while d_L and a_P do not significantly change with the water content. The linear relationship between d and $1/\phi_L$ holds, i.e. the bilayer

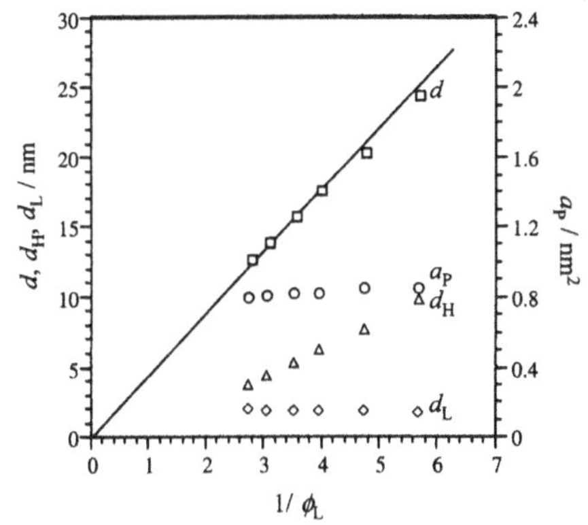

Fig. 2 Small-angle X-ray scattering data for the lamellar phase in the $Si_{14}C_3EO_{52}$-water system at 25 °C. Interlayer spacing, *d* (*squares*); effective cross-sectional area per copolymer molecule, a_P (*circles*), half of the hydrophilic thickness, d_H (*triangles*); half of the lipophilic thickness, d_L (*diamonds*)

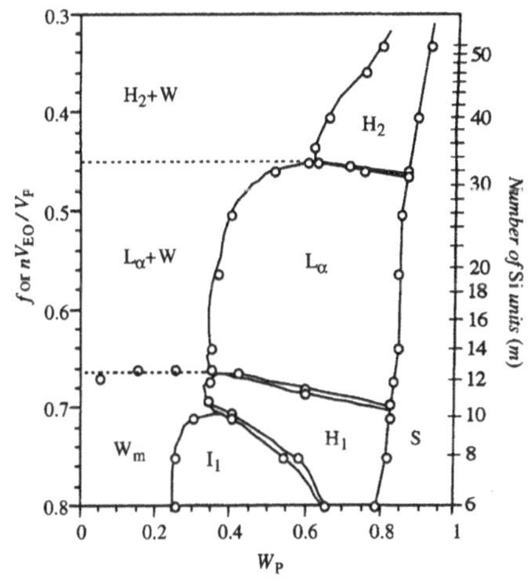

Fig. 3 Phase diagram of the binary water–$Si_mC_3EO_{52}$ system as a function of the poly(dimethylsiloxane) chain length, *m*) at 25 °C. *f* is the volume ratio of the poly(oxyethylene) chain to the copolymer. *Wm*, I_1, H_1, L_α and H_2 indicate aqueous micellar solution, normal micellar cubic, normal hexagonal, lamellar and reverse hexagonal phases, respectively. *W* is an excess water phase and *S* is a solid-present region

is just swollen by water and there is only little variation in the hydration of the POE chain. The PDMS chain is suppressed to about half (2.1 nm) of its extended length (3.9 nm); however, the layer curvature cannot be changed to positive (convex toward water) upon addition of water because of the large a_P (around 0.82 nm^2).

Phase diagram of the water–$Si_mC_3EO_{52}$ system at 25 °C

The phase diagram of the water–$Si_mC_3EO_{52}$ system constructed at 25 °C as a function of the volume fraction of the hydrophilic moiety of the copolymer, *f* (vertical axis), and the weight fraction of the copolymer in the system, W_P (horizontal axis), is presented in Fig. 3. The corresponding number of siloxane units, *m*, is also shown on the right-hand axis. A large solid-present region is observed in the concentrated region. The solid-present phase consists of semicrystalline POE and amorphous PDMS chains. Addition of water dissolves the POE chains and increases the segregation between water and PDMS, consequently the copolymers self-organize into different microstructures. When the content of the PDMS chain increases, the flexibility of the chain increases and hence the solid-present region shrinks to higher concentrated regions at high *m* values.

Short-PDMS-chain copolymers, $Si_{5.8-10}C_3EO_{52}$ ($f=0.8$–0.7) form successively aqueous micellar solution (W_m), discontinuous micellar cubic (I_1), and normal hexagonal (H_1) phases with increasing surfactant concentration. The I_1 and H_1 phases are thermally stable and they melt to a micellar solution phase at high temperature [10]. A large lamellar region is observed in the middle of the phase diagram. An L_α–H_1–W_m phase transition takes place at $m \approx 12$. The L_α phase swells a large amount of water and coexists with excess water (L_α + W) at $m > 13$. The samples look bluish in the very dilute region of the L_α + W phase. In this region, vesicles were observed by video-enhanced polarized microscopy. The phase boundary between W_m and L_α + W was identified by visual observation and by microscopy. Long-PDMS-chain copolymers $Si_{32-52}C_3EO_{52}$ form reverse hexagonal liquid crystal (H_2). The H_2 phase solubilizes a smaller amount of water compared to the L_α phase. Although the POE chain length of the copolymer is the same (EO_{52}), the solubilization of water decreases with the increase of *m* in the H_2 phase.

POE-type nonionic surfactants form bicontinuous cubic phases on both sides of the lamellar phases [12]. However, the bicontinuous cubic phases are absent in the present system because of the large conformational asymmetry of the POE–PDMS chains [7]. However, mixtures of this copolymer ($Si_{25}C_3EO_{52}$) and nonionic surfactant ($C_{12}EO_5$) form a bicontinuous cubic phase in water [11]. In the mixed system, both of the blocks of the copolymer have more freedom or less packing restriction because of the presence of short spacer molecules at the interface of the aggregates.

The microstructures of the different liquid crystals were identified by using SAXS. The reciprocal spacing of the typical SAXS peaks are found to be in the ratio $1:\sqrt{3}$ for the H_2 phase, 1:2 for the L_α phase and $1:\sqrt{3}:2$ for the H_1 phase. These anisotropic liquid crystals were also identified from their typical textures observed by video-enhanced microscopy. The four peaks resolved in the cubic phase showed spacing ratios of $\sqrt{3}:\sqrt{4}:\sqrt{8}:\sqrt{11}$. It is considered that these peaks represent diffractions from (111), (200), (220) and (311) planes, respectively, that correspond to a face-centered-cubic structure belonging to the *F*23 space group [13].

We measured the interlayer spacings, d, of the I_1, H_1, L_α and H_2 phases as a function of surfactant concentration for particular PDMS chain lengths ($m = 5.8$, 10, 12, 14, 25, 38, 52) at 25 °C. The d values changed almost linearly with copolymer concentration. If the shape of the aggregate is unchanged with copolymer concentration, the d values can be fitted to straight lines passing through the origin in d versus $(1/\phi_L)^{1/3}$, $(1/\phi_L)^{1/2}$, $1/\phi_L$ and $(1-\phi_L)^{1/2}/\phi_L$ plots for the I_1, H_1, L_α and H_2 phases, respectively, according to equations described in Ref. [8]. The effective cross-sectional area per copolymer molecule, a_P, was also calculated for each liquid crystal from the measured d values [8]. There is no significant variation in a_P with changing copolymer concentration. However, we also calculated the a_P from d values for 100% copolymer obtained by extrapolating the lines, for example, in Fig. 2; d for undiluted $S_{14}C_3EO_{52}$ is obtained (12.7 nm) from the line at $1/\phi_L = 2.7$. The a_P values are 0.80, 0.81, 0.81, 0.82, 1.02, 1.10 and 1.13 for the copolymer of $m = 5.8$, 10, 12, 14, 25, 38 and 52. The values for $m = 14$, 25, 38 and 52 are almost same as those of the corresponding copolymer forming the liquid crystals in the melt state and measured at 53 °C [7]. It is observed that a_P increases with increasing PDMS chain length irrespective of the types of liquid crystals. Hence the copolymer layer curvature changes from positive to negative (PDMS embedded in POE to POE embedded in PDMS matrix) and consequently I_1–H_1–L_α–H_2 phase transitions occur in the concentrated region with the increase of the PDMS chain. However, for $m = 5.8$–12, a_P does not increase significantly and is an analogue of the POE-type nonionic surfactant (C_mEO_n)-water systems in which the area per surfactant molecule does not change with composition and is almost the same for at least C_{12}–C_{18} at constant EO chain lengths [5, 14]. C_mEO_n surfactants show rich polymorphic phase behavior with varying concentration and the surfactant layer curvature in aggregates changes from negative to positive with water content. Similar properties were observed, for example, H_1–I_1–W_m, H_1–W_m or L_α–H_1–W_m phase transitions occur with an increase in water content for short-PDMS-chain copolymers (Fig. 3) because the PDMS chain up to this length ($m \sim 12$) can be elongated to a cone shape in normal types of aggregates.

In contrast, the long PDMS chain owing to its high flexibility and low cohesive energy becomes coiled and increases a_P. Hence it is hard to change the layer curvature toward positive because in this case the long PDMS chain has to be elongated to form a normal aggregate, which is energetically unfavorable. Hence, upon addition of water, L_α and H_2 phases coexist with water instead of forming an aqueous micellar solution, which is observed in the dilute region of short PDMS-chain copolymer.

Undiluted $Si_{5.8-12}C_3EO_{52}$ does not form an ordered morphology because of low segregation between short-PDMS-chain and POE, whereas for $m > 12$ either an L_α or an H_2 phase is formed in the melt state [7]. However, addition of water to $Si_{5.8-12}C_3EO_{52}$ increases the effective volume fraction of POE by hydration and enhances the segregation between POE and PDMS, where hydrogen bonds stiffen and cross-link the POE chains. Consequently, in hydrous conditions the ordered-state morphologies can be formed at lower M_W of the copolymer.

The short-PDMS-chain copolymers show surfactant-like behavior: i.e. they exhibit polymorphic behavior in water by changing composition (the segregation tendency is solvent-dependent), whereas polymorphic phase behavior is modulated by changing the relative volume ratio of the blocks in the long PDMS copolymers. In the latter case, the segregation tendency of POE and PDMS chains is more dominant to form a particular self-assembly than solvent (water) effects.

Conclusion

POE–PDMS copolymers successively form normal micellar cubic, normal hexagonal, lamellar and reverse hexagonal liquid crystals in water. The cubic phase belongs to a face-centered-cubic structure. With the increase of the PDMS chain of the copolymer, the chain becomes coiled and increases the effective cross-sectional area and hence the copolymer layer curvature changes from positive to negative, in which the PDMS chain has more spaces. Consequently, the I_1–H_1–L_α–H_2 phase transitions occur. In the short-PDMS-chain copolymers, the liquid-crystalline phases change to micellar solution phases upon dilution, whereas lamellar and reverse hexagonal phases formed in the long-PDMS-chain copolymer coexist with excess water, otherwise to form a normal micelle the long PDMS chain has to be elongated, which is energetically unfavorable. The relative sizes and the segregation of *A–B* blocks are more dominant for the polymorphic behavior than water effects.

Acknowledgements We thank Asao Harashima and Haruhiko Furukawa at Dow Corning Toray Silicone Co. Ltd., Japan, for supplying the silicone copolymers.

References

1. Hamley IW (1998) The physics of block copolymers. Oxford University Press, Oxford
2. Bates FS, Fredrickson GH (1999) Phys Today February 32
3. Seddon JM (1990) Biochim Biophys Acta 1031:1
4. Hajduk DA, Kossuth MB, Hillmyer MA, Bates FS (1998) J Phys Chem B 102:4269
5. Huang K-L, Shigeta K, Kunieda H (1998) Prog Colloid Polym Sci 110:171
6. Uddin MH, Yamashita Y, Furukawa H, Harashima A, Kunieda H Prog Colloid Polym Sci (in press)
7. Uddin MH, Rodriguez C, Lopez-Quintela A, Solans C, Esquena J, Kunieda H (2003) Macromolecules 36:1261
8. Kunieda H, Uddin MH, Horii M, Furukawa H, Harashima A (2001) J Phys Chem B 105:5419
9. Uddin MH, Rodriguez C, Watanabe K, Lopez-Quintela A, Kato T, Furukawa H, Harashima A, Kunieda H (2001) Langmuir 17:5169
10. Rodriguez C, Uddin MH, Furukawa H, Harashima A, Kunieda H (2001) Prog Colloid Polym Sci 118:53
11. Kunieda H, Uddin MH, Furukawa H, Harashima A (2001) Macromolecules 105:5419
12. Kunieda H, Shigeta K, Ozawa K, Suzuki M, (1997) J Phys Chem B 101:7952
13. Li X, Washenberger RM, Scriven LE, Davis HT, Hill RM (1999) Langmuir 15:2267
14. Kunieda H, Yamaguchi Y, Ozawa K, Suzuki M (1999) Colloids Surf A 160:15

Progr Colloid Polym Sci (2004) 126: 6–13
DOI 10.1007/b93988

C. Raffournier
M. Ollivon
P. Couvreur
C. Dubernet

Solubility of triacylglycerols or stearylamine in phospholipid vesicles

C. Raffournier (✉) · M. Ollivon
P. Couvreur · C. Dubernet
UMR CNRS 8612, Faculté de Pharmacie,
Université de Paris-XI, 5 rue J.B. Clément,
92296 Châtenay-Malabry, France
e-mail: chrisraf@club-internet.fr
Tel.: +33-1-46835584
Fax: +33-1-46619334

Abstract The solubilities of two types of lipids, triglycerides (TG), weakly polar compounds, and stearylamine (SA), a polar molecule, within phospholipid bilayers of vesicles are examined by time-resolved selective solubilisation and compared. Both TG and SA were added to phosphatidylcholine (Lipoid E80, PL) vesicles at different concentrations ranging from 0 to 20% (weight per weight of PL). TG insertion in lipid bilayers is a known phenomenon, which develops up to a limit of solubility beyond which emulsion droplets coexist with liposomes. The characterisation of the various colloids formed is obtained by selective solubilisation of the aggregates resulting from continuous detergent addition (Triton X-100). Turbidity monitoring allowed this identification thanks to the well-known "vesicle-to-micelle transition process". The low TG concentration only slightly modifies the solubilisation profiles with low turbidity values at the solubilisation completion corresponding to mixed-micelle formation. In contrast, the high TG concentration results in high residual turbidity values interpreted as the coexistence of mixed micelles and emulsion droplets. The amount of detergent used to reach solubilisation completion is proportional to both PL and TG concentrations. The process is dramatically different with SA, although high turbidity values are also observed at solubilisation completion. Indeed, when the SA proportion increases in the vesicles, the amount of detergent necessary to solubilise the vesicles is reduced. So, depending on the type and polarity of molecules added to conventional PL vesicles, the solubilisation can be delayed (TG) or promoted (SA).

Keywords Liposomes · Triton X-100 · Micelle · Emulsion · Solubility limit

Introduction

Preparation of an emulsion is a convenient way, known from ancient times, to disperse insoluble molecules not using organic solvents. The multiple applications of emulsions in many fields account for this interesting ability. For example, preparation of an oil-in-water (O/W) emulsion enhances solubilisation and bioavailability of hydrophobic molecules in a hydrophilic medium. While such emulsions are thermodynamically unstable systems, addition of charged emulsifiers or amphiphilic polymers, or both, allows their stabilisation for long periods of time and even their dilution. Among the major disadvantages of this dispersion system for therapeutic use are instability and incompatibility with intravenous administration partly owing to the size of the droplets. To remedy these inconveniences, submicronic cationic emulsions have been investigated. The presence of cationic lipids stabilises the O/W interface and favours the formation of droplets of small size (200 nm) as well as allows intravenous injection [1]. Formulation of such systems is characterised by the presence in excess [2] of three surface-active agents stabilising the oil droplets made of C_8 and C_{10} triglycerides (TG): egg phospholipids (Lipoid E80, PL), stearylamine (SA) (monoacyl chain cationic lipid) and poloxamer 188. In this context, cryogenic transmission electron microscopy experiments (cryo-TEM) were performed to observe the morphology of these emulsions. Cryo-TEM has revealed the existence of unexpected bicompartmental structures where a phospholipid membrane stabilises both an oily and a water phase [3].

In order to understand how such structures form, we decided to analyse interactions between the surface agents used and TG. In a first step, the system was simplified to build up miscibility models of PL/TG and PL/SA mixtures in a water phase. Insertion of TG in lipid bilayers is a well-known phenomenon. Hamilton and Small [4] have shown that up to 3% mol/mol of triolein (three acyl chain C_{18-1}) is incorporated in dipalmitoylphosphatidylcholine vesicles and 2.8% in pure egg phosphatidylcholine (PC). Dahim et al. [5] have shown that this solubility limit is extended to 3.8% for incorporation of TG ($C_{18-1}=60\%$ and mixture of different acyl chains) in pure egg PC. In the present work, the time-resolved selective solubilisation technique developed by Dahim et al. was used to characterise the population of particles formed upon mixing of TG or SA with PL as well as for the determination of the limits of solubility of these molecules into PL.

Materials and methods

Materials

Three types of lipids were used for vesicle preparation: PL ($M_r=707\pm3$) was purchased from Lipoid (Ludwigshafen, Germany), SA ($M_r=269.5$) from Sigma (l'Isle d'Abeau, France), and TG ($M_r=505\pm3$) from Société Industrielle des Oléagineux (St. Laurent Blangy, France). PL corresponds to a complex mixture of egg phospholipids with various acyl chains (mainly C_{16} and C_{18}) in which PC is the main constituent. The oil phase is a mixture of TG synthesised from two saturated fatty acids: capric acid (C10:0) and caprylic acid (C8:0). Triton X-100 (TX-100) ($M_r=624$) was purchased from Sigma (l'Isle d'Abeau, France). Glycerol was from Fluka (Sigma, l'Isle d'Abeau, France).

Vesicle preparation and characterisation

Lipid films were prepared by evaporation under vacuum of chloroformic solutions of either TG (0–5 mg) and PL (25 mg) or SA (0–5 mg) and PL (up to 25 mg). The amount of lipid added was determined exactly. The vesicle composition is shown in Table 1. These lipid films were rehydrated with an aqueous solution of glycerol (2.25% w/w) and then sonicated (Vibracell, Bioblock scientific) in an ice-cooled vial to reduce the size of the liposomes formed, that is to say to transform multilamellar vesicles into small unilamellar vesicles.

The vesicle suspensions were centrifuged (3K30, Sigma, Bioblock) at low speed to remove the titanium microparticles resulting from sonication and were used shortly after preparation (3 days).

The vesicle size was measured by quasielastic light scattering using a Nanosizer (N4Plus, Coultronics) and was found to be between 70 and 100 nm with a polydispersity index not greater than 0.5. The composition of the vesicles was checked by enzymatic analysis of choline (Biomérieux, France).

Solubilisation monitoring

Liposome solubilisation was followed at 25 °C by monitoring the turbidity at 400 nm with a double-beam UV–vis spectrophotometer (Lambda 2, PerkinElmer). The initial sonicated preparations were diluted (3/4, 1/2, 1/4, 1/10) to a total volume of 1.2 ml and placed in a glass cuvette (10x10 mm). At each of these vesicle suspensions a TX-100 solution was added continuously, at a constant rate, under stirring, using a syringe pump as previously described [6]. The experimental setup is described in Fig. 1. The TX-100 concentration was adapted to the lipid concentration: a stock solution at 62.2 mM was diluted as mentioned earlier (3/4, 1/2, 1/4, 1/10), in order to maintain the initial TX-100 to lipid concentration ratio constant and to get the same solubilisation duration whatever the initial lipid concentration. The influence of the detergent addition rate was determined (data not shown). Different rates (0.1157, 0.2135, 0.4630, 0.9260, 1.852 μl/s) were examined. The solubilisation profiles were found not to be significantly modified. This means that the combination of the TX-100 concentration and the addition rate used is low enough to maintain dynamic equilibrium in detergent/PL molecule rearrange-

Table 1 Compositions and molar ratios of triglyceride (*TG*)/Lipoid E80 (*PL*) and stearylamine (*SA*)/PL vesicles. *Left*: PL concentration was constant and TG was added from 0 to 20% w/w. *Right*: Total weight of PL and SA was kept constant so that SA represents 0–20% w/w

Composition		0% TG	2% TG	4% TG	6% TG	8% TG	10% TG	20% TG	0% SA	4% SA	10% SA	20% SA
[PL]	mg				25				25	24	22.5	20
	mM				5.44				5.44	5.22	4.9	4.35
%PL	% w/w	100	98	96	94	92	90	80	100	96	90	80
	% mol/mol	100	97.3	97.4	92.3	89.9	87.7	78.1	100	89.4	76.1	58.6
% TG	% w/w	0	2	4	6	8	10	20				
	% mol/mol	0	2.7	5.3	7.7	10.1	12.3	21.9				
% SA	% w/w								0	4	10	20
	% mol/mol								6	10.6	23.9	41.4

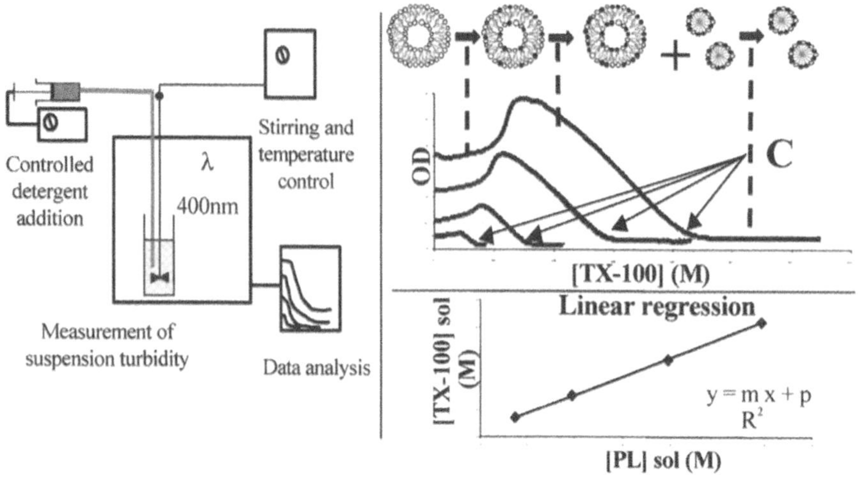

Fig. 1 *Left*: Schematic diagram of the setup used for turbidity monitoring. Triton X-100 (*TX-100*) is continuously added under stirring to the liposome suspension thanks to the syringe pump. *Right*: Solubilisation of liposomes followed by time-resolved turbidity as a function of detergent addition. *Top*: The vesicle/micelle transition. *Middle*: Solubilisation curves for pure Lipoid E80 (*PL*) vesicles at four different concentrations corresponding to dilutions 0.75, 0.5, 0.25 and 0.1 of initial the vesicle suspension. Breakpoint *C* corresponds to solubilisation completion (mixed-micelle formation is complete). *Bottom*: The amount of TX-100 needed to solubilise a given quantity of lipids at breakpoint C as a function of lipid concentration. The slope, *m*, and the ordinate value, *p*, at [PL] = 0 mM of the regression line, representing the number of moles of TX-100 needed to solubilise 1 mol of lipids and the concentration of monomer in equilibrium with the mixed micelles, respectively

ments occurring during TX-100 addition. The intermediate rate of 0.4630 μl/s was chosen for all the experiments.

The solubilisation curves, representing the evolution of turbidity during the surfactant addition, are divided into two main stages (Fig. 1). At first, addition of detergent solution to the liposomes causes an increase in the optical density (OD) corresponding to an aggregation/fusion process of vesicles. Further addition of TX-100 induces a progressive decrease in the OD until a low value corresponding to the formation of mixed micelles. Thus, the whole process corresponds to a vesicle/mixed-micelle transition. The solubilisation completion is indicated on the curve by a breakpoint called C. TX-100 concentrations at point C are reported as a function of PL concentrations in the liposome suspension at point C. The points obtained fall on a line (that more or less corresponds to a phase diagram boundary), the slope of which represents the number of moles of TX-100 necessary to solubilise one mole of PL (the ordinate at a PL concentration of 0 mM is negligible), in other words the slope corresponds to the molar ratio $[TX\text{-}100]_C/[PL]_C$.

The analysis of the solubilisation curves permits us to determine the solubility limit of TG in lipid bilayers according to Dahim et al. [5]. Different proportions of TG are added to the vesicles. At low TG proportions, TG is totally inserted in the PL bilayers and only TG-containing liposomes are present. When the TG concentration is greater than the TG insertion limit in the PL bilayer, the excess TG organises itself in emulsion droplets. Detergent addition, thanks to the selective time-resolved solubilisation recordings, allows us to detect the presence of emulsion droplets in TG/PL mixtures. Upon detergent addition, at the end of solubilisation, for low TG concenrations, liposomes are transformed into mixed micelles of low turbidity. Beyond the solubility limit of TG into vesicle bilayers, the excess TG forms emulsion droplets that are not significantly modified by TX-100. At the completion of vesicle solubilisation, both mixed micelles and oil droplets coexist in the suspension at breakpoint C. The presence of unsolubilised emulsion droplets dramatically increases the turbidity of the dispersions at this point. In order to compare different preparations, the OD at point C can be normalised as follow: $OD_{Ci}/[TX\text{-}100]_{Ci}$ for each of the four dilutions *i* of the vesicle suspension. The specific OD is the average of the four values obtained for a given formulation. The specific turbidity increases linearly with the TG concentration but two regimes, which correspond to the two populations described earlier, are observed in the turbidity dependence of point C (Fig. 2). Two straight lines, the intersection of which allows the determination of the solubility limit of TG into vesicle bilayers, best fit these two regimes.

Results and discussion

Profiles of solubilisation curves

The vesicle suspensions containing different proportions of TG or SA were diluted 10, 4, 2 and 4/3 times (volume per volume), corresponding, respectively, to vesicles concentrations of 0.1, 0.25, 0.5 and 0.75 of the mother solution. The solubilisations were achieved by addition of TX-100 solutions the concentrations of which, $[TX\text{-}100]_0$, were proportional to the initial lipid concentration, $[lipid]_0$. For each solubilisation series $[TX\text{-}100]_0/[lipid]_0$ was a constant value (11.4).

The solubilisation curves obtained for four suspensions of vesicles containing 0, 4, 10 and 20% of TG are shown in Fig. 3. For a TG concentration of 0, that is to say with pure PL vesicles, the initial OD is low (0.29 or less). Two steps are observed in the initial OD increase during TX-100 addition. The first step corresponds to TX-100 insertion into vesicles, while the step with a sharp increase is interpreted as a vesicle fusion/aggregation process. Further TX-100 addition also results in two steps, an important OD decrease until low OD values (0.14) of mixed micelles and a slight decrease related to their dilution.

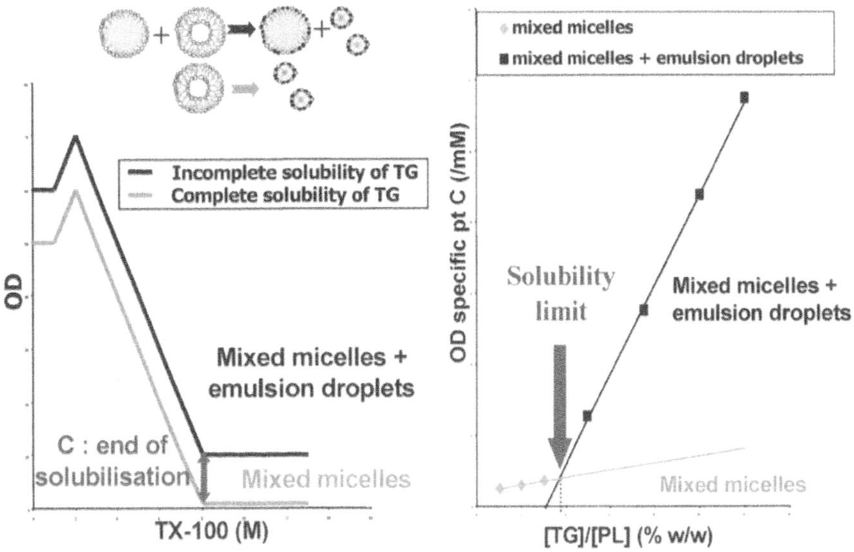

Fig. 2 Schematic description of the method used for the determination of solubility limits of triglycerides (*TG*) in liposome suspensions. *Left*: Solubilisation curves for liposomes (*lower curve*) and emulsions (*upper curve*). It is worth noting the difference in the optical density (*OD*) for complete and incomplete solubility of TG in bilayers. *Right*: $OD_{specific}$ versus $[TG]_{tot}/[PL]_{tot}$ ratio allowing the determination of the solubility limit of TG in lipid bilayers

Along with the increase in the proportion of TG in the vesicle preparations the initial OD value becomes higher (0.39 for 4% of TG, 0.51 for 10% of TG and 0.70 for 20% of TG), the second step of the OD increase becomes less marked and the OD at point C also increases (0.16 for 4% of TG, 0.22 for 10% of TG and 0.36 for 20% of TG).

The turbidity (OD) depends on the number, the size and the type of the particles observed by light scattering [7]. The turbidity increase observed at high TG concentrations in the initial preparations (0.7 at 20% of TG against 0.29 at 0%) is related to the presence of emulsion droplets in the suspensions. Indeed, it has been shown that excess of TG that is not solubilised in phospholipid bilayers self-aggregates in emulsion droplets [4, 5]. The observation of the same series of events in the solubilisation profiles, whatever the TG concentration, attests to the presence of vesicles. Then, the two populations of aggregates coexist beyond a certain TG-to-PL concentration ratio. Consequently the solubilisation profiles observed in the presence of TG and their evolution as a function of the TG-toPL concentration ratio are interpreted as the selective solubilisation of vesicles. In comparison with 0% TG curves, the higher turbidity observed at breakpoint C for 4, 10 and 20% of TG corresponds to the presence of oil droplets and demonstrates that these objects are not significantly modified by detergent addition. Then, the increase in the OD during the fusion/aggregation process, less and less well defined as the proportion of TG increases, corresponds to the presence of emulsion droplets masking vesicle events.

The solubilisation curves obtained for four suspensions of vesicles containing 0, 4, 10 and 20% of SA are shown in Fig. 4. The set of turbidity profiles observed for 0% of SA, which is identical to that of Fig. 3 (pure PL vesicles), is shown for comparison with SA-containing

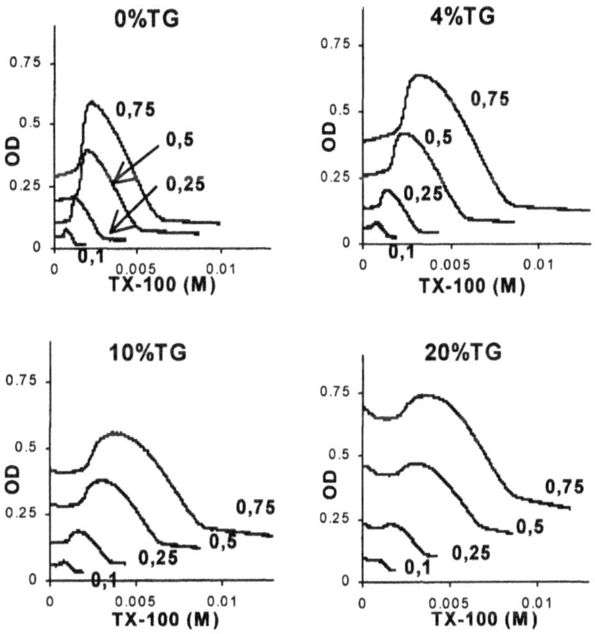

Fig. 3 Evolution of solubilisation curves for suspensions (liposome with and without emulsion droplets) as a function of both PL and TG contents. The four solubilisation plots shown correspond to 0, 4, 10 and 20% of oil. The four different dilutions examined are directly indicated on the plot

ones. Whatever the SA concentration, the solubilisation process is different from that of TG-containing aggregates. At low SA concentrations (4%) only a single OD increase is observed, while at SA concentrations of 10% and above the increase step vanishes and is replaced by a slow decrease. The duration of this last step diminishes as a function of SA content. As for TG, the similitude of the profiles and the high residual turbidity attests to the presence of two populations of aggregates; vesicles

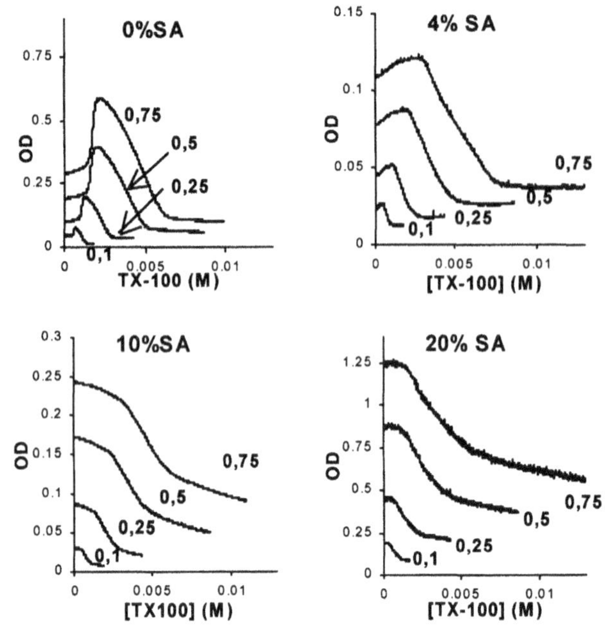

Fig. 4 Evolution of solubilisation curves as a function of both lipid and stearylamine (*SA*) contents. The four solubilisation plots shown correspond to 0, 4, 10 and 20% of SA and the four different dilutions examined are directly indicated on the plot

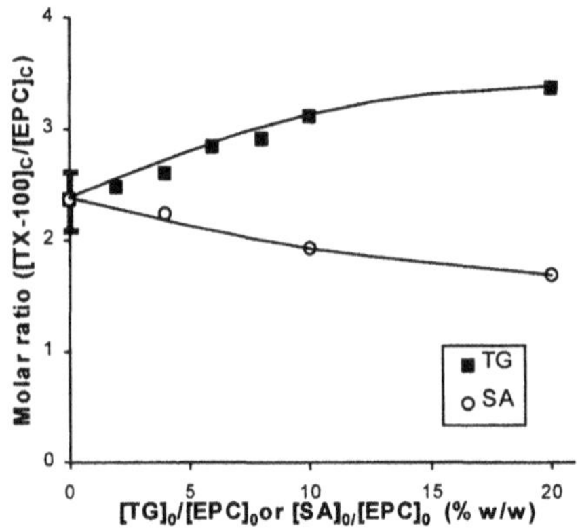

Fig. 5 Evolution of the amount of TX-100 needed for liposome solubilisation at breakpoint C expressed in $[\text{TX-100}]_C/[\text{PL}]_C$ molar ratio in the mixed micelles as a function of TG or SA content. The ratios correspond to the aggregate composition at C for the different preparations solubilised: 0, 2, 4, 6, 8, 10 and 20% of TG or 0, 4, 10 and 20% of SA

coexist with another type of highly turbid aggregate. It is remarkable that the overall turbidities recorded for 4% of SA, initial as well as final ones, are less than observed for the other concentrations, including 0%.

Influence of solubilised molecule type on the amount of TX-100 needed for solubilisation completion

The amount of TX-100 needed to reach breakpoint C has been plotted as a function of PL content at this point for each TG and SA concentration. The 11 linear regression plots (seven for TG and four for SA, from 0 to 20%) are not shown but the molar ratios ($[\text{TX-100}]_C/[\text{PL}]_C$) deduced from the slope *m* (Fig. 1) of the plot are reported in Fig. 5 as a function of TG or SA content. Indeed, the slope of the linear regression plots, which corresponds to a phase boundary, allows us to deduce the composition of the aggregates. The mean value obtained from four different experiments involving four independent batches of pure PL vesicles is 2.4 ± 0.3, meaning that about 2.5 mol of TX-100 is needed to solubilise 1 mol PL.

The evolutions of the molar ratios versus the proportion TG or SA are dramatically different in relation to the presence of TG or SA. While the proportion of TG increases, the molar ratio increases almost linearly from 0 to 10% and then, from 10 to 20%, it increases more slowly and seems to reach a plateau. In contrast, the reverse process is observed for SA: an increasing proportion of SA reduces the molar ratio in a practically linear way from 0 to 10% and in a rather asymptotic way from 10 to 20%.

This divergence in molar ratio evolution is due to the amphiphilic properties of the lipids used. TGs are weakly polar compounds and more TX-100 is necessary for their solubilisation in comparison with PL molecules. Moreover, in the emulsion droplets, interpenetration of TG acyl chains of the core and PL monolayer could result in rigidifying of its surface and modifying the solubilisation process [8]. Consequently, these events contribute towards an increase of the molar ratio with increasing proportion of TG.

In contrast, the presence of charged lipids induces disorder in lipid bilayers [9] which can help the insertion of molecules or interbilayer water swelling [10]. In addition to that, SA is a monoacyl chain aminolipid and has been shown, in some cases, to permeabilise lysosomal membrane; this implies a SA tensioactive property [11]. Thus, the amphiphilic character of SA could help TX-100 insertion in the bilayer. So, TX-100 and SA behave in a synergic way in the solubilisation of vesicles. This explains why the molar ratio decreases when the proportion of SA increases in the liposome preparations.

Determination of solubility limits

To determine the solubility limits of TG and SA in lipid bilayers, the specific OD ($OD_{specific}$ = average value of $OD_{Ci}/[TX]_{Ci}$), calculated from the data of Figs. 4 and 5, is reported in Figs. 6 and 7 as a function of the proportion of TG or SA in the vesicles ($[TG]_0$ or $[SA]_0/[PL]_0$ % w/w) [5]. Dahim et al. showed that depending on the relative proportion of TG in the

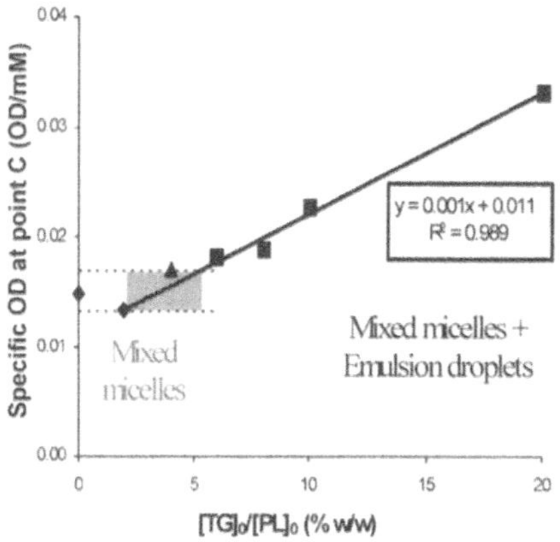

Fig. 6 Determination of TG solubility limit. The specific OD calculated from the data in Fig. 3 allows us to calculate a regression equation and to delimit a zone (in *grey*) corresponding to the TG solubility limit in the vesicles

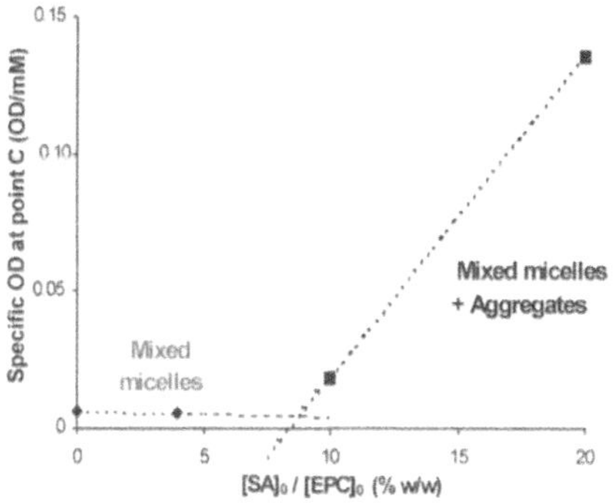

Fig. 7 Determination of SA solubility limit. The specific OD calculated from Fig. 4 does not allow us to calculate regression lines but suggests a limit between mixed micelles and the coexistence of mixed micelles with aggregates

vesicles, at breakpoint C one or two aggregate populations are formed (mixed micelles only or mixed micelles coexisting with emulsion droplets). As shown in Fig. 2, the dependence of the specific turbidity upon TG concentration allows the determination of the limit of solubility of TG into the vesicles. The intercept of the regression line corresponding to coexistence of emulsion droplets and mixed micelles with that corresponding to the pure mixed micelles yields such a solubility threshold. The solubility limit of SA in lipid bilayers has not been determined previously to our knowledge.

Figure 6 represents the determination of the TG solubility limit into PL vesicles. The seven specific OD values reported correspond to the average value of two experiments. At high TG concentrations, specific turbidity is measured in a domain in which two populations coexist. Both mixed micelles and emulsion droplets contribute to the scattering of light and then to the higher turbidity. In this domain, the experimental points are numerous enough to calculate a regression equation ($R^2=0.989$), which represents the specific OD evolution. At low TG concentrations, only the mixed-micelle population is formed upon solubilisation, the OD of which is lower. Only a few measurements were recorded in this domain and it was not possible do draw a precise linear regression in order to get an intercept. Taking into account that the OD is not expected to change very much in this domain as shown by Dahim et al., we estimated a solubility limit between 3 and 5% ($[TG]_0/[PL]_0$ w/w), equivalent to 4.2 and 7% mol/mol, respectively. This result is to be compared with that found by Dahim et al. and Hamilton and Small. Dahim et al. used C_{18} TG and pure PC and found 3.8% mol/mol. Hamilton and Small using triolein and egg lecithin found a 2.7% mol/mol limit. Our experiments with C_8/C_{10} TG and PL should lead to a higher solubility limit because acyl chains are shorter and insert more easily in the PC bilayers of the vesicles. However, as the values obtained with these different products are similar, it could be supposed that this easier insertion has a limited influence on the limit value.

A plot for the determination of the SA solubility limit into PL is shown in Fig. 7. The experimental points (average of two experiments) are not numerous enough to calculate regression lines but it can be supposed that they should follow the dotted lines drawn. The solubility limit of SA in PL bilayers cannot be precisely determined yet but X-ray experiments (data not shown) confirmed a change in the supramolecular structure of PL/SA mixtures around 8% w/w corresponding to 18.6% mol/mol. Moreover, we deduced that the amount of SA by weight that can be inserted in lipid bilayers seems to be greater than that of TG. Indeed, SA is a lipid with only one acyl chain (TG has three) and an important polar group with cationic charge. These properties could facilitate SA insertion in PL bilayers. Indeed, Dahim et al. observed that the insertion of methyl esters, resulting from esterification of TG, displayed an insertion rate within bilayers which is about 1 order of magnitude larger than that shown by the original TG (Dahim et al., private communication).

While the specific OD values observed at high SA and TG concentrations are similar, the comparison of specific OD values observed at high SA and TG concentrations indicates that much larger or more numerous aggregates (or both) are formed with SA.

Conclusion

The results show that addition of TG or SA to the initial composition of PL vesicles induces modifications of these aggregates up to limits that can be determined by the selective solubilisation technique of continuous detergent addition. At low concentrations of TG or SA, these molecules are solubilised in the initial vesicles and do not really modify these aggregates. The changes locally induced by the presence of molecules of packing parameters different from that of PL are not sufficient to modify the whole curvature of these aggregates. Beyond a certain concentration of TG or SA, a limit of solubility is observed and a second population of aggregates is formed next to the vesicles. This second population of excess aggregates, which is very probably emulsion droplets in the case of TG, has not yet been characterised for SA. Thanks to the high specific turbidity of this second population and to the fact it is not solubilised first by TX-100, its existence is easily evidenced by the time-resolved selective solubilisation technique. In addition, this selective solubilisation technique yields the molecular composition of the aggregates formed at boundary limits during the solubilisation of TG- or SA-containing PL bilayers. Indeed, the apparent solubility limits found might not correspond to absolute solubility limits found at equilibrium since vesicles are not aggregates in this thermodynamic state. The determination of molecular compositions of intermediate aggregates observed during the solubilization process of vesicles is THE specific interest of this time-resolved solubilisation technique, which works even for complex mixtures and in dynamic equilibrium conditions [6, 12].

Opposite behaviour was found for both additives: more detergent is needed to solubilise additional TG molecules of the vesicles (addition of 10% of TG to the lipid composition of the vesicles results in an increase of the molar ratio ($[\text{TX-100}]_C/[\text{PL}]_C$) from 2.4 to 3.1); less detergent is required to form mixed micelles in the presence of SA (the ratio drops from 2.4 to 1.9 in similar conditions). In other words, TG molecules can be considered as vesicle-structuring molecules while SA molecules are rather destructuring ones. The addition of a long-chain charged amphiphile while stabilising the vesicle suspension helps its solubilisation thanks to the surfactant character of SA [11, 13].

Urbaneja et al. [14] have already examined the solubilisation of SA containing egg PC vesicles by TX-100. It was observed that the presence of 10% of SA induced a decrease in the detergent-to-lipid concentration ratio at breakpoint C from 4.0 to 3.0 in agreement with our data. Both decreases can be compared taking into account the difference in phospholipid composition between the work of Urbaneja et al. and our work (pure egg PC and PL).

The shape of the time-resolved solubility profiles of the TG-containing aggregates is dominated by that of the vesicle solubilisation which superimposes on the turbidity of the emulsion droplets, increasing with their proportion but not displaying the accident of that of the former, as shown in Fig. 3. This dominant effect induces a large similarity of the solubilisation profiles whatever the PL and TG concentrations.

Such behaviour is not observed for SA-containing aggregates, which clearly exhibit a turbidity minimum at a SA-to-PL concentration ration of $8 \pm 1\%$ w/w and very different profiles as a function of SA concentration, as shown in Fig. 4. The exact origin of this minimum has not yet been determined. Beyond this minimum, a large turbidity increase is observed in the SA-containing aggregates populations that can be related to a change in size and different processes. The size increase could result from the aggregation/fusion process of SA-containing aggregates induced by lipid complex formation with mixed micelles. Indeed, Casals et al. [15] have shown that at 20% mol/mol, corresponding to almost 8% w/w, micelles of pure SA coexist with Lipoid S-100/SA liposomes. These positively charged micelles could aggregate with mixed micelles by interacting with an anionic lipid present in PL and form the highly turbid aggregates observed. Nevertheless, Korner et al. [16] have shown by measurements of surface properties that PL and SA strongly interact at their polar groups. This is not in favour with the presence of pure SA micelles separated from PC/SA vesicles. Moreover, SA is frequently used as a charged additive to PL for the stabilisation of liposome preparations thanks to the repulsion forces that it exerts [12, 17]. This is in contradiction with the aggregation phenomenon. In fact, except for very low concentrations, the exact location of SA in the lipid vesicles is not known. X-ray diffraction experiments are now under way to characterise PL/SA aggregates as well as the location and interactions with lipoid molecules.

References

1. Wehrle P, Korner D, Benita S (1996) Pharm Dev Techno 1:97–111
2. Ollivon M, Potier-Gumery L (1998) In:Grossiord JL, Seillier M (eds) Multiples emulsions:structures, properties and applications, Les éditions de la Santé, Paris, pp 195–222
3. Teixeira H, Dubernet C, Rosilio V, Benita S, Lepault J, Erk I, Couvreur P (2000) Pharm Res 17:1329–32
4. Hamilton JA, Small DM (1981) Biophysics 78:6878–6882
5. Dahim M., Gustafsson J, Puisieux F, Ollivon M (1998) Chem Phys Lipids 97:1–14
6. Ollivon M, Walter A, Blumenthal R (1988) Biochemistry 27:1695–1703
7. Kerker M (1969) The scattering of light. Academic, London
8. Mims M, Guyton JR, Morrisett JD, (1986) Biochemistry 35:474–483
9. Faure C, Tranchant JF, Dufourc EJ (1996) Biophys J 70:1380–1390

10. Hauser H, Gains N, Eibl HJ (1986) Biochemistry 25:2126–2134
11. Jonas AJ, Speller RJ (1989) Biochim Biophys Acta 984:257–261
12 Paternostre M, Meyer O, Grabielle-Madelmont C, Lesieur S, Ghanam M, Ollivon M (1995) Biophys J, 69: 2476–2488
13. McIntosh TJ, Magid AD, Simon SA (1990) Biophys J 57:1187–1197
14. Urbaneja MA, Alonso A, Gonzalez-Mañas JM, Goñi FM, Partearroyo MA, Tribout M, Paredes S (1990) Biochem J 270:305–308
15. Casals E, Soler M, Gallardo M, Estelrich J (1998) Langmuir 114:7522–7526
16. Korner D, Benita S, Albrecht G, Baszkin A (1994) Colloids Surf B 3:101–109
17. Armengol X, Estelrich J (1995) J Microencapsul 12:525–535

Progr Colloid Polym Sci (2004) 126: 14–20
DOI 10.1007/b93970

Klaus Lunkenheimer
Simon Schrödle
Werner Kunz

Dowanol DPnB in water as an example of a solvo-surfactant system: adsorption and foam properties

K. Lunkenheimer (✉)
MPI für Kolloid- und Grenzflächenforschung,
14424 Potsdam/Golm, Germany
e-mail: klaus.lunkenheimer@mpikg-golm.mpg.de
Tel.: +49-331-5679445
Fax: +49-331-5679502

S. Schrödle · W. Kunz
Institut für Physikalische und Theoretische Chemie, Universität Regensburg,
93040 Regensburg, Germany

Abstract Dowanols being chemically ethers of *n*-alkanols with bis(isopropylene glycol) represent characteristic properties of hydrotropic molecules or "solvo-surfactants", i.e. they have some surfactant properties in addition to typical solvent properties. The dynamic surface tension behaviour of aqueous Dowanol solutions ($c \leq 10^{-1}$ M) is a consequence of different surface-active components owing to the presence of different isomers and surface-active contaminants. By applying a special purification method ("flounder technique") we removed additional (non-Dowanol) surface-active trace impurities by aspirating them from the highly compressed adsorption layers. The various sucked off fractions reveal peculiar surface properties that are quite different from those of the original and of the "surface-chemically" pure solutions. Whereas the former show only short-time foam stability of a few seconds, the sucked-off fractions formed (meta)stabile foams lasting for a few hours. It is concluded that the Dowanol's solvo-surfactant characteristic is caused by the synergistic effect of various surface-active species rather than by the special structure of an individual component.

Keywords Solvo-surfactant · Hydrotrope · Dowanol · Purity · Adsorption at air/water interface

Motivation

Solvo-surfactants or hydrotropic detergents are interesting molecules because they combine properties of solvents with those of surfactants. For example, they allow one to solubilise hydrophobic organic compounds in water, partly by acting as a solvent and leading to a molecular solubilisation, partly by organising the liquid structure, leading to micelles, microemulsions and similar structures. Solvo-surfactants can play the role of solvents, cosolvents, cosurfactants or true surfactants, depending on their molecular composition and on temperature.

As solvo-surfactants have found many interesting applications, for example, they are very good cleaning agents, we would like to know if their properties are really brought about by their particular chemical structure or whether they might also be caused by a synergetic effect of different compounds.

Until now, mostly *n*-alkyl polyglycol ethers (C*i*E*j*) were studied in some detail [1], in particular, in mixtures with water and *n*-alkanes [2, 3, 4]. Their structures, especially of 2-butoxyethanol, were investigated by various scattering techniques [5]. In the present contribution our interest is focussed on a special type of solvo-surfactants, taken from the Dowanol P series (C*i*P*j*), the so-called DPnB, which is a bis(isopropylene glycol) butyl ether [6]:

$$C_4H_9O[CH_2CH(CH_3)O]_2H. \quad (1)$$

Note that C_iP_j are considered to be much less toxic than short-chain C_iE_j, so they will gain increasing interest in the future. DPnB is partially miscible with water at room temperature, the mixtures showing a lower critical point below 0 °C. The corresponding value of the shorter homologue DPnP (the propyl ether) is about 13 °C. Such a phase separation is typical for mixtures of hydrotropes with water. In some cases even closed loops with lower and upper critical points were found [5].

Experimental

High-performance purification

In order to investigate the surface-active properties of DPnB, we removed surface-active contaminants by applying automatic high-performance surfactant purification ("flounder technique" [7]). By this procedure the surface-active contaminants are removed from the surfactant solutions by repeatedly aspirating their highly compressed adsorption layers. The volume fractions aspirated were collected and the dynamic surface tension and the foam behaviour of the original, the "surface-chemically pure" (scp) solutions and the sucked-off fractions were investigated. The principle of the surface-chemical purification is illustrated in Fig. 1. In position 1 the solution is kept in the flat glass container such that its surface area is comparatively large (about 215 cm^2). This position is maintained until the solution has practically established its equilibrium surface tension value. Thereafter the apparatus is tilted into position 2. Because of this the solution's surface is drastically reduced to the area of the small neck of the container (about 7 cm^2). The thus compressed adsorption layer is aspirated by means of a small platinum capillary. Having done this, we turned the apparatus back into position 1. This cyclic procedure is repeated automatically until the surface-active contaminants have been completely removed from the original solution, i.e., the state of "surface-chemical purity" has been achieved.

Surface tension measurement

Surface tension measurements were performed with the du Noüy ring tensiometer taking into account necessary modifications for the measurement of surfactant solutions [8]. The measurements were performed at 295 K.

Foamability measurement

Foamability measurements were performed by a simple Bartsch test. This procedure consists of gently shaking equal amounts of solution (7 ml) in an upright glass cylinder (15-cm height, 1.5-cm inner diameter) and measuring the height of foam with respect to time.

Theoretical evaluation

The experimental equilibrium surface tension, σ_e, versus concentration, c, isotherms were evaluated by the two-state approach to the surface equation of state put forward in Ref. [9, 10]. The basic idea of this approach is that the amphiphile's surface conformation does not remain unaltered during adsorption, but changes gradually between (at least two) different surface conformations. At low surface coverages conformations

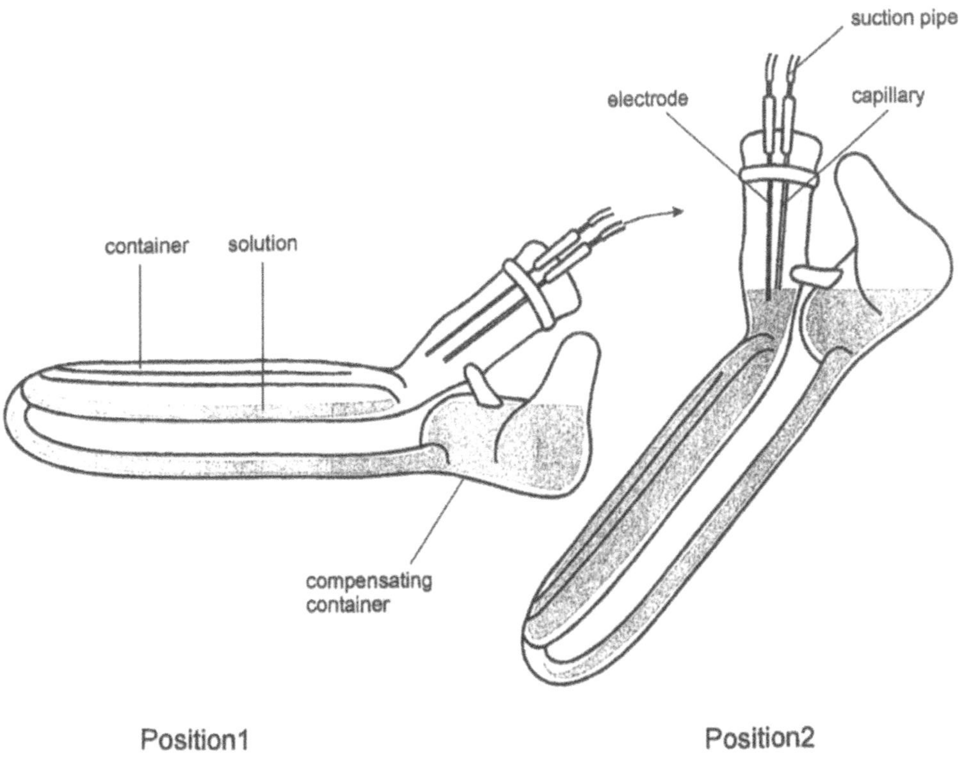

Fig. 1 Schematics of the high performance purification apparatus for surfactant solutions a) Position of the container in the state of adsorption: large surface area, b) container position when the adsorption layer is sucked off by a fine capillary: highly compressed adsorption layer

requiring larger surface areas are favoured, whereas at higher concentrations smaller cross-sectional area conformations are favoured. Thus, the experimental isotherm $\sigma_e(c)$ is described as

$$\Delta\sigma_e = \alpha\Delta\sigma_{e,1} + (1 - \alpha)\Delta\sigma_{e,2}, \quad (2)$$

where $\Delta\sigma_e = \sigma_W - \sigma_e$ stands for surface pressure with σ_W being the surface tension of pure water and σ_e being the equilibrium surface tension of the surfactant solution. $\alpha(c)$ describes the transition function changing monotonically between $0 \le \alpha \le 1$ [9, 10]. $\Delta\sigma_{e_i}$ denotes the adsorption equation of Henry, Langmuir (*i* "ideal surface behaviour") and Frumkin ("regular surface behaviour"), respectively.

Results and discussion

The dynamic surface tension of the original solutions is characterised by a long time-dependence and periods with certain induction intervals (Fig. 4). A detailed gas chromatography/mass spectrometry analysis reveals the presence of eight isomers as shown in Fig. 2 for DPnP. A corresponding analysis of DPnB shows a very similar composition. So the commercial product is quite pure in so far as only Dowanol molecules are present, but the different isomers can indeed have quite different surface properties, owing to their slightly different geometry and polarity.

Figure 3 shows the purification characteristics of the flounder technique, i.e., the dependence of the equilibrium surface tension value, σ_e, on the number, j, of purification cycles of three different apparatus. In agreement with the thermodynamic requirement, the surface tension values of the scp solutions are greater than those of the original solutions, being identical for the different apparatus. The slope $(d\sigma_e/dj)_{p,T} = 0$ proves the surface-chemical purity of the solutions [11].

The dynamic surface tension behaviour of the sucked-off fractions at increasing stages of purity is presented in Fig. 4. The solutions of the fractions $j = 100$, 200 and 300 have relatively low surface tension values, indicating enrichment of strongly surface active material. The considerably higher surface tension values and the shorter dynamic surface tension change for the final solution ($j = 377$) prove that surface-active impurities had been removed. Interestingly, within the dynamic surface tension, $\sigma(t)$, intervals of different characteristic courses are observed for $j \le 300$ cycles, hinting at the presence of not only one but of several trace impurity components.

The most remarkable feature in the dynamic surface tension is found for the solutions of successively increasing purity. There are also characteristic periods like those observed with the sucked-off fractions, but in these solutions the surface tension always increases with time up to 1 mN/m. This dynamic surface tension behaviour is shown in Fig. 5. In this figure the dynamic surface tension of the original, as-received solution is also shown. Its behaviour is opposite to that of the former solutions.

Evaluating the entire σ_e versus log c isotherm (Fig. 6), we achieved a good fit only if we applied the two-state approach to the surface equation of state (see earlier). However, using the two-state approach resulted in the same quality of the best fit whether it was applied as a Henry–Langmuir or as a Langmuir I–Langmuir II set (standard deviation $s = \pm 0.18$ mN/m). The corresponding cross-sectional areas of the adsorbate's surface conformations differ slightly, amounting to 56.0 and 52.4 Å^2/molecule, respectively. The difference of their standard energy of adsorption is 1.7 kJ/mol. This difference is characteristic of amphipiles' conformational changes [12]. The interesting feature of the curve-matching is that the transitional interval is very broad, covering almost the entire σ_e versus log$c$ isotherm. Behaviour like this was only observed for definite, SCP n-alkyloxypropylene oligomers [13].

Concerning the adsorption kinetics of the SCP Dowanol solutions, it should be noted that the surface tension increases slightly with time, and that the equilibrium is established after a rather long adsorption time only. Usually the adsorption kinetics of ordinary surfactants obeys a diffusion-controlled mechanism, i.e., for surfactant concentrations $c \ge 10^{-3}$ M the adsorption equilibrium ought to be established in the order of magnitude of milliseconds [14, 15].

Thus, summarizing, there are two unusual features of the scp solutions of the solvo-surfactant Dowanol DPnB: first, an unsual increase of dynamic surface tension during adsorption – although a small effect only – and, second, the comparatively slow establishment of the adsorption equilibrium. The reason for this is probably the presence of the different isomers. The isomers' surface parameters will differ from each other only slightly. Thus, for example, it is imaginable that the main isomer having a smaller cross-sectional area, which is adsorbed first, can be substituted to a certain extent by another isomer present in rather low bulk concentration but having slightly stronger surface activity and a larger cross-sectional area. This might result in a small increase of the solution's equilibrium surface tension.

It is also imaginable that some conformational changes of the main component in the adsorption layer are responsible for an effect like this. However, the relatively long adsorption time for establishing the equilibrium of the scp solution hints rather at an effect similar to that of a trace impurity.

In Fig. 7 the height of the foam is plotted against time for the original, a sucked-off fraction and the purified solution. The foam behaviour of these solutions is quite different. Thus, the original (as-received) and the scp solutions showed some short-time

Fig. 2 Eight isomers of bis-isopropyleneglycol butyl ether (DPnP), identified by GC/MS analysis

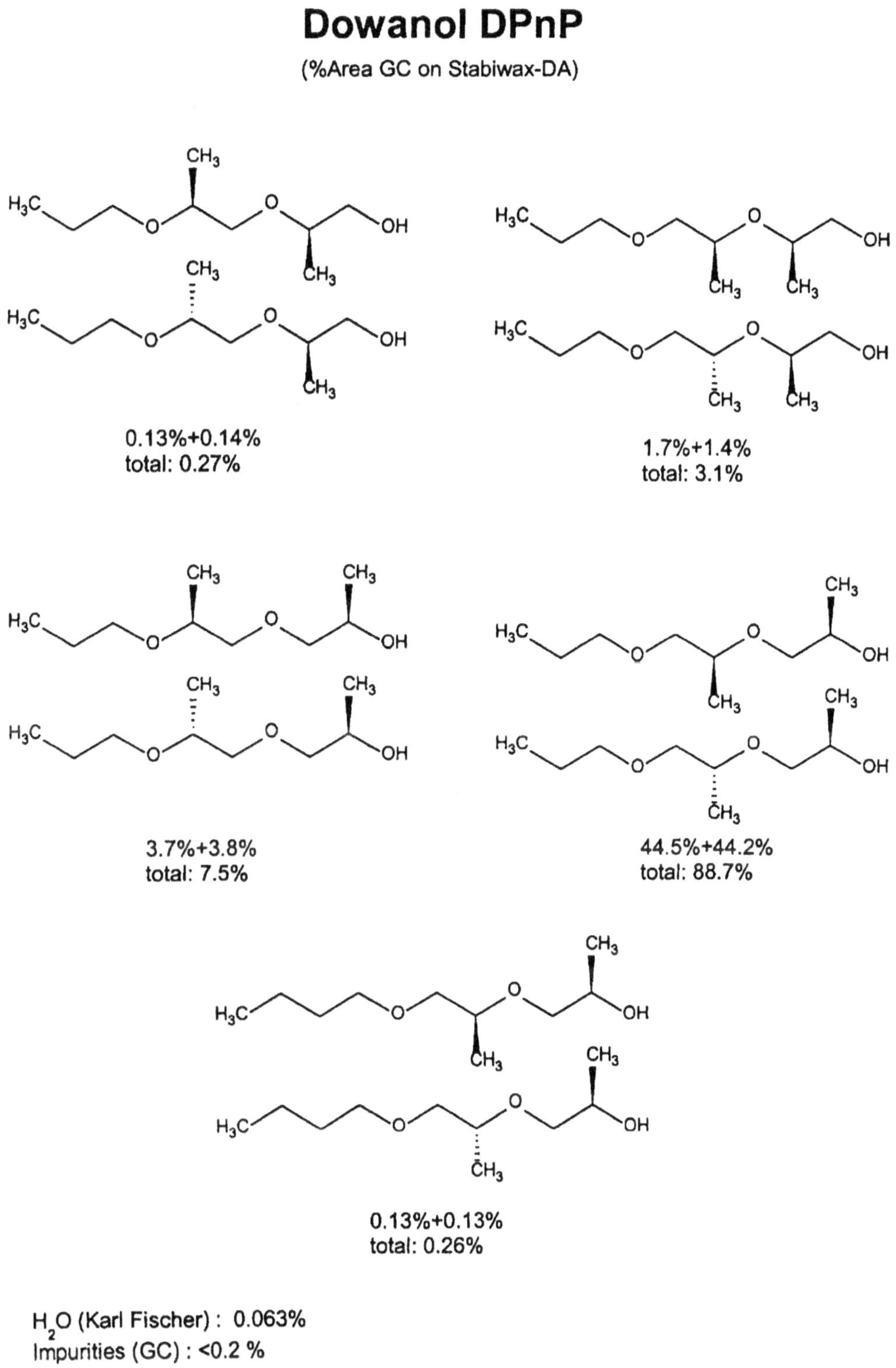

foamability although the residual foam of the scp solution persists a little longer (note the logarithmic time scale). Behaviour like this is characteristic of "wet foam" systems, for example, of aqueous solutions of soluble *n*-alkanols or *n*-alkanoic acids [16]. But unlike these, the sucked-off solutions revealed long foam stability lasting 2 h and longer. Such behaviour is characteristic of "dry foams" obtained with typical surfactant solutions forming stable foams [16]. The quite different foam behaviour between these solutions is so big that it cannot be explained by the presence of related isomers of the main component. Thus, the

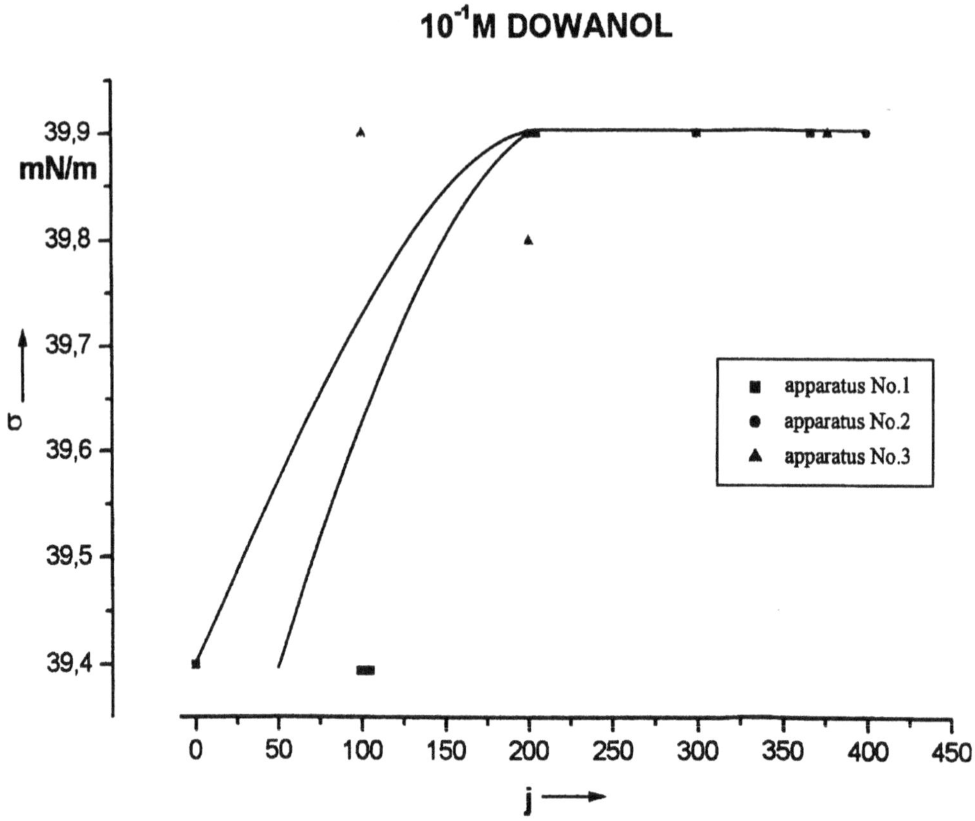

Fig. 3 Purification characteristics of the purification apparatus ("flounder technique"): equilibrium surface tension value, σ_e, in dependence on the number j of the purification cycles of three different apparatus; stock solution: 1×10^{-1} M Dowanol

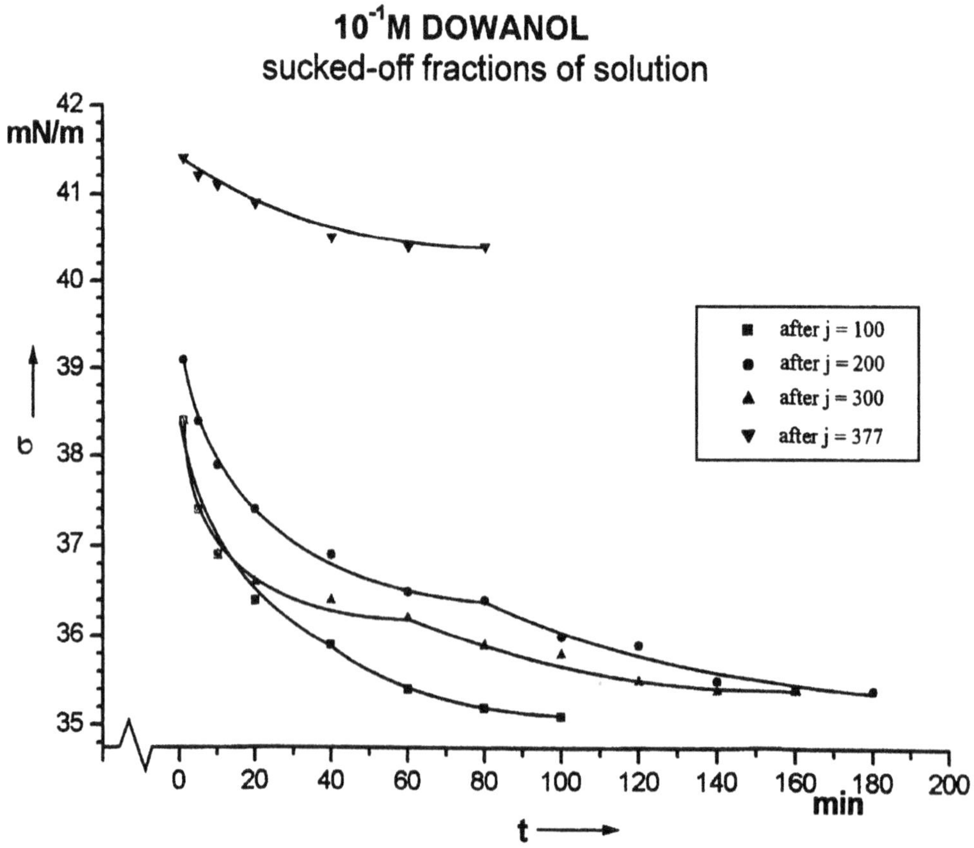

Fig. 4 Dynamic surface tension behaviour of the sucked-off fractions of 1×10^{-1} M Dowanol solution at increasing grade of purity. j – number of automatic purification cycles

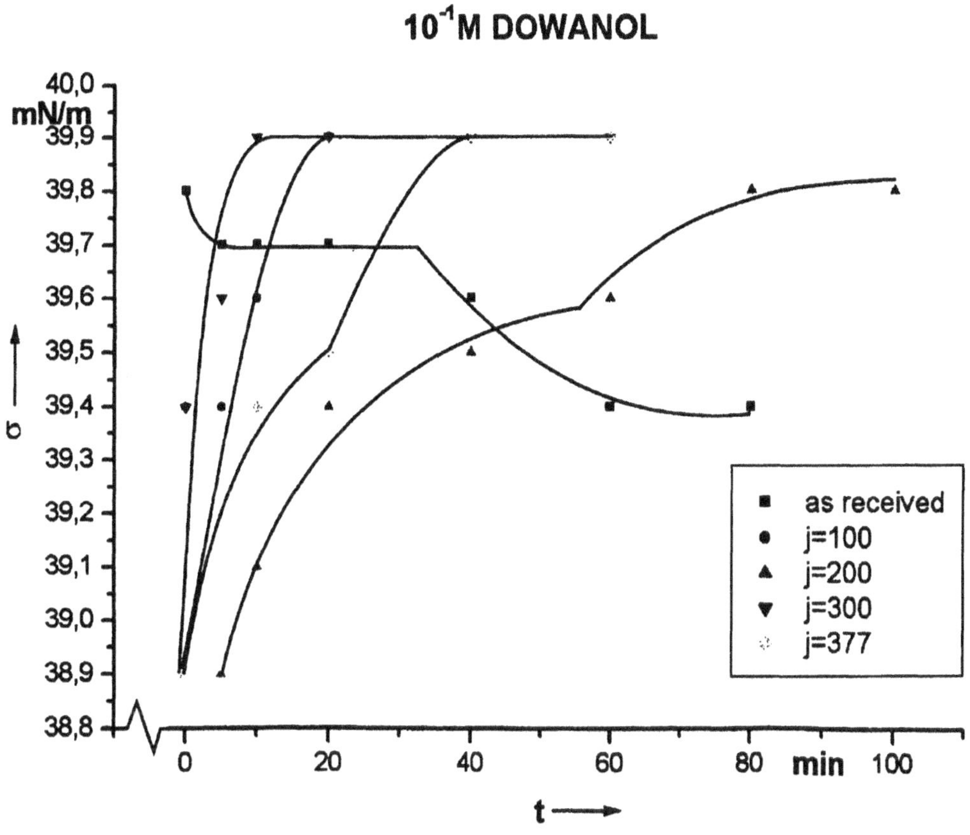

Fig. 5 Dynamic surface tension behaviour of the Dowanol solutions under purification at successively increasing states of purity. Dynamic tension of the original "as-received" solution is also given. ■ ■ ■ as received; ● ● ● after $j = 100$; ▲ ▲ ▲ after $j = 200$; ▼ ▼ ▼ after $j = 300$; ◆ ◆ ◆ after $j = 377$

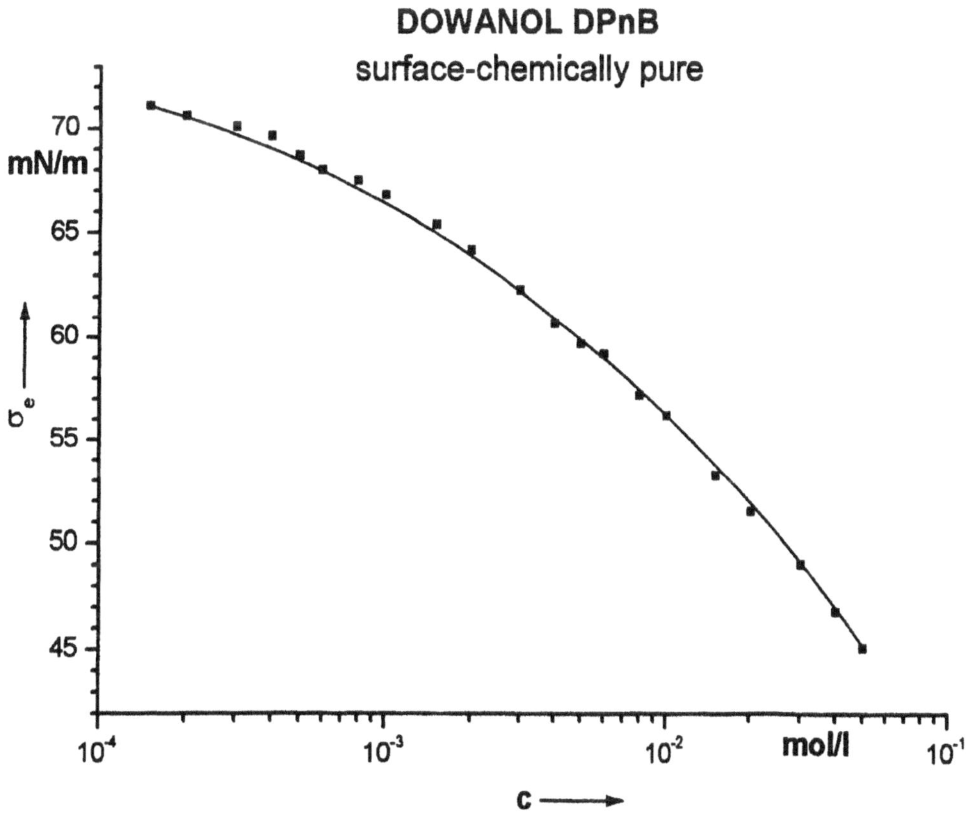

Fig. 6 Equilibrium surface tension vs. concentration isotherm of the surface-chemically pure solution of Dowanol DPnB

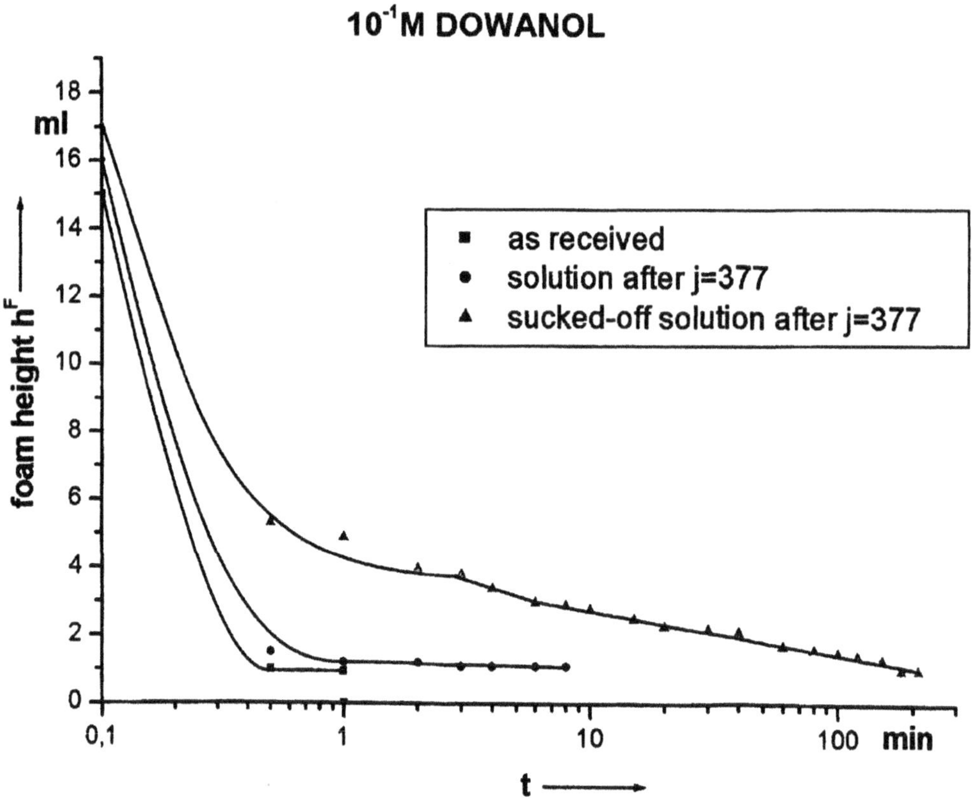

Fig. 7 Height of foam column in dependence on logarithm of time for the stock solution of Dowanol DPnB: a) original, b) a sucked-off fraction, and c) the surface-chemically pure solution

original Dowanol DnPB must also contain (a) typical surfactant(s) of relatively strong surface activity, at least in trace concentration(s), which provide(s) the sucked-off solutions in which it (they) is (are) preferentially enriched with typical (dry foam) behaviour. However, for the moment it was not possible to identify this substance.

Conclusions

The aqueous solutions of the solvo-surfactant Dowanol DnPB possess certain interesting adsorption properties. Although we do not yet understand all peculiarities well, these results indicate that the label Dowanol contains (at least) traces of various surface-active substances of different properties. It seems that these impurities together with the presence of eight different isomers give rise to the versatile properties of this hydrotrope in addition to the chemical composition of DPnB.

Acknowledgements The authors thank Dr. E. Eibler and the staff of the central analytical facilities (Regensburg University) for the mass spectra of the Dowanols. S.S. is thankful to the "Verband der chemischen Industrie e.V. (VCI)" for his grant.

References

1. Schulz J, Bowers GH, Findenegg GH (2001) J Phys Chem B 105:6956
2. Burauer S, Sachert T, Sottmann T, Strey R (1999) Phys Chem Chem Phys 1:4299
3. Mitrinovic DM, Williams SM, Schlossmann ML (2001) Phys Rev E 63:1
4. Yamashita F, Perron G, Desnoyers JE, Kwak JCT (1986) ACS Symp Ser 311:79
5. Quirion F, Magid LJ, Drifford M (1990) Langmuir 6:244
6. Product Information Dowanol DPnB of the Dow Chemical Company, Midland, Michigan 48674 USA of April 1997
7. Lunkenheimer K, Pergande H-J, Krüger H (1987) Rev Sci Instrum 58:2313
8. Lunkenheimer K, Wantke K (1981) Colloid Polym Sci 259:354
9. Lunkenheimer K, Hirte R (1992) J Phys Chem 96:8683
10. Hirte K, Lunkenheimer K (1996) J Phys Chem 100:13786
11. Lunkenheimer K, Miller R (1987) J Colloid Interface Sci 120:176
12. Vold MJ (1984) J Colloid Interface Sci 100:224
13. Lunkenheimer K, Holzbauer H-R, Hirte R (1994) Prog Colloid Polym Sci 97:116
14. Fang J-P, Wantke K-D, Lunkenheimer K (1995), J Phys Chem 99:4632
15. Chang C-H, Franses EI (1995) Colloids Surf A 100:1
16. Exerowa D, Kruglyakov PM (1998) Foam and foam films: theory, experiment, application. Elsevier, Amsterdam

Progr Colloid Polym Sci (2004) 126:21–24
DOI 10.1007/b93966

N.M. Gosselet
A.M. Layre
V. Wintgens
C. Amiel

Physicochemical study of modified dextrans with a β-cyclodextrin polymer

N.M. Gosselet (✉) · A.M. Layre
V. Wintgens · C. Amiel
CNRS, U.M. C7581, Laboratoire de Recherche sur les Polymères,
2–8 rue Henri Dunant,
94320 Thiais, France
e-mail: gosselet@glvt-cnrs.fr,
Tel.: +33-1-49781229,
Fax: +33-1-49781208

Abstract The aim of this work was the design of a pH-sensitive system consisting of an amphiphilic polymer and a β-cyclodextrin coepichlorhydrin water-soluble polymer (pβCD/EP). Different modified dextrans were synthesized from reaction of dextran ($M_w = 40{,}000$) with phthalic anhydride, 2-phenylglutaric anhydride and 1,2-cyclohexane dicarboxylic anhydride with grafting ratio around 2%. The viscometric study of modified dextrans with pendant cyclohexane carboxylic acid groups mixed with pβCD/EP was performed at pH 1 and 7 and showed a small increase of reduced viscosity compared to mixtures of dextran and pβCD/EP. The phase diagram shows association phase-separation at low pH, whereas monophasic mixtures were obtained at pH higher than the pK_a. Furthermore this phase-separation is reversible according to the pH. The use of 8-anilinonaphtalene-1-sulfonic acid, as a fluorescent probe, allowed us to determine the inclusion constants of the modified dextrans. Complexation constants higher by 1 order of magnitude for the most hydrophobically modified dextrans were obtained at pH 2.5 compared to pH 7.

Keywords Associative polymers · β-Cyclodextrin polymer · Inclusion · Dextran

Introduction

Associating polymers can bind together through interactions such as hydrogen bonding [1], ionic associations [2] or hydrophobic interactions [3]. These associating polymers can lead to high viscosities. A new concept of associating polymers was recently developed [4] by using complementary cyclodextrin polymers and hydrophilic guest polymers bearing hydrophobic groups (1–5%). The aim of this work was to elaborate a pH-dependant system by incorporating on the dextran chains hydrophobic moieties whose affinity for a β-cyclodextrin (βCD) cavity can be triggered by the pH. In this paper, the first part is devoted to the modification of dextran with phthalic anhydride (DM2), 2-phenyl glutaric anhydride (DM7) and 1,2-cyclohexane dicarboxylic anhydride (DM4) and the second part is devoted to a physicochemical study of these modified dextrans alone or in the presence of βCD coepichlorhydrin water-soluble polymer (pβCD/EP).

Experimental

Dextran modification

Dry dextran T40 (Table 1) (5 g) was first solubilized in dimethylformamide (125 ml) in the presence of dry LiCl (1.25 g) at 80°C for 1.5 h. An excess of anhydride (1.1×10^{-2} mol) and an equimolar quantity of pyridine were added. The reaction occurred at 50°C for 4 h. Modified dextran was then precipitated with 2-propanol and dried. It was then purified by dialysis with a regenerated cellulose membrane of 1,000 cutoff for 48 h. The solution was then freeze-dried.

Table 1 Characteristics of polymers determined by size-exclusion chromatography and light scattering. β-Cyclodextrin coepichlorhydrin water-soluble polymer (*pβCD/EP*)

Sample	M_n	M_w	Cavities (wt%)
Dextran T40	29,000	42,800	
pβCD/EP1	110,000	160,500	59
pβCD/EP2	6,130	14,920	76.4

pβCD/EP synthesis

The pβCD/EP was synthesized by polycondensation with epichlorhydrin in alkaline media, according to the method of Renard et al. [5]. The polymer was then purified by ultrafiltration on a regenerated cellulose membrane of 30,000 or 1,000 cutoff according to the sample and freeze-dried. The characteristics are reported in Table 1.

Experimental measurements

Fluorescence measurements are done with an SLM Aminco 8100 fluorimeter. The excitation and emission wavelengths were fixed at 340 and 460 nm, respectively. IR spectra were recorded using a PerkinElmer 1760 IR Fourier transform spectrometer. ^{1}H NMR spectra were recorded using a Brucker 200 MHz instrument. Viscometry measurements were done with a Ubbelhode viscometer at 25±0.1 °C. The polymer solutions were prepared and stirred 24 h before the measurements. The solutions were filtered prior to the measurements.

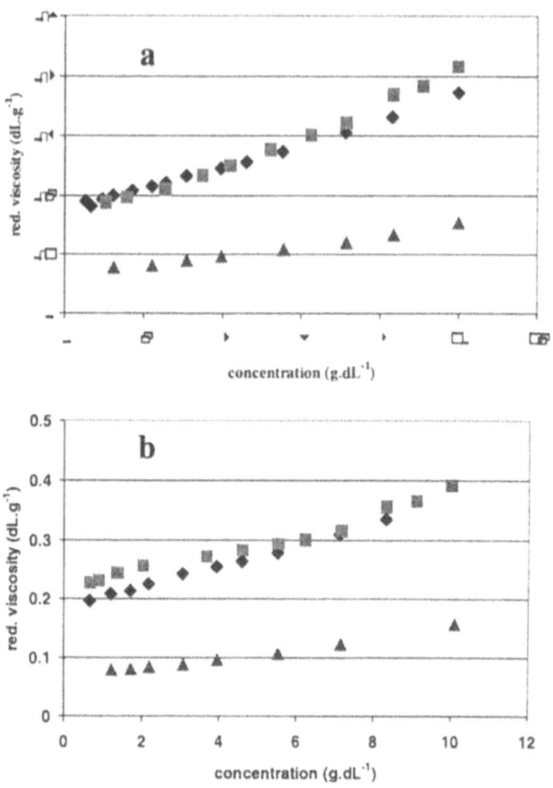

Fig. 1 Variation of reduced viscosity of dextran (*diamonds*), dextran modified with 1,2-cyclohexane dicarboxylic anhydride (*DM4*) (*squares*) and β-cyclodextrin coepichlorhydrin water-soluble polymer (*pβCD/EP1*) (*triangles*) versus concentration at **a** pH 1 and **b** pH 7

Results

The IR spectra of modified dextrans show an absorption band characteristic of the ester function at 1,715 cm^{-1} for DM2, at 1,725 cm^{-1} for DM4 and at 1,722 cm^{-1} for DM7. Modification of dextran with anhydride results in carboxylic pendant groups on the dextran chains. The pK_a values are 3.9, 4.8 and 4.77 for DM2, DM4 and DM 7, respectively. The ^{1}H NMR spectra allowed the determination of the modification rates of these dextrans, which are around 3, 2 and 2.25% (mole) for DM2, DM4 and DM7, respectively.

The viscometric properties of dextran, DM4 and pβCD/EP1 at pH 1 are reported in Fig. 1a. DM4 shows a viscosity similar to that of unmodified dextran. This is an indication that the grafted groups do not have a strong tendency to associate in hydrophobic microdomains. At pH 7 (Fig. 1b) the negative charges carried by DM4 lead to a reduced viscosity higher than that of dextran at low concentration. At higher concentration, this polyelectrolytic behavior disappears as a consequence of the overlapping of chains. The viscometric study of mixtures of DM4 and pβCD/EP, 80/20 w/w was possible in the domain 0–30 g l^{-1} total concentration without separation of phases at all pH. In these mixtures, the ratio of dextran pendant groups to cyclodextrin cavities is close to 1. The results at pH 1 (Fig. 2a) show slightly higher reduced viscosities for mixtures of DM4 and pβCD/EP1 than for mixtures of dextran and pβCD/EP, 80/20, in the same domain of concentrations, as a consequence of association properties between both polymers through the inclusion of hydrophobic groups in the βCD cavities. Interaction between the two polymers may lead to higher viscosities in the concentration range close to the overlap concentration of the chains (60 g l^{-1}); however, viscosity measurements were not performed in this range since the mixtures phase-separate (see later). The values of the reduced viscosity of mixtures of DM4 and pβCD/EP1 at pH 7 (Fig. 2b) are higher than those obtained for mixtures of dextran and pβCD/EP1 owing to the addition of two different effects: low viscosity increase due to polymer–polymer associations and swelling due to the presence of charges. However the observed effects are weak probably because of complexation constants of the hydrophobic groups in the βCD cavities that are too low. At low pH, the polymers are neutral and the viscosity behavior can be mainly attributed to the inclusion complexes. At pH 7, the DM4 polymer is negatively charged and the

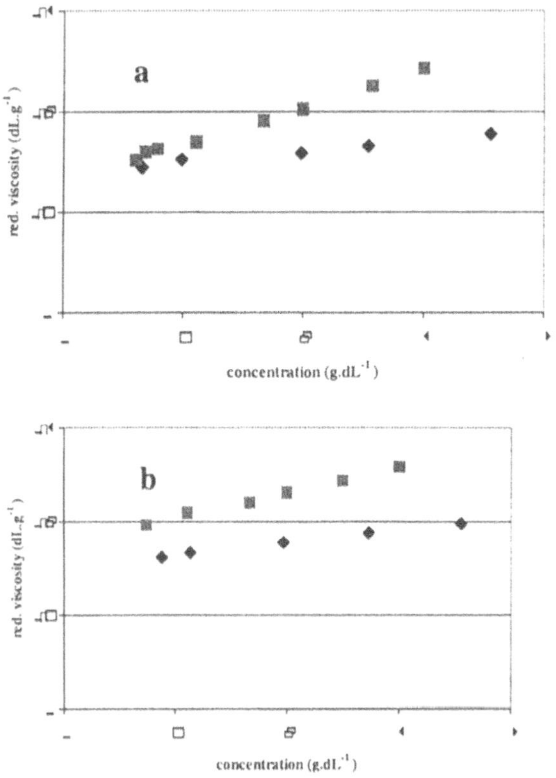

Fig. 2 Variation of reduced viscosity of mixtures of dextran (*diamonds*) or DM4 (*squares*) and pβCD/EP1 in the proportion 80/20 (w/w) at **a** pH 1 and **b** pH 7

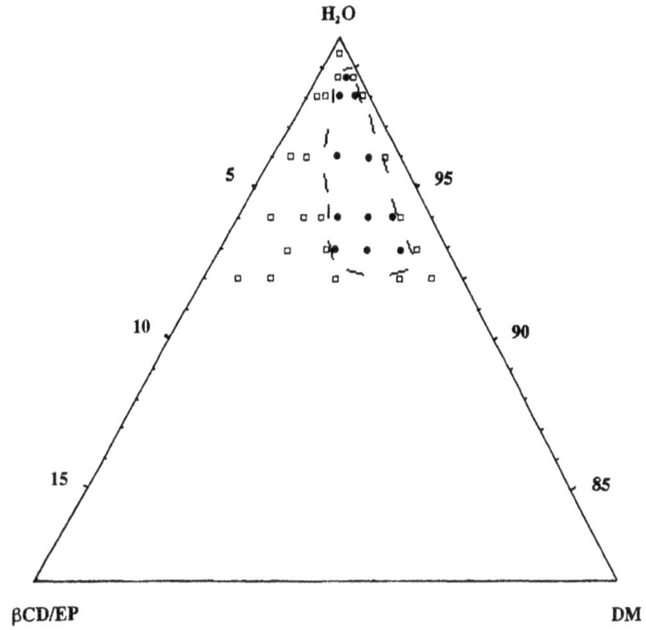

Fig. 3 Phase diagram of mixtures of DM4 and pβCD/EP1 at pH 1: one phase (*squares*); two phases (*circles*)

Table 2 Values of inclusion constants in pβCD/EP2

Sample	K at pH 2.5	K at pH 7
ANS	2,155	3,100
DM2	100	50
DM4	1,575	95
DM7	2,515	245

swelling of the polymer–polymer complexes can compensate for lower affinity of the hydrophobic groups for the cavities.

We have observed phase separations at acidic pH, until pH 5.8 for mixtures of DM4 and pβCD/EP1. At higher pH, the mixtures are monophasic. This phase separation is reversible according to the pH: increasing the pH to values higher than 6 in phase-separated mixtures results immediately in clear monophasic solutions. The phase diagram at pH 1 is shown in Fig. 3 and indicates a closed domain of phase separation, present at low polymer concentration and located around the stoichiometry one dextran pendant group for one βCD cavity. This diagram is typical of an associative phase separation where the two polymers are experiencing attractive interactions. These are the result of the inclusion of hydrophobic groups in the βCD cavities.

The attractive interactions between the modified dextrans and the pβCD/EP were also characterized by fluorescence spectroscopy, leading to the determination of the complexation constants between the βCD cavities and the hydrophobic groups. We used a method of competition for the inclusion in the βCD cavities, the competitor 8-anilinonaphtalene-1-sulfonic acid (ANS) being the fluorescent probe. After determining the association constant, K_1, of ANS with pβCD/EP2 using the Benesi–Hildebrand [6] relation, the association constants of DM2, DM4 and DM7 were deduced from the decrease of the fluorescence intensity by adding DM2, DM4 or DM7 to a solution of βCD/EP and ANS at fixed concentration (4.64×10^{-4} mol cavities l^{-1} and 2×10^{-5} mol l^{-1}, respectively). Low molecular weight pβCD/EP2 was used in these experiments in order to reduce the light scattered from the samples. The measurements were done at two pH: 2.5 and 7, which are well below and above respectively the pK_as. The results are reported in Table 2. In all cases the inclusion constants are higher at acidic pH than at neutral pH, in good correlation with the associative phase separation observed at low pH only. DM2 does not show a high affinity for pβCD/EP at all pH. For DM4 and DM7, the association constant at low pH is about 10 times the association constants at low pH. This finding is in good agreement with literature results [7] for the inclusion of nonpolymeric guests into cyclodextrins: ratios of the complexation constants of the protonated/unprotonated guest of the order of 10 are generally reported in the literature.

Conclusion

In this study, amphiphilic dextrans were designed in order to trigger their interaction with βCD polymers as a function of pH. Three kinds of hydrophobic groups containing carboxylic acid function were grafted on dextrans. These modified dextrans associate with βCD polymers by making inclusion complexes between the dextran hydrophobic moieties and the βCD cavities. DM2 interacts only weakly with pβCD/EP. At low pH and for DM4, these attractive interactions are characterized by a small thickening behavior at low concentration and by an associative phase separation at higher concentration. Increasing the pH above the pK_a of the modified dextrans results in monophasic mixtures, owing to the reduced interactions between the two polymers. The charged hydrophobic moieties have an affinity for βCD cavities about 10 times lower than at low pH as shown by fluorescence measurements.

References

1. Leibler L, Rubinstein M, Colby RH (1991) Macromolecules 24:4701.
2. Semenov AN, Nyrkova L, Khokhlov A (1996) In: Schlik S (ed) Ionomers: characterization, theory and applications. CRC, Boca Raton, pp 251–279
3. Glass JE (ed) (1996) Hydrophilic polymers: performance with environmental acceptance. ACS advances in chemistry series 248. American Chemical Society, Washington, DC
4. Amiel C, Moine L, Sandier A, Brown W, David C, Hauss F, Renard E, Gosselet NM, Sébille B (2001) In: Mc Cormick CL (ed) ACS symposium series 780. American Chemical Society, Washington, DC, pp 58–81
5. Renard E, Deratani A, Volet G, Sébille B (1997) Eur Polym J 33:49
6. Benesi HA, Hildebrand J (1949) J Am Chem Soc 71:2703
7. Junquera E, Aicart E (1997) J Inclusion Phenom 29:119

Progr Colloid Polym Sci (2004) 126: 25–29
DOI 10.1007/b93969

Durga P. Acharya
Eiko Oshimura
Kazutami Sakamoto
Hironobu Kunieda

The lyotropic ribbon phase and its transformation to the lamellar phase in the potassium *N*-dodecanoyl-DL-alaninate/water/decanol system

D. P. Acharya · H. Kunieda (✉)
Graduate School of Environment and Information Sciences,
Yokohama National University,
79-7 Tokiwadai, Hodogaya-ku,
Yokohama 240-8501, Japan
e-mail: kunieda@ynu.ac.jp
Tel.: +81-45-3394190;
Fax: +81-45-3394190

E. Oshimura · K. Sakamoto
Aminoscience Lab., Ajinomoto Co.,
Inc., 1-1 Suzuki-cho,
Kawasaki 210-8681, Japan

Abstract We have studied the variation in the aggregate size, shape and the unit cell of the rectangular ribbon (R_1) phase formed between the hexagonal (H_1) and the lamellar phase with water defects ($L_\alpha{}^H$ phase) in the potassium *N*-dodecanoyl-DL-alaninate/water/1-decanol system. With an increase in surfactant concentration at roughly constant decanol concentration, the aspect ratio, ρ, of the aggregate increases, whereas the rectangular unit cell shrinks, suggesting a decreasing separation between the ribbonlike aggregates. Similarly increasing decanol concentration at fixed surfactant-to-water ratio increases the unit cell width and ρ. The small-angle X-ray scattering results establish an epitaxial relationship between the (11) plane of the R_1 phase and the (10) plane of the $L_\alpha{}^H$ phase. This suggests that with successive addition of decanol, the ribbons of the (11) plane of the R_1 phase come closer and partially merge along their length, forming a surfactant bilayer pierced with water defects ($L_\alpha{}^H$ phase).

Keywords Potassium *N*-dodecanoyl-DL-alaninate · Acyl amino acid surfactant · Rectangular ribbon phase · Defective lamellar phase · Lyotrophic liquid crystal

Introduction

Long-chain alcohols are well-known for their ability to modify the interfacial curvature in the aggregates of aqueous ionic surfactants. Addition of decanol to the hexagonal (H_1) phase of anionic surfactant/water systems decreases the interfacial curvature and induces the formation of lamellar (L_α) phases [1, 2], sometimes via other liquid-crystal phases [3, 4]. When the surfactant molecules in the cylindrical aggregates in the H_1 phase are progressively substituted by decanol molecules, the circular cross-section of the cylinders undergoes a structural deformation, giving rise to a new phase consisting of rodlike aggregates with somewhat elliptical cross-section with no uniform interfacial curvature. In this liquid-crystal phase, called the rectangular ribbon (R_1) phase, these "ribbonlike" aggregates are organized in a two-dimensional rectangular lattice.

Recently, we observed the R_1 phase located between the hexagonal and lamellar phases in the phase diagram of potassium *N*-dodecanoyl-DL-alaninate (DL-KDA)/water/decanol systems at 25 °C, as shown in the phase diagram in Fig. 1. There are several reports of the formation of the R_1 phase in surfactant/water binary systems [5, 6]; however, we are not aware of any report of the R_1 phase in other surfactant/water/decanol systems except in the system of sodium dodecyl sulphate [4, 7].

With successive addition of decanol to the R_1 phase in the DL-KDA/water/1-decanol system, a defective lamellar ($L_\alpha{}^H$) phase having surfactant bilayers pierced by water holes or "defects" is formed, as shown in the phase diagram (Fig. 1). In this phase, the surfactant bilayers are not completely flat, rather they have depressions due to water defects [8] which are not correlated within single as well as within neighbouring bilayers. With successive addition of decanol, the water holes progressively

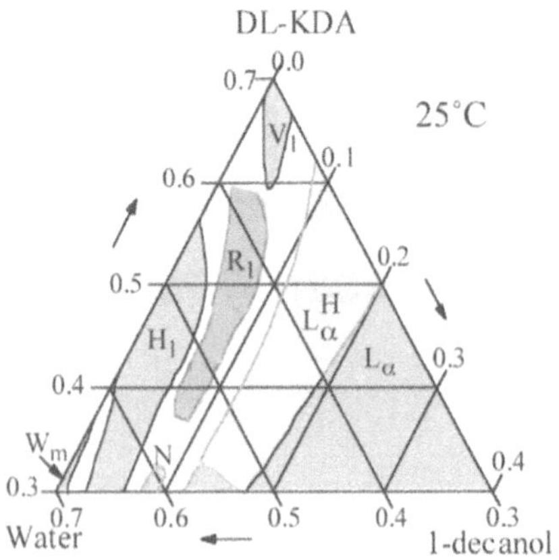

Fig. 1 Partial phase diagram of the potassium *N*-dodecanoyl-DL-alaninate (*DL-KDA*)/water/1-decanol system at 25 °C. The phase symbols are as follows: micellar (W_m), hexagonal (H_I), bicontinuous cubic (V_I), nematic (*N*), defective lamellar (L_α^H) and lamellar (L_α)

disappear and the conventional L_α phase is formed, as shown in the phase diagram (Fig. 1). The $L_\alpha{}^H$ phase has been observed in sodium dodecyl sulphate/water/decanol systems [9, 10] and in a number of binary systems of ionic and nonionic surfactants [6].

In this paper we discuss the variation of size and aspect of the rectangular unit cell and the aggregates of the R_I phase with composition in the DL-KDA/water/decanol system. Calculation of the aspect (or axial) ratio of the aggregates is based on the "hexagon-rod model" proposed by Hagslatt et al. [5]. We also present a brief discussion on the $R_I \rightarrow L_\alpha^H$ phase transformation induced by decanol in the DL-KDA system on the basis of the variation of aggregate size and the small-angle X-ray scattering (SAXS) patterns.

Experimental

Materials

N-dodecanoyl-DL-alanine (purity greater than 99%) was received from Ajinomoto Co. Its potassium salt, abbreviated as KDA, was used as a surfactant in this study. Potassium hydroxide (general reagent) and standard (5 M) hydrochloric acid (Junsei Chemical) and 1-decanol (99%) (Aldrich) were also used. All the reagents were used as received, without further purification. Carbon dioxide free Millipore-filtered water was used to prepare samples and solutions.

Preparation of samples

Aqueous KOH, standardized by titration against 5M HCl, was used to neutralize the amino acid. To ensure the complete neutralization of amino acid, a slight excess (about 2% of the stoichiometric amount) of KOH was taken. Required amounts of reagents were taken in glass ampoules with narrow constriction. The ampoules were flame-sealed and the sample mixtures were homogenized by repeated centrifugation. The samples were left in a water bath at 25 °C for several weeks.

SAXS measurement

SAXS measurements at 50 kV, 300 mA, (18 kW Rigaku rotating anode generator RINT 2500) were carried out to determine the interlayer spacing of liquid crystals. For SAXS measurement, the samples were covered with plastic films (Mylar method).

Results and discussion

Variation of the aggregate shape and size with composition

The first two peaks of the SAXS pattern of the R_I phase are very intense, and are followed by weak peaks [3, 5, 11]. In the present system, the most probable symmetry seems to be *pgg* with a rectangular unit cell. However, the intensities of the peaks observed for higher *q* (scattering vector) values are not sufficient to completely rule out the *cmm* symmetry [3]. The unit cell dimensions *a* and *b* are derived from *q* values of the first two peaks ($hk = 02$ and 11) of the SAXS pattern using Eq. (1):

$$q_{hk} = 2\pi\left(\frac{h^2}{a^2} + \frac{k^2}{b^2}\right)^{1/2}. \tag{1}$$

Since the unit cell parameters are the same whether the structure has *pgga* or *cmm* symmetry, uncertainty in the exact symmetry does not introduce any error in the calculation of the size of the aggregates. The values of *a* and *b* obtained from the SAXS data and the volume fraction of the lipophilic part of the aggregate, ϕ_{hc}, give the total cross-sectional area of the lipophilic core of the aggregate, *A*, according to the relation

$$A = \phi_{hc}ab/2. \tag{2}$$

Since there are two ribbons in a unit cell, the factor 1/2 appears on the right side of Eq. (2). The quantity ϕ_{hc} is determined from the composition of the components using relations given elsewhere [2].

To calculate the aggregate size, we applied hexagon-rod model of the R_I phase [5]. In this model, the cross-section of the aggregate is assumed to be like a hexagon, having all of its internal angles of 120°, to obtain the minimum interfacial area, and a constant water thickness between the neighbouring aggregates is assumed. The hexagon model of the ribbon phase is illustrated in Fig. 2a. Although the interface of the ribbon is not likely to have corners as in a hexagon, the latter is still close to the actual shape of the ribbon and the axial ratio of the ribbon can be calculated easily with this model. The length of the flat part, *L*, of the lipophilic region (shaded area), the smallest thickness of the lipophilic core across the flat part, $2R$, of the aggregate, and the thickness of the aqueous layer (including surfactant headgroups) around the lipophilic core, d_w, are illustrated in Fig. 2b. *L* is determined from Eq. (3), obtained from the geometrical consideration of the hexagons. Simple geometry allows us to express the area of the lipophilic part of the aggregate in terms of L, which is then combined with Eq. (2) to obtain Eq. (4) for R:

$$L = \frac{1}{2}\left(b - a\sqrt{3}\right). \quad (3)$$

$$R = -\frac{L}{2\sqrt{3}} + \sqrt{\frac{L^2 + A2\sqrt{3}}{12}}. \quad (4)$$

Now, the axial ratio of the hydrocarbon core of the aggregate, ρ, and the thickness of aqueous layer, d_w, (including surfactant head groups) can be calculated using the equations

$$\rho = (2R + L)/2R \quad (5)$$

and

$$d_w = 0.5(a - 2R) \quad (6)$$

The rodlike aggregate with circular cross-section has $\rho = 1$. The variation of the unit cell dimensions and, L, R and ρ with composition is shown in Table 1.

On increasing the surfactant concentrations, keeping the decanol concentration approximately constant, the height of the unit cell, a, decreases with the thickness of the water layer between the aggregates. The width of the unit cell, b, is insensitive to the increase in surfactant concentration in the lower surfactant concentration region, but at higher surfactant concentration it decreases. On the other hand, the length of the lamellar part of the ribbon, L, and the axial ratio, ρ, increase with surfactant concentration. This suggests that the added surfactant is incorporated in the flat part of the ribbon. With the increase in the length of the longer axis of the ribbon ($\approx L + 2R$), b should have increased. Most probably, the decrease in the water layer thickness compensates this effect and b remains constant with an increase in decanol concentration. These changes in the size of the unit cell and the aggregate cross-section imply that the ribbons come closer with increasing concentration of the surfactant. If we assume the thickness of the layer of surfactant headgroups around the lipophilic core to be 0.5 nm, the separation between the charged layers of neighbouring aggregates in the R_1 phase ranges roughly from 0.5 nm (for high surfactant concentration) to 1.0 nm (for low surfactant concentration).

On increasing the decanol concentration at fixed DL-KDA-to-water ratio, the height of the unit cell, a, shows a small decrease, although the interlayer spacing between the (11) planes increases!. The width, b, increases in the lower decanol concentration region, and then attains a nearly constant value in the higher decanol concentration region within the R_1 phase. With increasing concentration of decanol, L also increases. With decreasing a and constant b in the decanol-rich region of the R_1 phase, the increase in L (and ρ) of the ribbon would reduce the separation between the aggregates of adjacent (02) planes. The thickness of the aqueous layer (including surfactant head group), d_w, also decreases with an increase in decanol concentration (Table 1).

This type of variation in the aggregate structure in the R_1 phase is believed to be caused by microsegregation of surfactant and decanol molecules in the aggregate, with the decanol molecules lying in the flatter part and surfactant molecules in the curved part of the aggregate [13]. Gelbart et al. [14] and Dubois et al. [15] have shown theoretically that to produce a stable aggregate the cosurfactant molecules must reside in the flatter part of the aggregate, and hence this part grows at the expense of the curved edges, leading to an increase in the aggregate size and phase stability. Our result shows that R in the R_1 phase is much smaller than the radius of the circular cross-section of the aggregate in the H_1 phase ($R_{H1} = 1.39$ nm) at the same surfactant-to-water ratio

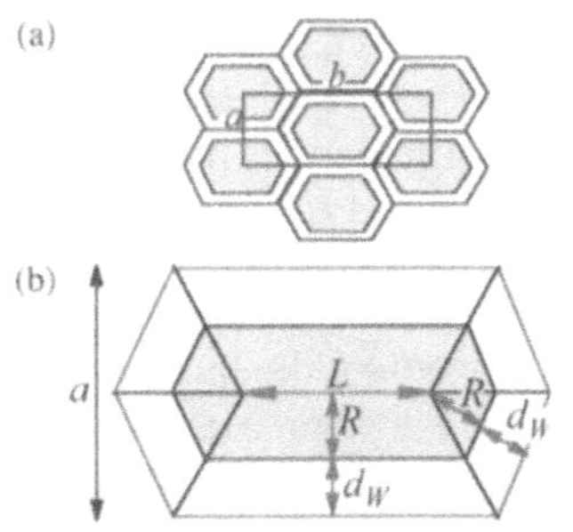

Fig. 2a,b Schematic model of the structure of the R_1 phase based on the hexagon-rod model. The *shaded area* is the lipophilic region of the aggregate of thickness $2R$ and length L at the flat part of the aggregate surrounded by a constant thickness d_w of the water layer including a headgroup

Table 1 Variation of unit cell dimension, and aggregate size parameters (for symbols see Fig. 2 and text) of the R_1 phase with surfactant composition. W_s is the weight percent of surfactant in surfactant and water, and X_o is the weight percent of decanol in the whole system

Composition	a/nm	b/nm	L/nm	r/nm	ρ	d_w/nm
$W_s = 44.9$, $X_o = 7.6$	4.63	10.90	1.44	1.28	1.56	1.04
$W_s = 49.3$, $X_o = 6.7$	4.40	10.70	1.54	1.25	1.62	0.95
$W_s = 50.0$, $X_o = 7.4$	4.40	10.83	1.61	1.26	1.64	0.96
$X_o = 8.34$	4.39	10.90	1.65	1.27	1.65	0.92
$W_s = 52.4$, $X_o = 5.0$	4.31	9.97	1.25	1.26	1.50	0.90
$X_o = 8.0$	4.28	10.77	1.68	1.26	1.66	0.88
$W_s = 60.2$, $X_o = 5.21$	4.04	10.21	1.60	1.25	1.64	0.77

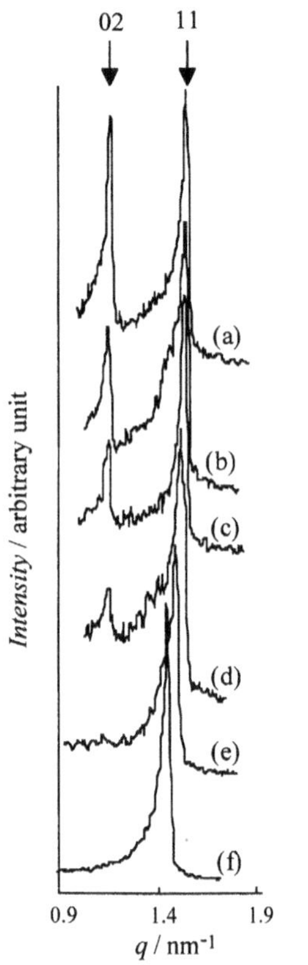

Fig. 3 Development of the small-angle X-ray scattering patterns from the R_1 phase to the $L_\alpha{}^H$ phase with successive addition of decanol at fixed DL-KDA-to-water ratio, W_s (1:1 by weight). Intensities are not drawn to scale. The decanol concentrations in *a–f* are 7.37, 8.34, 9.94, 10.78, 11.79 and 13.00, respectively

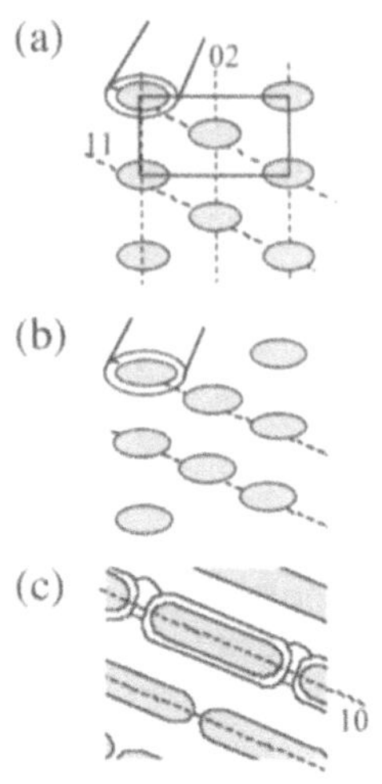

Fig. 4a–c Schematic model of the $R_1 \rightarrow L_\alpha^H$ phase transformation with successive addition of decanol. With successive addition of decanol, the rectangular symmetry of the R_1 phase (**a**) is broken (**b**) and the $L_\alpha{}^H$ phase (**c**) is formed

in the absence of decanol. This indicates that the interfacial curvature at the edge (curved part) of the aggregate is less than the curvature the circular interface of the H_1 phase would assume at that composition, and, therefore, the free energy of the curved part is expected to be higher than that of the flat part of the ribbon. This result supports the theoretical prediction of Gelbart et al.

$R_1 \rightarrow L_\alpha^H$ phase transformation

The development in the SAXS pattern from the R_1 to the $L_\alpha{}^H$ phase with successive addition of decanol is shown in Fig. 3. The intensity of the peak corresponding to the (02) plane decreases gradually and disappears completely in the $L_\alpha{}^H$ phase, whereas the (11) plane of the R_1 phase shows a gradual increase in interlayer spacing and eventually transforms into the (10) plane of the $L_\alpha{}^H$ phase. This suggests that with the increase in the flatness of the ribbons (or decanol concentration), the ribbons come closer along the (11) plane and the rectangular symmetry is gradually broken, as shown schematically in Fig. 4. Although the R_1 phase in Fig. 4 is shown to have *cmm* symmetry for the sake of simplicity, the scheme is expected to be applicable for *pgg* symmetry as well, since these two symmetries of the R_1 phase differ only in that in the latter symmetry the planes of the ribbons are slightly tilted to the longer axis of the rectangular unit cell. With further addition of decanol, the ribbons along the (11) planes merge laterally with the adjacent ribbons to form a surfactant bilayer extending "infinitely". However, the formation of a lamellar-like interface may not take place uniformly and gradually [16], rather the aggregates are locally joined at small segments to form patches of surfactant bilayers and water "holes" or "defects" of the $L_\alpha{}^H$ phase, as shown in Fig. 4. With successive addition of the cosurfactant, the number density of the water defects decreases and the lamellar-like interface is promoted. This restores the amount of water in the interlamellar space and increases the interlayer spacing. This explains the gradual shift of the most intense SAXS peaks toward lower q (or higher interlayer spacing) in Fig. 3 at higher decanol concentration.

Conclusion

The R_1 phase formed between the classical H_1 and L_α phases in the DL-KDA/water/1-decanol system shows a variation in aggregate size and unit cell dimension. With increase in surfactant concentration the aspect ratio, ρ, of the aggregate increases, whereas the rectangular unit cell shrinks, bringing the ribbonlike aggregates closer to each other. Similarly, increasing decanol concentration increases the unit cell width, b, and also ρ. The SAXS results show that an epitaxial relationship exists between the (11) plane of the R_1 phase and the (10) plane of the L_α^H phase, suggesting that with successive addition of decanol the ribbons of the (11) plane of the R_1 phase come closer and partially merge along their length, forming a surfactant bilayer of the L_α^H phase pierced with water defects.

Acknowledgement We are thankful to A. M. Lopez-Quintela, Department of Physical Chemistry, University of Santiago Compostela, Santiago de Compostela, Spain, for his valuable comments and suggestions.

References

1. Ekwall P (1975) In: Brown GH (ed) Advances in liquid crystals, vol 1. Academic, New York, pp 1–142
2. Jonsson B, Wennerstrom H (1987) J Phys Chem 91:338
3. Acharya DP, Lopez-Quintela AM, Kunieda H, Oshimura E, Sakamoto K Langmuir (2003) J Oleo Sci 52:407
4. Hendrikx Y, Charvolin J, Rawiso M (1981) J Phys 42:1427
5. Hagslätt H, Söderman O, Jönsson B (1992) Liq Cryst 12:667, and references therein
6. Holmes MC (1998) Curr Opin Colloid Interface Sci 3:485
7. Gustafsson S, Quist P, Halle B (1995) Liq Cryst 18:545
8. Fairhurst CE, Holmes MC, Leaver MS (1997) Langmuir 13:4964
9. Hendrikx Y, Pansu B (1996) J Phys II 6:33
10. Kekicheff P, Cabane B (1987) J Phys 48:1571
11. Gustafsson S, Quist P -O (1996) J Colloid Interface Sci 18:564
12. Kunieda H, Horii M, Koyama M, Sakamoto K (2001) J Colloid Interface Sci 236:78
13. Hendrikx Y, Charvolin J, Rawiso M (1984) J Colloid Interface Sci 100:597
14. Gelbart WM, McMullen WE, Masters A, Ben-Shaul A (1985) Langmuir 1:101
15. Dubois M, Belloni L, Zemb TH, Deme B, Gulik-Krzywicki T (2000) Prog Colloid Polym Sci 115:238
16. Leaver MS, Holmes MC (1993) J Phys II 3:105

Progr Colloid Polym Sci (2004) 126: 30–34
DOI 10.1007/b93968

V. Abou
H. Benalla
M. J. Meziani
J. Zajac

The effect of phenol on the critical micelle concentration and the enthalpy of micellisation of cationic dimeric surfactants in aqueous solutions

V. Abou · H. Benalla · M. J. Meziani
J. Zajac (✉)
Laboratoire des Agrégats Moléculaires
et Matériaux Inorganiques,
UMR 5072, Université Montpellier 2,
Place E. Bataillon,
34095 Montpellier Cedex 5, France
e-mail: zajac@univ-montp2.fr

Abstract The effect of phenol addition to the aqueous phase on the critical micelle concentration and molar enthalpy of micellisation of alkanediyl-α,ω-bis(dodecyldimethylammonium bromide) Geminis (referred to as $C_{12}C_SC_{12}$, where *S* is the spacer length) has been studied at 298 K. For $C_{12}C_2C_{12}$, specific conductivities of aqueous surfactant solutions were measured over concentration ranges about the respective critical micelle concentrations at three selected phenol contents. UV absorbances of phenol, taken at a wavelength of 269.5 nm, were compared for different phenol and $C_{12}C_2C_{12}$ concentrations. Titration calorimetry was used to determine the molar dilution enthalpy of a micellar stock solution, in pure water and in a 1.5 mmol kg^{-1} solution of phenol in water, as a function of the equilibrium surfactant concentration for three Geminis, $C_{12}C_2C_{12}$, $C_{12}C_6C_{12}$ and $C_{12}C_{12}C_{12}$. Depression of the critical micelle concentration and the enhanced thermal effect of micellisation in the presence of phenol were inferred from the experimental results. The location of polarisable phenol molecules in the outer portion of the surfactant micelles is postulated.

Keywords Phenol solubilisation · Cationic Gemini surfactants · Titration calorimetry · Specific conductivity · UV absorbance

Introduction

During the last decade, the behaviour of Gemini surfactants in aqueous solution and at interfaces has been intensively studied by a variety of experimental techniques [1, 2, 3, 4, 5, 6, 7, 8]. By far the most often investigated Geminis belong to the family of cationic alkanediyl-α,ω-bis(dodecyldimethylammonium bromide) compounds for which the simplified notation $C_nC_SC_n$ is used in the present paper, with *n* and *S* denoting the number of carbon atoms in each of the hydrophobic tails and the spacer, respectively. Owing to the increased difficulty in packing the surfactant hydrophobic tails into spherical micelles, short-spacer homologues form linear threadlike aggregates and long-spacer ones produce a mixture of spheroids and treelike micelles [1]. Micellisation of cationic Geminis is much more exothermic than that of their single-chain counterparts and the thermal effect is very sensitive to changes in the spacer length and the temperature [7, 8]. The critical micelle concentration (cmc) as a function of the spacer length reaches its maximum at *S* ranging between 4 and 6. This maximum cmc value corresponds to a strong minimum in the magnitude of the exothermic enthalpy of micellisation.

On account of unusual micelle morphologies and low cmcs, these compounds may be used in many potential applications, such as enhanced oil recovery, soil remediation, waste stream purification or drug delivery. The present paper reports some aspects of the micellisation of $C_{12}C_SC_{12}$ Geminis, with alkanediyl spacer groups, C_2H_4, C_6H_{12} and $C_{12}H_{20}$, in the presence of phenol (PhOH) in an aqueous medium. PhOH species of considerable toxicity and low biodegradability are undesirable components of aqueous streams and may thus be eliminated by using micellar solubilisation [9].

Experimental

Materials

The three alkanediyl-α,ω-bis(dodecyldimethylammonium bromide) Geminis used, $C_{12}C_2C_{12}$, $C_{12}C_6C_{12}$ and $C_{12}C_{12}C_{12}$, were kind gifts from R. Zana, Institut Charles Sadron, CNRS, Strasbourg, France. The synthesis and purification procedures were described in Ref. [10]. PhOH was Sigma high-performance liquid chromatography grade material with purity exceeding 99%. The water used throughout all the experiments was deionised and purified with a Millipore Super Q system. It had a pH value of 6 and a conductivity which varied between 0.05 and 0.1 $\mu S\ cm^{-1}$.

Methods

The measurements of the specific conductivities of surfactant aqueous solutions at some selected PhOH concentrations were performed using a Tacussel CDM 210 conductimeter. Each concentration series was made up by progressive injections of a micellar stock solution into the cell. Steady readings were obtained approximately 10 min after each addition of solution to the cell.

UV spectra of PhOH dissolved in water and dodecane were recorded using a Varian Cary 3E spectrophotometer with a 1-mm cell. In a series of experiments in the presence of Gemini $C_{12}C_2C_{12}$, an aqueous solution of the same surfactant concentration was used as a reference.

The enthalpy changes accompanying the successive dilutions of a micellar surfactant solution were monitored using a Montcal microcalorimeter [8, 11]. The stock solution of molality 10 times greater than the cmc of a given surfactant in a given environment was injected, in small steps of around 0.1 g min^{-1}, into the calorimetric cell containing initially 8 g water. PhOH was added to some systems in such a way to keep its concentration both in the cell and in the stock solution constant. For each injection step, heat absorption ceased after about 30 min and the recorder returned to its baseline. Calibration of the areas under the thermal peaks was performed using a Joule calibration probe immersed in the homogenised solution.

Results and discussion

The exact location of a given solubilisate in a surfactant micelle reflects the type of forces operating between both species [12, 13, 14, 15]. In the case of monomeric cationic surfactants, polarisable PhOH molecules are thought to be located in the polar mantle of micellar aggregates, or adsorbed at the micelle-solution interface, the π-electron cloud interacting with the electric field produced by the cationic headgroups of the micelle [13, 14, 15]. In the present work, the effect of PhOH on the micelle formation of $C_{12}C_2C_{12}$ was first investigated in the conductivity measurements. The results are presented in Fig. 1, where the concentration of the solutions is stated in molality. Commonly the specific conductivity versus

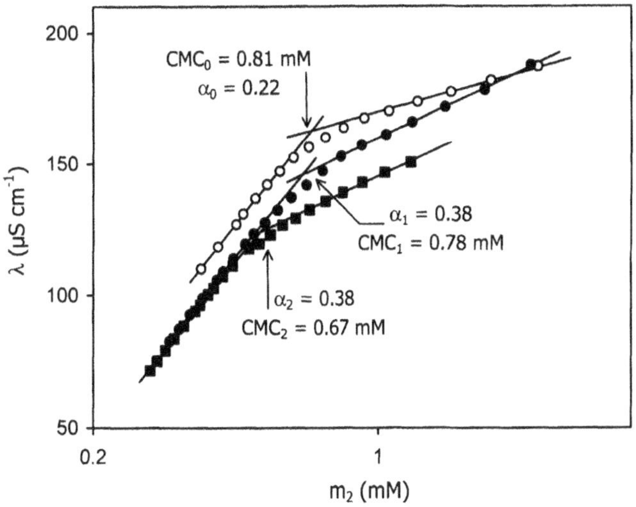

Fig. 1 The specific conductivity of $C_{12}C_2C_{12}$ in pure water (*open circles*), in a 1.5 mmol kg^{-1} phenol solution (*solid circles*) and in a 9.6 mmol kg^{-1} phenol solution (*squares*) as a function of the surfactant molality at 298 K. The critical micelle concentration (*cmc*) and the micellar degree of ionisation, α, are determined respectively from the point of intersection and from the slopes of the two straight lines [10]

concentration plots show an abrupt change of slope in the region of the cmc. The data in Fig. 1 were fitted with a pair of straight lines and the experimental points in Fig. 1 clearly indicate a smooth transition between the two linear portions. The effect is more pronounced for higher PhOH concentrations in the aqueous phase. This probably means that transition from the unassociated Gemini surfactant to the ultimate micellar state occurs over a quite wide range of molality, producing polydispersed intermediate aggregates. Determination of the cmc from the change in slope indicated by the linear portions is therefore less accurate than usual. The appropriate values of the cmc and the micellar degree of ionisation, α, are reported in Fig. 1. The cmc decreases steadily with increasing PhOH content in the aqueous phase. Although α is increased by addition of PhOH, it seems to be independent of the PhOH concentration. It is clear that PhOH units interact with the micellised $C_{12}C_2C_{12}$ species and this interaction affects mainly the charge of the surfactant micelles. Since the pH of the aqueous medium is lower than the pK_a of C_6H_5OH, the molecular form is predominant in the system. The nature of the changes observed suggests that polarisable PhOH molecules may be incorporated into the surfactant micelle, mainly in its outer portion.

Additional proof of the PhOH location was sought in UV spectroscopy measurements, on the basis of clear qualitative differences between the UV spectra of PhOH in water and in dodecane (not shown here). For different submicellar and supermicellar concentrations of $C_{12}C_2C_{12}$, the UV spectra of PhOH in surfactant solutions were identical with the UV pattern in pure

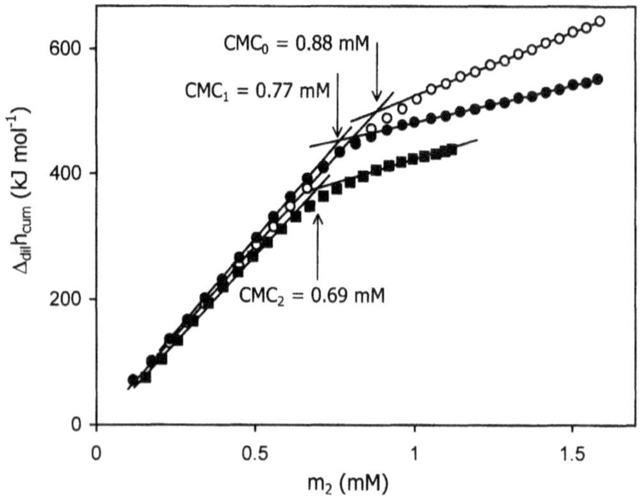

Fig. 2 The cumulative molar enthalpy of successive dilutions of a micellar stock solution of $C_{12}C_2C_{12}$ in pure water (*open circles*), in a 1.5 mmol kg^{-1} phenol solution (*solid circles*) and in a 9.6 mmol kg^{-1} phenol solution (*squares*) as a function of the equilibrium surfactant molality at 298 K. The point of intersection between the two straight lines is taken as the cmc

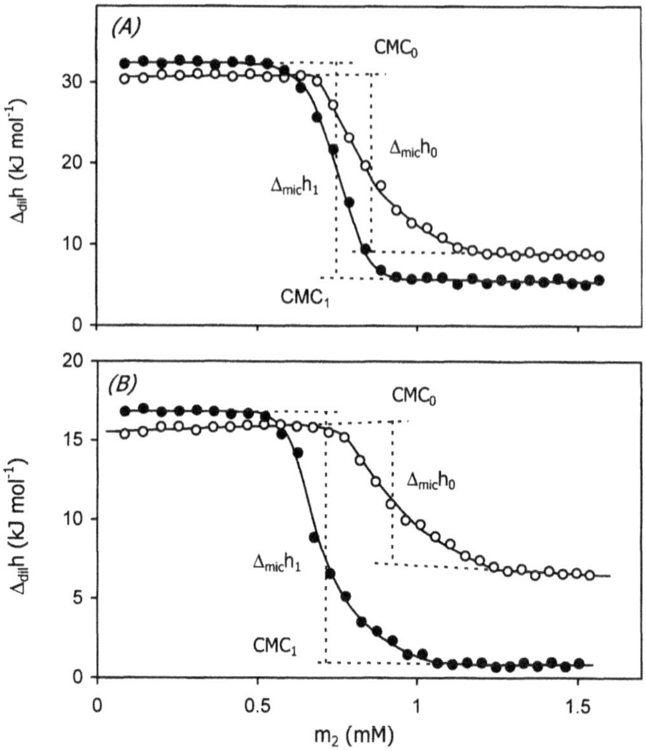

Fig. 3 The differential molar enthalpy of successive dilutions of a micellar stock solution of **a** $C_{12}C_2C_{12}$ and **b** $C_{12}C_6C_{12}$ in pure water (*open circles*) and in a 1.5 mmol kg^{-1} phenol solution (*solid circles*) against the equilibrium surfactant molality at 298 K. The so-called Maxwell geometrical construction is based on extrapolation of the premicellar and postmicellar linear parts to a hypothetical cmc and replacement of the real dilution profile by a step function [11] (the model dilution profiles are represented by the *dashed lines*). The value of the cmc is chosen in such a way that the two areas, bounded by the curve of $\Delta_{dil}h$ versus m_2 and the three model straight lines, compensate for one another

water, the intensity of electromagnetic radiation transmitted through a sample at a given wavenumber being only a function of the overall PhOH concentration in the solution. The plots of UV absorbance, taken arbitrarily at a wavelength of 269.5 nm, against the PhOH concentration gave a unique straight line, irrespective of the actual surfactant content. Even though it was inserted into the surfactant micelle, PhOH appears to keep its polar microenvironment. It should thus be located close to the micelle–solution interface and its main effect is to modify the nature of the electrical double layer surrounding the ionic micelle. For example, it can moderate the mutual repulsion between the dimeric headgroup cations or induce some subtle changes in the degree of binding of the bromide counterion to the micelle.

The cmc may also be evaluated from direct calorimetric measurements of the enthalpy of dilution by plotting the values of the cumulative molar enthalpy of dilution, $\Delta_{dil}h_{cum}$, as a function of the equilibrium molality, m_2. Examples of the cumulative dilution curve for Gemini $C_{12}C_2C_{12}$ in the absence and in the presence of PhOH are presented in Fig. 2. Each set of cumulative enthalpy data can be represented as a pair of straight lines with different slopes, the cmc being obtained easily from the intersection point. Nevertheless, as in the case of specific conductivity data, the change in slope is rather gradual, which increases the uncertainty in the cmc determination. For pure Gemini surfactant, the cmc_0 value inferred from the calorimetric data differs from that obtained on the basis of the specific conductivity versus molality plots by about 9%. In the presence of PhOH, there is generally a small difference in the value of the cmc obtained by the two methods.

Alternatively, the results of dilution calorimetric measurements can be processed to plot the molar enthalpy of dilution, $\Delta_{dil}h$, as a function of the surfactant concentration, m_2, in the equilibrium solution [11]. Direct calorimetric measurements were carried out for three Gemini surfactants, $C_{12}C_2C_{12}$, $C_{12}C_6C_{12}$ and $C_{12}C_{12}C_{12}$, first from pure deionised water (without PhOH) and then from a 1.5 mmol kg^{-1} solution of PhOH in water. Examples are shown in Fig. 3. The low PhOH content in the system allows the problem to be treated on the basis of the pseudo-phase transition model for both the micellisation and the PhOH distribution between the aqueous phase and the micelles [16]. The aromatic component is therefore assumed to behave as a probe not affecting the structure of dimeric micelles. PhOH is included in the mean solvent and the dilution enthalpy values are expressed per unit amount of the surfactant.

In all cases, $\Delta_{dil}h$ is almost a constant function of m_2 in the premicellar region, in which the enthalpy change on

Table 1 Critical micelle concentration (*cmc*, mmol kg^{-1}) and enthalpy change ($\Delta_{mic}h$, kJ mol^{-1}) accompanying the micellisation of Gemini $C_{12}C_SC_{12}$ in binary water–surfactant or ternary water–surfactant–phenol systems, as inferred from direct calorimetric measurements of the molar enthalpy of dilution and the Maxwell geometrical construction

m_{PhOH}	Surfactant					
	$C_{12}C_2C_{12}$		$C_{12}C_6C_{12}$		$C_{12}C_{12}C_{12}$	
	cmc	$\Delta_{mic}h$	cmc	$\Delta_{mic}h$	cmc	$\Delta_{mic}h$
No PhOH	0.86 ± 0.05^a	-21.9 ± 1.5^a	0.92 ± 0.05	-9.0 ± 0.8	0.39 ± 0.03	-13.1 ± 0.9
1.5 mM	0.75 ± 0.04	-26.6 ± 1.7	0.71 ± 0.04	-15.9 ± 1.3	0.38 ± 0.03	-19.7 ± 1.4

[a]Maximum relative uncertainties of 4 and 2.5% were estimated for the molar enthalpy of dilution in the premicellar and postmicellar region, respectively. The relative uncertainty of the equilibrium molality did not exceed 1.5%

injection is ascribed to destruction of micelles and dilution of the resulting monomers, and in the postmicellar region, where the coexisting micelles and monomers are diluted. Dilution of micelles is evidently an endothermic process both in the presence and in the absence of PhOH. The invariance of the enthalpy effect above the cmc indicates that the size and the shape of the surfactant aggregates remain essentially unchanged and dilution of a concentrated stock solution in this concentration range is accompanied mainly by changes in intermicellar interactions. The so-called Maxwell geometrical construction [11] can be used to determine both the cmc and the enthalpy change, $\Delta_{mic}h$, accompanying the micellisation of dimeric $C_{12}C_SC_{12}$ in binary water–surfactant or ternary water–surfactant-PhOH systems. The micellisation parameters obtained from the differential enthalpy data are given in Table 1. Note that the cmc values evaluated in this manner are equal, within experimental error, to that obtained from the cumulative molar enthalpy, thus pointing out the reliability of the Maxwell geometrical construction. For $C_{12}C_SC_{12}$ in pure water, somewhat different values of cmc_0 and $\Delta_{mic}h_0$ have been reported by Bai et al. [7] and Grosmaire et al. [8]. Such discrepancies are quite common in view of the dependence of both micellisation parameters on the underlying theoretical models for micellisation. $\Delta_{mic}h$ obtained simply from the difference between "somehow selected upper and lower enthalpy values" may be poorly reproducible, especially when the enthalpy effects in the premicellar and postmicellar region are not perfectly constant. The geometrical construction described earlier is well justified by the pseudo-phase transition model; however, this procedure appears very sensitive to the accuracy of the calorimetric measurement.

The detailed analysis of Table 1 reveals that the presence of PhOH in the system makes the enthalpy change on micellisation more exothermic. If one considers that incorporation of small amounts of PhOH into a Gemini micelle affects mainly intermicellar interactions, without greatly modifying the micelle structure, the absolute increase in the enthalpy, i.e., $\Delta_{mic}h$–$\Delta_{mic}h_0$, per mole of surfactant can be taken as a rough estimate of the enthalpy effect of PhOH transfer from the aqueous phase to the outer portions of the micelle: $C_{12}C_2C_{12}$ 4.7 kJ mol^{-1}; $C_{12}C_6C_{12}$ and $C_{12}C_{12}C_{12}$ 6.6 kJ mol^{-1}. The enthalpy effect is smaller for $C_{12}C_2C_{12}$, probably because fewer PhOH molecules are solubilised in micellar structures with more densely packed cationic headgroups. It is also interesting to note that the enthalpy difference $\Delta_{mic}h$–$\Delta_{mic}h_0$ is approximately twice that obtained for monomeric dodecyltrimethylammonium bromide with the same PhOH content [15].

Conclusion

The combined conductivity, UV spectroscopy and calorimetric studies clearly indicate that polarisable PhOH molecules are located in the outer portions of micellar aggregates of Gemini $C_{12}C_SC_{12}$ surfactants under the action of the electric field induced by the dimeric headgroup cations of the micelle. The specific conductivity and cumulative enthalpy of dilution data can be represented as a pair of straight lines to about the same approximation for the two methods but there is generally some difference in the cmc values thus obtained owing to a gradual change in slope in the region of the cmc. The cmc of a given surfactant is lowered in the presence of PhOH. The transfer of PhOH from the aqueous phase to the micellised species is exothermic and it markedly contributes to the enthalpy gain on micelle formation.

References

1. Zana R (1997) In: Robb ID (ed) Specialist surfactant. Chapman and Hall, London, p 81, and references therein
2. Rosen M J, Zhu ZH, Hua XY (1992) J Am Oil Chem Soc 69:30
3. Esumi K, Goino M, Koide Y (1996) J Colloid Interface Sci 183:539
4. Manne S, Schäffer TE, Huo Q, Hansma PK, Morse DE, Stucky GD, Aksay IA (1997) Langmuir 13:6382
5. (a) Chorro C, Chorro M, Dolladille O, Partyka S, Zana R (1998) J Colloid Interface Sci 199:169; (b) Chorro C, Chorro M, Dolladille O, Partyka S, Zana R (1999) J Colloid Interface Sci 210:134
6. Meziani MJ, Benalla H, Zajac J, Partyka S (2000) Prog Colloid Polym Sci 115:72
7. Bai GY, Yan HK, Thomas RK (2001) Langmuir 17:4501
8. Grosmaire L, Chorro M, Chorro C, Partyka S, Lagerge S (2001) Thermochim Acta 379:255
9. Syamal M, De S, Bhattacharya PK (1997) J Membr Sci 137:99
10. Zana R, Benrraou M, Rueff R (1991) Langmuir 7:1072
11. Zajac J (2000) Colloids Surf 167:3
12. Christian SD, Scamehorn JF (eds) (1995) Solubilisation in surfactant aggregates. Dekker, New York
13. Hirose C, Sepulveda L (1981) J Phys Chem 85:3689
14. Shah SS, Awan MA, Idris SA (1995) Colloids Surf 105:319
15. Benalla H, Zajac J, Partyka S, Rozière J (2002) Colloids Surf 203:259
16. De Lisi R, Miliot S (1995) In: Christian SD, Scamehorn JF (eds) (1995) Solubilisation in surfactant aggregates. Dekker, New York, chap 3

Progr Colloid Polym Sci (2004) 126: 35–39
DOI 10.1007/b94007

Mingfu Zhang
Pierre Teissier
Marina Krekhova
Valérie Cabuil
Axel H.E. Müller

Polychelates of amphiphilic cylindrical core–shell polymer brushes with iron cations

M. Zhang · A.H.E. Müller (✉)
Makromolekulare Chemie II and
Bayreuther Zentrum für Kolloide und
Grenzflächen, Universität Bayreuth,
95440 Bayreuth, Germany
e-mail: axel.mueller@uni-bayreuth.de
Tel.: +49-921-553399
Fax: +49-921-553393

P. Teissier · V. Cabuil
UMR 7612 (LI2C) – Équipe Colloïdes
Inorganiques, Université Pierre et Marie
Curie, 75252 Paris Cedex 05, France

M. Krekhova
Anorganische Chemie I,
Universität Bayreuth,
95440 Bayreuth, Germany

Abstract Well-defined amphiphilic cylindrical polymer brushes with a poly(acrylic acid) core and a poly(*n*-butyl acrylate) shell were synthesized via atom transfer radical polymerization. Unimolecular polymer cylinders were directly visualized on mica by scanning force microscopy. The hydrophilic cores of the brushes, poly(acrylic acid), were neutralized by NaOH and afterwards reacted with Fe^{3+} and Fe^{2+} cations. The formation of polychelates of polymer brushes and iron cations was confirmed and characterized by Fourier transform IR spectroscopy, UV–vis spectroscopy, transmission electron microscopy and scanning force microscopy.

Keywords Cylindrical polymer brush · Unimolecular micelle · Polychelate · Iron cation · Complex

Introduction

Cylindrical polymer brushes, which have one side chain per main chain monomer unit, have attracted extensive interest, owing to the possibility of forming extended chain conformations and their peculiar properties [1–6].

Polymerization of macromonomers [1–3] as well as the "grafting-from" technique [4–6] have been used to synthesize cylindrical polymer brushes. In the latter method, the backbone of the brush is first synthesized and then modified to carry an initiating group at each monomer unit. The subsequent polymerization initiated by the backbone forms side chains. Using this method well-defined polymer brushes with high grafting density and narrow distributions of both backbone and side chain can be obtained.

With amphiphilic block copolymers as side chains, polymer brushes have a core–shell structure and can be regarded as unimolecular micelles. Very recently, we succeeded in the synthesis of core–shell cylindrical brushes with amphiphilic block copolymer side chains, such as poly(acrylic acid)-*block*-polystyrene (PAA-*b*-PS) or PAA-*block*-poly(*n*-butyl acrylate) (PAA-*b*-P*n*BA), by atom transfer radical polymerization (ATRP) [5, 6]. The living character of ATRP [7] enables the control of the distribution of the side chain, so the resulting amphiphilic brushes have well-defined sizes.

As part of our continuous effort for the synthesis and applications of amphiphilic core–shell cylindrical brushes, we report here the synthesis and characterization of polychelates of amphiphilic brushes and iron cations. Further reactions such as alkalization or reduction of the loaded iron cations are expected to produce superparamagnetic nanoparticles inside the cores of polymer brushes, forming a new class of hybrid nanoparticles.

Experimental

Materials

$FeCl_3$ (Merck, anhydrous, for synthesis), $FeCl_2 \cdot 4H_2O$ (Fluka, analytical grade) and NaOH (Merck, analytical grade) were used as received. All organic solvents were of analytical grade and used without further purification. Water was ultrapure and was obtained from a Milli-Q plus system (Millipore). The synthesis of amphiphilic core–shell cylindrical polymer brushes was reported in our previous papers [5,6].

Polymer characterization

The monomer conversion in the polymerization was determined by gas chromatography using a polymethylsiloxane capillary column. ^{1}H NMR spectra were recorded using a Bruker AC-250 spectrometer. The apparent molecular weights of the polymer brushes were measured by gel permeation chromatography (GPC) using tetrahydrofuran as the eluent at a flow rate of 1.0 ml/min at room temperature. The column set was 5 μm SDV gel, 10^5, 10^4, 10^3, 10^2 Å, 30 cm each (PSS, Germany) and the detectors were a Waters 410 differential refractometer and photodiode array detector operated at 254 nm. PS standards (PSS) were used for the calibration of the column set. Membrane osmometry was performed at 35 °C in toluene to determine the true molecular weight of the backbone of the polymer brush using a cellulose triacetate membrane with a Gonotec 090 osmometer. The samples for scanning force microscopy (SFM) measurements were prepared by dip-coating from dilute solutions of polymer brushes or polychelates onto a freshly cleaved mica surface. The SFM images were taken with a Digital Instruments Dimension 3100 microscope operated in tapping mode.

Preparation and characterization of polychelates

The typical procedure was as follows. Amphiphilic polymer brushes were dissolved in a mixture of methanol and chloroform (3/5–1/1 v/v). Then a calculated amount of 1.0 M NaOH (80% of the molar amount of the COOH group in polymer) was added to the polymer solution, followed by stirring overnight. Afterwards a threefold excess of $FeCl_3$ (1.0 M aqueous solution) was added and the mixture was stirred for 1 day. The unassociated ferric cations were removed by dialysis against the solvent mixture which was used to dissolve the polymer. Dialysis was carried out for 1 week (and the solvent was changed once a day) using a regenerated cellulose membrane tube with a molecular weight cutoff of 6–8,000.

The polychelates synthesized were characterized by Fourier transform (FT) IR spectroscopy (Bio-Rad, USA), UV–vis spectroscopy (Lambda 15, PerkinElmer), SFM and transmission electron microscopy (TEM). A small drop of a dilute solution of polychelate was placed on a copper grid coated with a Formvar–carbon film. After drying, the grid was used for TEM measurements using a Zeiss CEM 902 electron microscope operated at 80 kV.

Results and discussion

Synthesis and characterization of amphiphilic polymer brushes

The synthesis was carried out as described in detail previously [5, 6]. The structure of the amphiphilic core–shell polymer brush, PAA-*b*-P*n*BA, is illustrated in Scheme 1. Owing to the core–shell cylindrical structure it can be regarded as a unimolecular micelle.

The backbone of the polymer brush was synthesized via anionic polymerization. GPC measurements showed that the backbone had a very narrow distribution ($M_w/M_n = 1.08$), and its degree of polymerization was determined as $DP_n = 1{,}500$ from its absolute molecular weight obtained by membrane osmometry measurements.

Core–shell cylindrical brushes with poly(*t*-butyl acrylate)-*block*-P*n*BA (P*t*BA-*b*-P*n*BA) side chains were formed via sequential ATRPs. Selective hydrolysis of the core block of the side chain, P*t*BA, into PAA resulted in amphiphilic brushes. The results of the characterization of the polymer brushes used for the synthesis of polychelates are shown in Table 1.

The morphology of the amphiphilic brush was characterized by SFM. A typical SFM image of brush 1, dip-coated from dilute $CHCl_3/CH_3OH$ (1/1 v/v) solution onto mica is shown in Fig. 1. Wormlike polymer cylinders can be clearly visualized. The average dimensions of the cylinders are length 180 nm, core diameter 25 nm and height 2.5 nm, obtained from statistical analysis over 250 separated cylinders.

Formation and characterization of polychelates of amphiphilic polymer brushes and iron cations

First FTIR spectroscopy was used to verify the ion exchange between iron cations and carboxylic acid groups in polymer brushes. The FTIR spectra of brush 2 without and with addition of $FeCl_2$ are shown in Fig. 2. Besides the peak at 1,735 cm^{-1} corresponding to carbonyl groups of –COOH and $-COO(CH_2)_3CH_3$ in the polymer brush, a new peak at 1,595 cm^{-1} appeared when brush 2 was mixed with $FeCl_2$. This peak is associated

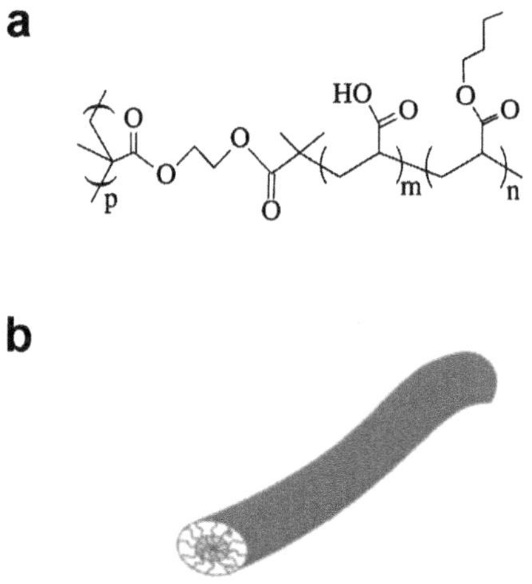

Scheme 1. Structure of amphiphilic core–shell cylindrical polymer brush with poly(acrylic acid)-*block*-poly(*n*-butyl acrylate) side chains: **a** chemical formula; **b** schematic 3D structure

with the asymmetric stretching of carboxylate salts, clearly indicating the coordination of Fe^{2+} to COOH groups. $FeCl_3$ has similar behavior when it reacts with carboxylic acid groups of the polymer [8].

It has been reported that conversion of carboxylic acid groups to sodium carboxylate results in a large increase in both the rate and the extent of transition-metal ion uptake from metal chlorides [9]. Thus the carboxylic acid groups in amphiphilic polymer brushes were neutralized by NaOH aiming to load more iron cations into the polymer core. SFM measurements showed there were no changes in morphology and size before and after neutralization. Then an excess amount of iron salts such as ferric chloride were added. Dialysis was used to remove the unreacted ferric cations. The complete removal of unreacted ferric cations was confirmed by SFM, since unreacted $FeCl_3$ formed spherical particles on mica. A typical SFM image of the polychelate of brush 1 and $FeCl_3$ after purification by dialysis is depicted in Fig. 3. Comparison of images in Figs. 1 and 3 indicates that the morphologies and sizes of the pure polymer brush and the polychelate differ significantly. In contrast to the smooth contour of brush 1, polychelates have periodic undulations in height (and also diameter) along the backbone, as seen in Fig. 3b. We speculate that this change in morphology may be due to the inter-side chain coordination of multivalent ferric cations with monovalent carboxylate groups from different side chains, which introduces the aggregation

Table 1 Characterization results of polymer brushes with a poly (acrylic acid) core and a poly(*n*-butyl acrylate) shell. The degree of polymerization (*DP*) of the side chain block was calculated according to the initial monomer-to-initiator ratio and the monomer conversion

Polymer[a]	$10^{-5}\times M_{n,GPC}$[b]	$M_w/M_{n,GPC}$[b]	DP_{core}	DP_{shell}
Brush 1	9.34	1.40	37	48
Brush 2	7.72	1.40	29	32

[a] $DP_{backbone} = 1{,}500$
[b] Number-average molecular weight and polydispersity of unhydrolyzed brushes, with a poly(*t*-butyl acrylate) core and a poly (*n*-butyl acrylate) shell, obtained from gel permeation chromatography measurements

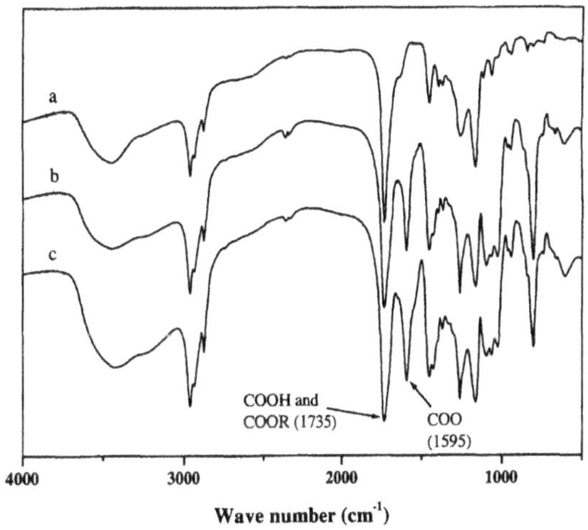

Fig. 2 Fourier transform IR spectra of *a* brush 2, and complexes of brush 2 with $FeCl_2$: *b* $Fe^{2+}/COOH = 0.25$, *c* $Fe^{2+}/COOH = 0.5$

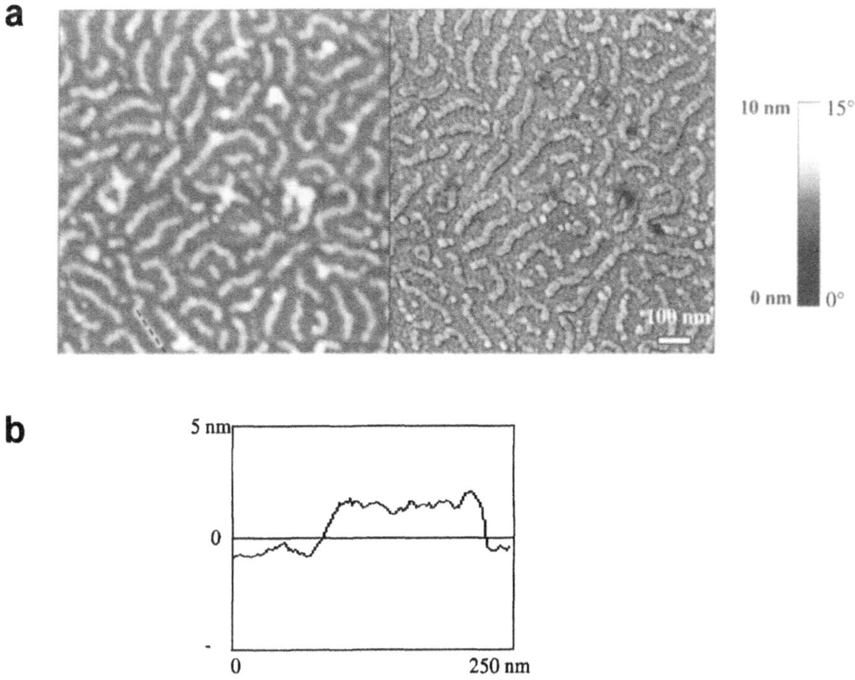

Fig. 1 Scanning force microscopy (*SFM*) image of brush 1: **a** height (*left*) and phase (*right*); **b** cross-section analysis of one cylinder (marked by a *dashed line* in **a**)

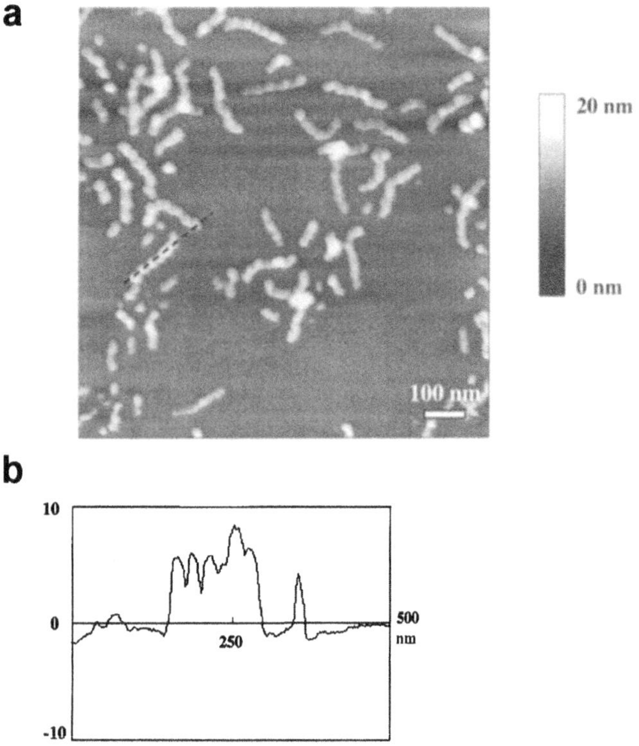

Fig. 3 SFM image of the polychelate of brush 1 and $FeCl_3$ after dialysis: **a** height image and **b** cross-section analysis of one cylinder (marked by a *dashed line* in **a**)

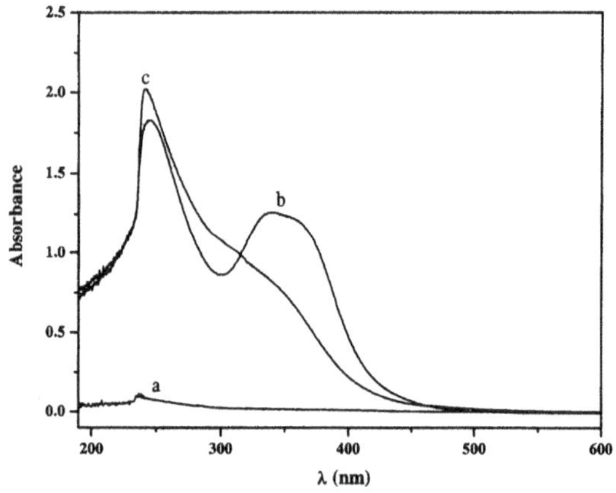

Fig. 4 UV–vis spectra of *a* brush 1 neutralized with NaOH (=80% COOH, polymer concentration 0.25 g/l), *b* $FeCl_3$, 0.042 g/l and *c* polychelate of brush 1 and $FeCl_3$ (after dialysis) with a polymer concentration of 0.25 g/l. Solvent $CHCl_3/CH_3OH$ (5/3 v/v)

of side chains. Moreover, the height of the polychelate is much greater than that of the polymer brush, indicating that the loading of ferric cations into the polymer core stiffens the polymer cylinder dramatically.

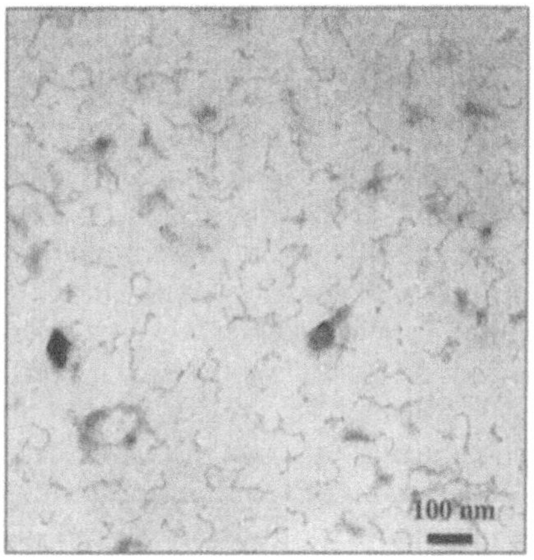

Fig. 5 Transmission electron microscopy image of the polychelate of brush 1 and $FeCl_3$ (after dialysis)

UV–vis measurements also confirmed the formation of polychelates. The UV–vis spectra of neutralized brush 1, $FeCl_3$ and the corresponding polychelate are shown in Fig. 4. With the same polymer concentration, the polychelate has a much stronger absorption than the polymer. Comparison between the spectra of the polychelate and $FeCl_3$ indicates that the absorption of polychelate mainly originates from the coordinated ferric cations.

Although the polymer brush itself is not visible by TEM, the core of the polychelate should be visible owing to the high electron density of ferric cations. A TEM image of the polychelate of brush 1 and $FeCl_3$ is depicted in Fig. 5. As expected, thin wormlike cylinders are clearly seen. Ferric cations worked as a staining agent in this case. The cylindrical contour shown in Fig. 5 indicates again the well-defined structure of the polymer brush.

Conclusion

Amphiphilic cylindrical polymer brushes with a PAA core and a P*n*BA shell were synthesized via ATRP. The core of the polymer brush was first partially neutralized by NaOH, then reacted with iron cations via ion exchange. The formation of polychelates of the polymer brush and ferric (and ferrous) chlorides was confirmed by various methods. The conversion of the polychelates to magnetic nanocylinders via either reduction or alkalization of iron cations inside the polymer core is in progress and will be published elsewhere.

Acknowledgements The authors thank A. Göpfert for TEM measurements. This work was supported by the Deutsche Forschungsgemeinschaft and the Bavarian–French University Center.

References

1. Tsukahara Y, Mizuno K, Segawa A, Yamashita Y (1989) Macromolecules 22:1546–1552
2. Wintermantel M, Gerle M, Fischer K, Schmidt M, Wataoka I, Urakawa H, Kajiwara K, Tsukahara Y (1996) Macromolecules 29:978–983
3. Gerle M, Fischer K, Roos S, Müller AHE, Schmidt M, Sheiko SS, Prokhorova S, Möller M (1999) Macromolecules 32:2629–2637
4. Beers KL, Gaynor SG, Matyjaszewski K, Sheiko SS, Möller M (1998) Macromolecules 31:9413–9415
5. Cheng G, Boeker A, Zhang M, Krausch G, Müller AHE (2001) Macromolecules 34:6883–6888
6. Zhang M, Breiner T, Mori H, Müller AHE (2003) Polymer 44:1449–1458
7. Wang J-S, Matyjaszewski K (1995) J Am Chem Soc 117:5614–5615
8. Sohn BH, Cohen RE (1997) Chem Mater 9:264–269
9. Clay RT, Cohen RE (1998) Supermol Sci 5:41–48

Progr Colloid Polym Sci (2004) 126: 40–43
DOI 10.1007/b94006

Hideharu Mori
Delphine Chan Seng
Mingfu Zhang
Axel H.E. Müller

Hybrid silica nanoparticles with hyperbranched polymer and polyelectrolyte shells

Abstract We present the synthesis of hyperbranched polymer–silica hybrid nanoparticles by self-condensing vinyl polymerization (SCVP) via atom transfer radical polymerization (ATRP) from silica surfaces. ATRP initiators were covalently linked to the surface of silica particles, followed by SCVP of an initiator-monomer ("inimer") which has both a polymerizable acrylic group and an initiating group in the same molecule. Well-defined polymer chains were grown from the surface to yield hybrid nanoparticles composed of silica cores and hyperbranched polymer shells having multifunctional bromoester end groups. Self-condensing vinyl copolymerization of the inimer and *tert*-butyl acrylate from the functionalized silica nanoparticles created branched poly(*tert*-butyl acrylate)–silica nanoparticles. Novel hybrid nanoparticles with branched polyelectrolytes, poly(acrylic acid)–silica, were obtained after hydrolysis of linear segments of the branched poly(*tert*-butyl acrylate).

Keywords Hyperbranched polymer · Hybrid nanoparticle · Polyelectrolyte · Self-condensing vinyl polymerization · Atom transfer radical polymerization

H. Mori · D. Chan Seng · M. Zhang
A.H.E. Müller (✉)
Makromolekulare Chemie II and
Bayreuther Zentrum für Kolloide
und Grenzflächen,
Universität Bayreuth,
95440 Bayreuth, Germany
e-mail: axel.mueller@uni-bayreuth.de
Tel.: +49-921-553399;
Fax: +49-921-553393

Introduction

The well-controlled synthesis of nanocomposite materials consisting of a polymer with incorporated inorganic nanoparticles is an area of increasing research activity. Owing to wide scientific and industrial interest, several processes have been developed to prepare polymer–silica particle nanocomposites. In recent years, much attention has been paid to the use of controlled/"living" polymerizations from surfaces [1, 2], because this allows better control over the target molecular weight and molecular weight distribution of the target polymer. Highly branched polymers play an increasingly important role in interface and surface sciences, since their distinctive chemical and physical properties can be used advantageously as functional surfaces and as interfacial materials.

Here we describe a novel synthetic approach to prepare hybrid silica nanoparticles with hyperbranched polymer and branched polyelectrolyte shells [3]. Surface-initiated self-condensing vinyl (co)polymerization (SCV(C)P) was applied as a new method for the synthesis of highly branched polymers grafted from spherical surfaces. The resulting hybrid nanoparticles were characterized using various techniques.

Experimental

The preparation of the α-bromoester initiator attached to silica particles was conducted by the reaction of the crude trichlorosilyl derivative with silica particles in a manner similar to that reported previously [4, 5]. The hydrophilic silica gel (Aerosil 200, Degussa) used in this study had a mean particle diameter of 16 nm. The elemental analysis indicated that about 0.4 initiator molecules per nanometre squared are immobilized on the silica particles, corresponding to about 320 initiator molecules per silica particle. Synthesis of an acrylic AB^* inimer, 2-(2-bromopropionyloxy)ethyl acrylate (BPEA), was conducted by the reaction of an α-bromoacid halide with 2-hydroxyethyl acrylate in the presence of pyridine as reported previously [5, 6]. All polymerizations were carried out in a round-bottom flask sealed with a plastic cap under nitrogen. After the polymerization, any absorbed polymer formed was removed from the silica particles by Soxhlet extraction in tetrahydrofuran for 24 h. The branched structures were evaluated by characterization of the soluble polymers using conventional gel permeation chromatography (GPC), GPC/viscosity, and ^{1}H NMR measurements. Morphology and average particle sizes of the polymer–silica hybrid particles were characterized using transmission electron microscopy (TEM), field emission scanning electron microscopy (FE-SEM), scanning force microscopy (SFM), and dynamic light scattering (DLS). The detailed procedures were reported elsewhere [3].

Results and discussion

Hybrid silica nanoparticles with hyperbranched polymer shells were synthesized in two steps: (1) attachment of atom transfer radical polymerization (ATRP) initiators on the surface of silica nanoparticles and (2) SCVP of an initiator–monomer ("AB^* inimer", BPEA), as shown in Fig. 1. The formation of a 2-bromoisobutyryl fragment (B^*) layer on the silica surface was conducted by the reaction of the trichlorosilyl compound with the silica particles. The second step (SCVP) is based on an initiator–monomer ("inimer") of the general structure AB^*, where the double bond is designated A and B^* is a group capable of initiating the polymerization of vinyl groups. Because both the AB^* inimer and the functionalized silica particles have groups capable of initiating the polymerization, the chain growth can be started from both B^* in the initiators immobilized on the silica particles and a B^* group in the inimer. The surface-initiated SCVP from a flat silicon surface has been reported recently [5].

The bulk polymerization of BPEA with the functionalized silica particles was carried out in the presence of CuBr/N,N,N',N'',N''-pentamethyldiethylenetriamine (PMDETA) at a ratio of $[\text{BPEA}]_0:[\text{CuBr}]_0:[\text{PMDETA}]_0=125:1:1$. Under that condition, full conversion was reached after 2 h at 30 °C. The weight fraction of the surface-attached poly(BPEA) chains obtained after Soxhlet extraction was around 3.2 g polymer/g silica.

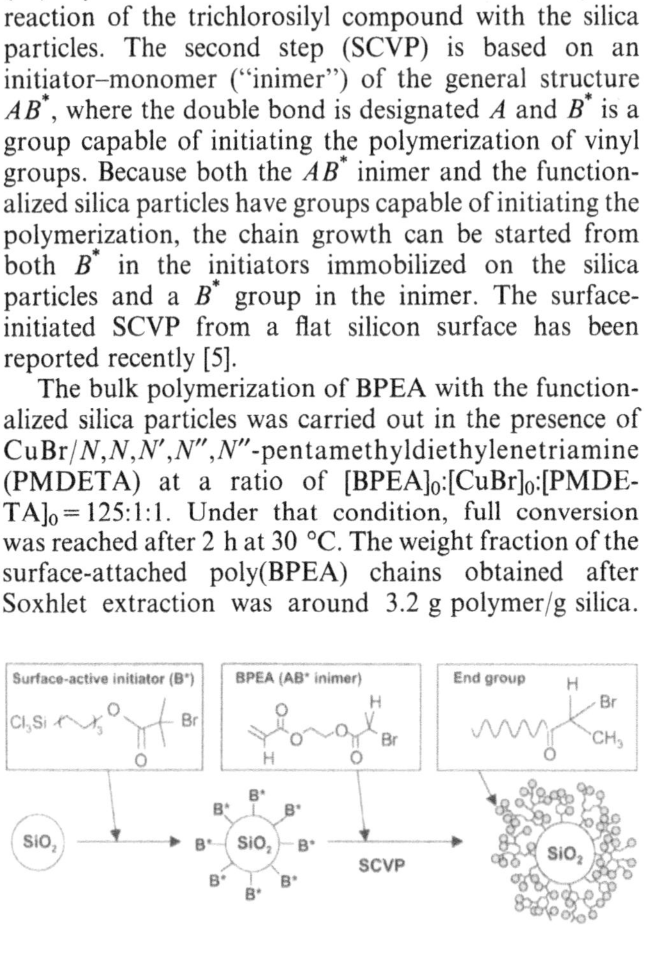

Fig. 1 Synthetic route for a silica particle with hyperbranched polymer shells

The number of the inimer (BPEA) units per silica particle was about 33,000, which can be calculated from the weight fraction and the number of the silica particles per gram (2.3×10^{17}/g silica). The value corresponds to the number of end groups per silica particle which may be distributed on the outermost surface of the hybrid as well as inside the polymer chains. Assuming complete initiator efficiency, which is roughly estimated from the grafting densities of the polymer chain and the initiator (about 0.4 chains/nm^2 in both cases), it can be derived that about 100 inimer units are attached to an initiating group on the surface, and the calculated molecular weight of the surface-grafted polymer is 25,000. The value is in agreement with the absolute molecular weight determined by GPC/viscosity ($M_n=30{,}800$). The surface coverage calculated by the elemental analysis and specific surface area of the bare silica substrate (200 m^2/g) was 9.9–21 mg/m^2, which is dependent on the polymerization conditions. Calculation from the weight fraction shows that about 10–20% of BPEA polymerizes from the surface of the silica nanoparticles under the conditions used in this study. The successful preparation of hyperbranched polymer–silica hybrid nanoparticles was confirmed by Fourier transform (FT) IR measurements (strong C=O vibration at 1,740 cm^{-1} and broad Si–O stretching vibration around 1,100 cm^{-1}) and elemental analyses (carbon content 28.0–32.9% in the products obtained by the bulk polymerization). The relative atomic composition of the organic part (C, 28.0; H, 3.5; Br, 18.8 in the product obtained by SCVP at $[\text{BPEA}]_0:[\text{CuBr}]_0=50$) is in agreement with the value (C, 25.1; H, 2.9; Br, 20.8) calculated from the weight fraction of the grafted polymer. These experimental data suggest that SCVP of AB^* inimer from the surface of the silica nanoparticle yields hybrid particles composed of a silica core and grafted hyperbranched polymer layers having multifunctional bromoester end groups. Such surface multifunctionality is ideally independent of the surface curvature of the core particle and the layer thickness of the polymer shell, which could not be achieved by linear polymers.

The synthesis of hybrid silica nanoparticles with branched polyelectrolyte shells was achieved by SCVCP of the AB^* inimer and a protected monomer (*tert*-butyl acrylate, tBuA), followed by hydrolysis, as shown in Fig. 2. Bulk copolymerization of BPEA with tBuA was conducted with CuBr/PMDETA in the presence of the functionalized silica particles. When the copolymerization was carried out at a comonomer ratio $[\text{tBuA}]_0:[\text{BPEA}]_0=\gamma=1.1$ or 6.1, the surface coverage was about 15–20 mg/m^2. The functionality of the end groups on the surface and the chemical composition as well as the structure of the branched polymers grafted on the silica nanoparticles could be controlled by the composition in the feed during the SCVCP, as confirmed by elemental analysis and FT-IR measurement.

The FE-SEM micrograph of the functionalized silica particles shows that spherical particles exist on the silicon substrate without aggregation, as can be seen in Fig. 3a. The diameter of the particles is in the range 40–150 nm, which is comparable to that determined by DLS. The larger size compared to the bare silica particles suggests that several functionalized silica particles may glue together into a spherical aggregate. The branched poly(*tert*-butyl acrylate) (PtBuA)–silica hybrid particles obtained by SCVCP at $\gamma = 6.1$ showed two different types of images. Spherical particles of about 40–160 nm in diameter are observed (Fig. 3b), corresponding to isolated particles. Aggregated forms are also seen (Fig. 3c) which are composed of many small particles. These isolated and aggregated forms are considered to be derived from the larger and smaller species of the functionalized silica particles (R_h obtained from DLS = 85 ± 3 and 20 ± 2 nm), respectively. Characteristic micrometer-sized dendric structures are seen occasionally (Fig. 3d), which may be formed by a crystallization process on the surface.

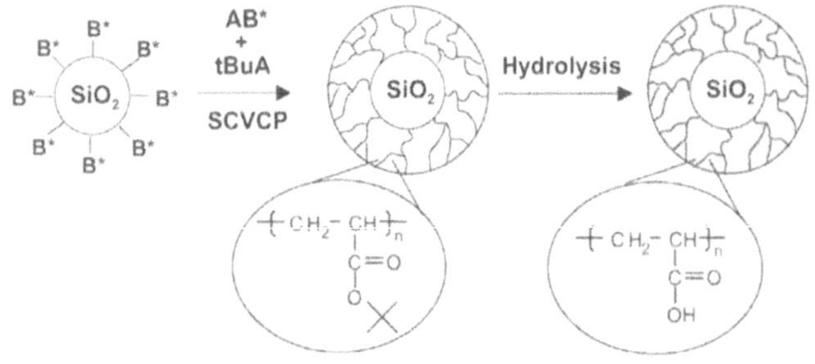

Fig. 2 Synthetic route for a silica particle with branched polyelectrolyte shells

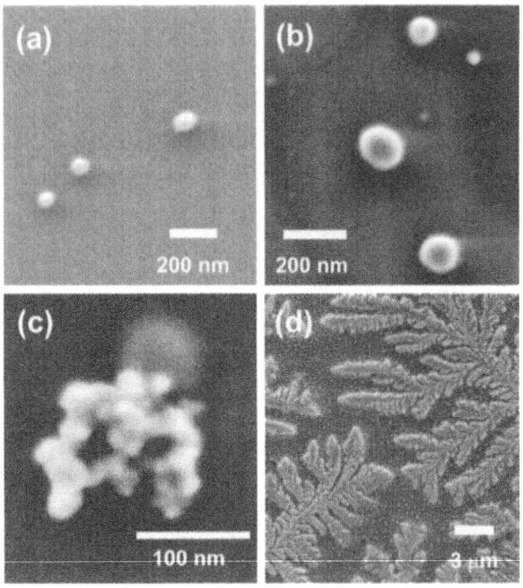

Fig. 3 Representative field emission scanning electron microscopy (*FE-SEM*) images of **a** the functionalized silica particles and **b–d** the branched poly(*tert*-butyl acrylate) (*PtBuA*)–silica hybrid particles obtained by self-condensing vinyl copolymerization (*SCVCP*) at $\gamma = 6.1$

Fig. 4 Representative **a–c** FE-SEM, **d** transmission electron microscopy, and **e** scanning force microscopy (phase 50° and cross-section taken at the position indicated by the *dotted line*) images of the branched PtBuA–silica hybrid particles obtained by SCVCP at $\gamma = 1.1$

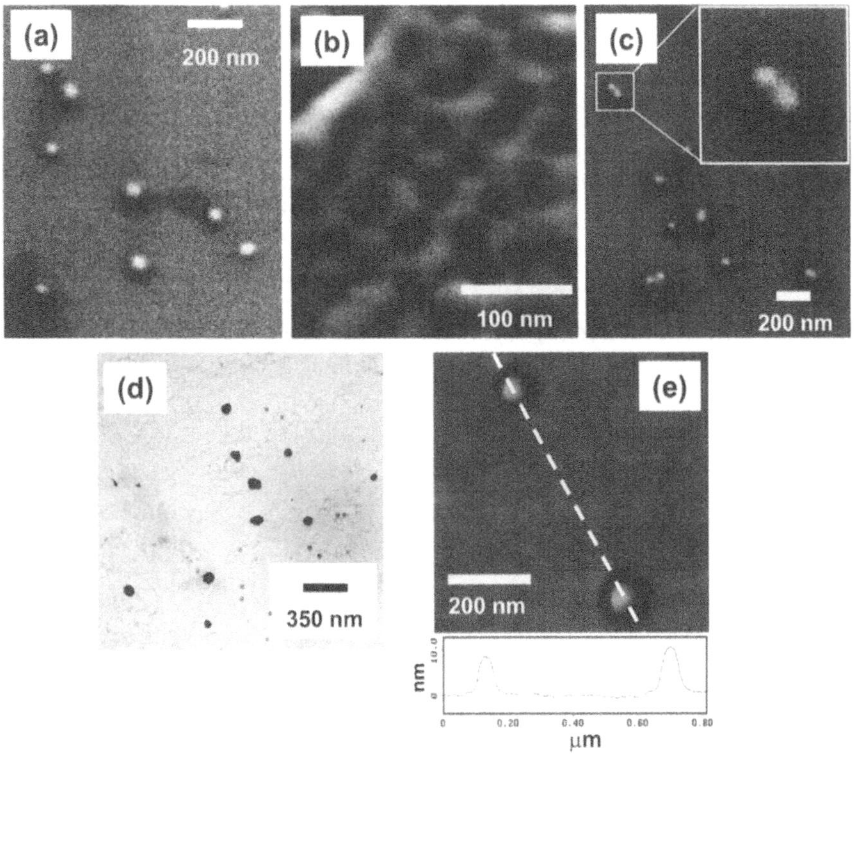

The small particles, which correspond to isolated forms, are seen in FE-SEM, TEM, and SFM measurements of the branched polymer–silica hybrid particles obtained at $\gamma = 1.1$ (Fig. 4a, d, e). Figure 4b shows a characteristic aggregation form containing continuous layers of silica particles in which the silica particles are arranged within a hybrid film on the silicon surface. In some isolated forms (Fig. 4c), bright spherical spots and dark halos around cores are observed. The dark halos were not observed in the functionalized silica particles before the polymerization, suggesting that the isolated form may consist of 1–4 silica particles as a core and the grafted copolymer shell existing around the core. These results indicate that the hybrid nanoparticles composed of the silica core and the hyperbranched polymer shell exist as isolated and aggregated forms, and the morphology can be modified by the polymerization conditions.

The *tert*-butyl ester groups in the branched PtBuA grafted on silica particles were hydrolyzed using trifluoroacetic acid. FT-IR measurements and elemental analyses indicate that the copolymerization of BPEA with tBuA, followed by hydrolysis of linear segments, yields characteristic silica hybrid nanoparticles with the tethered branched poly(acrylic acid) (PAA) chains as a shell in a controlled fashion. These polymers grafted on the nanoparticles can be designed to have a fairly open structure, allowing the functional materials, such as metal ions, to penetrate the film more easily than in conventional linear polymer layers. SCVCP was also applied for the synthesis of branched PAAs having different molecular weights and degrees of branching [7]. Hence, a well-controlled synthesis for these materials can lead to the creation of novel core–shell hybrids that are controllable on a nanoscopic scale and that have chemically sensitive interfaces.

Conclusions

Hyperbranched polymer–silica hybrid nanoparticles were synthesized by surface-initiated SCVP of an acrylic AB^* inimer from silica nanoparticles functionalized with monolayers of ATRP initiators. The SCVCP in the presence of the functionalized silica particles was a reliable method to prepare functional branched polymer–silica hybrid nanoparticles in which one can tune the architecture, chemical and physical properties, and particle morphology by the choice of comonomers and their composition in the feed. The present work can be applied to a wide range of inorganic materials having different size and shape for surface-initiated SCVP and SCVCP to allow the preparation of new functional branched polymer–metal hybrid nanoparticles.

Acknowledgements The authors thank C. Abetz (BIMF) and A. Göpfert for FE-SEM and TEM measurements. This work was supported by the Deutsche Forschungsgemeinschaft.

Reference

1. Zhao B, Brittain WJ (2000) Prog Polym Sci 25:677
2. Pyun J, Matyjaszewski K (2001) Chem Mater 13:3436
3. Mori H, Chan Seng D, Zhang M, Müller AHE (2002) Langmuir 18:3682
4. Husseman M, Malmström EE, McNamara M, Mate M, Mecerreyes D, Benoit DG, Hedrick JL, Mansky P, Huang E, Russell TP, Hawker CJ (1999) Macromolecules 32:1424
5. Mori H, Boeker A, Krausch G, Müller AHE (2001) Macromolecules 34:6871
6. Matyjaszewski K, Gaynor SG, Kulfan A, Podwika M (1997) Macromolecules 30:5192
7. Mori H, Chan Seng D, Lechner H, Zhang M, Müller AHE (2002) Macromolecules 35:9270

Progr Colloid Polym Sci (2004) 126: 44–46
DOI 10.1007/b93967

Catherine Amiel
Céline Galant
Loic Auvray

Ternary complexes involving a β-cyclodextrin polymer, a cationic surfactant and an anionic polymer

C. Amiel (✉) · C. Galant
Laboratoire de Recherche sur les Polymères, CNRS UMR 7581, 2–8 rue H. Dunant, 94320 Thiais, France
e-mail: amiel@glvt-cnrs.fr

L. Auvray
Laboratoire Matériaux Polymères aux Interfaces, Université d'Evry, Batiment des Sciences, rue du Père Jarlan, 91025 Evry Cedex, France

Abstract Interpolyelectrolyte complexes have been prepared by mixing in aqueous solutions an anionic polymer, a neutral β-cyclodextrin polymer and a cationic surfactant dodecyltrimethylammonium chloride. In this ternary system, electrostatic interactions are involved between the opposite charges of the surfactant and the anionic polymer, whereas inclusion complex interactions link the surfactant to the neutral β-cyclodextrin polymer. The occurrence of these ternary complexes has been proved for three different anionic polymers: sodium polystyrene sulfonate, sodium dextran sulfate and herring sperm DNA fragments.

Keywords β-Cyclodextrin polymer · Polyelectrolyte complexes · Cationic surfactant · Small-angle neutron scattering · Viscometry

Introduction

Association of a β-cyclodextrin polymer (p-β-CD), a cationic surfactant and an anionic polymer may lead to ternary complexes involving both inclusion complex and electrostatic interactions (Fig. 1). Use of DNA as the anionic polymer in this system provides an original gene carrier which has several interesting properties compared to the actual models of gene carriers [1, 2, 3]: DNA compaction can be realized by a simple control of the ratio of cationic surfactant to negative charges (DNA) in the medium. Moreover, inclusion complex interactions between the CD cavities and the surfactant are expected to be reversible and to improve dissociation of polymer/DNA complexes to a much higher degree than ionic interactions alone.

The aim of this study is to demonstrate the occurrence of these ternary complexes, using a p-β-CD with a branched structure, *n*-dodecyltrimethylammonium chloride (DTAC) as the cationic surfactant and anionic polymers of different rigidities. Sodium polystyrene sulfonate (NaPSS) was used as a model of a flexible polyanion, sodium dextran sulfate (NaDxS) was a model of a polyanion of intermediate stiffness, whereas DNA fragments have higher rigidity. The structures of the complexes were studied by viscometry and small-angle neutron scattering.

Material and methods

The p-β-CD was prepared by polycondensation with epichlorohydrin in a strongly alkaline medium [4]. The sample used had a weight-average molecular weight of 160,000 g mol^{-1} and a β-CD content of 59% w/w. The cationic surfactant DTAC was been supplied by Acros Organics, France, and was used as is. In all the solutions prepared for this study, the DTAC concentration was below the critical micelle concentration: 2×10^{-2} mol l^{-1}. NaDxS, supplied by Sigma, France, had a charge density corresponding to an average of two sulfate groups per glucose unit and a weight-average molecular weight of 10,000 g mol^{-1}. NaPSS, supplied by Alfa Aesar, Germany, had about one charge per monomer unit and a weight-average molecular weight of 70,000 g mol^{-1}. The herring sperm DNA fragments, supplied by Sigma, France, had molecular weights in the range 10,000–30,000 g mol^{-1}.

The viscosity measurements were performed with a Ubbelohde viscometer at 25 ± 0.1 °C. The small-angle neutron scattering experiments were carried out with the PACE spectrometer at the Laboratoire Léon Brillouin, Saclay, France. The experimental q range was $10^{-2} < q < 10^{-1}$ Å^{-1}.

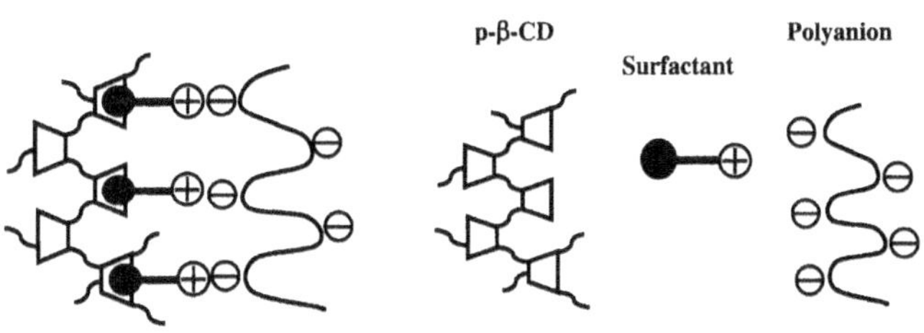

Fig. 1 Interaction scheme between the β-cyclodextrin polymer (p-β-CD), the cationic surfactant and the polyanion in the ternary complex

Results and discussion

The p-β-CD shows the regular behavior of a neutral polymer, characterized by a monotonic increase of the reduced viscosity as a function of the concentration (Fig. 2a) and a monotonic decrease of the scattering intensity as a function of the scattering wavevector (Fig. 3a). The DTAC molecules interact with p-β-CD by making inclusion complexes between the alkyl chains of the surfactants and the β-CD cavities. The complexation constant, 1,500 l mol^{-1}, is large enough to ensure the complexation of more than 95% of the surfactant in the concentration range of the experiments in this paper. Thus addition of DTAC to p-β-CD results in charged polymer systems. This polyelectrolyte behavior is clearly shown for p-β-CD/DTAC mixtures of stoichiometry 14/6 in Figs. 2a and 3a: the reduced viscosity decreases as a function of the concentration and the scattered intensity shows a correlation peak where q^*, the scattering wavevector at the maximum intensity, scales as $(\text{concentration})^{1/3}$ [5].

Addition of NaDxS to p-β-CD/DTAC solutions may result in a p-β-CD/DTAC/NaDxS ternary complex or in a mixture of free p-β-CD and DTAC/NaDxS binary complex. The obtaining of clear solutions in the concentration range explored is the first proof of the occurrence of the ternary complex because the DTAC/NaDxS binary aqueous mixtures precipitate in the same concentration range. Moreover, the electrostatic interactions were screened by interaction with the polyanion, as shown in Fig. 2b for the reduced viscosity, which increases as a function of the concentration, and in Fig. 3b, where the correlation peak has disappeared. The higher values obtained for the scattered intensity at low q are also indicative of the presence of several chain aggregates. The two interaction mechanisms, inclusion complexes and electrostatic interactions, thus participate in the formation of the p-β-CD/DTAC/NaDxS ternary complex.

Use of the two other polyanions, NaPSS and DNA, leads to comparable viscosity and small-angle neutron scattering behavior when mixed with p-β-CD and DTAC in aqueous solutions (Figs. 2b, 3b). However, these results need to be compared to the behavior of the DTAC/polyanion binary mixtures in order to be evidence of ternary complexation. As was observed with NaDxS, the DTAC/DNA binary mixture precipitates at the same concentration as in Fig. 2. The obtaining of a clear solution and the observation of large and less charged aggregates prove that p-β-CD/DTAC/DNA ternary complexes are formed. In contrast, DTAC/NaPSS forms a clear solution at the same concentration as in Fig. 2. Their scattering intensities at low q (shown in Fig. 3) are much lower than the p-β-CD/DTAC/NaPSS ones. This provides evidence of interpolymer complexation mediated by DTAC between p-β-CD and NaPSS. Thus, the formation of ternary complexes does not depend on the chemical nature nor on the rigidity of the three polyanions studied (DNA is more rigid than NaDxS, which is more rigid than NaPSS), given that they all have high charge densities.

The surfactant molecules connect the different chains in the aggregates. Depending on their relative molar concentration compared to the molar CD and anionic group concentrations, aggregates of different compactness are expected. Indeed, high connector densities should induce high compactness of the aggregates. The

Fig. 2 Reduced viscosities of the **a** p-β-CD solutions and a p-β-CD/dodecyltrimethylammonium chloride (*DTAC*) binary mixture and **b** p-β-CD/DTAC/polyanion ternary mixtures as a function of the p-β-CD concentration. In all the mixtures, the molar ratio of DTAC to CD concentration is kept constant and equal to 6/14 and, in the ternary mixtures, the number of negative charges is equal to the number of CD cavities

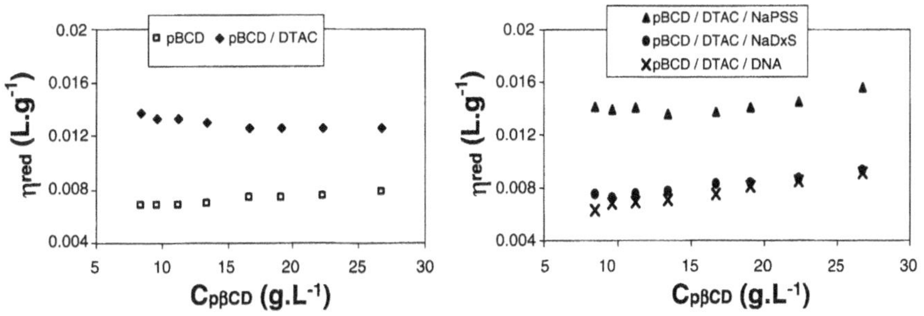

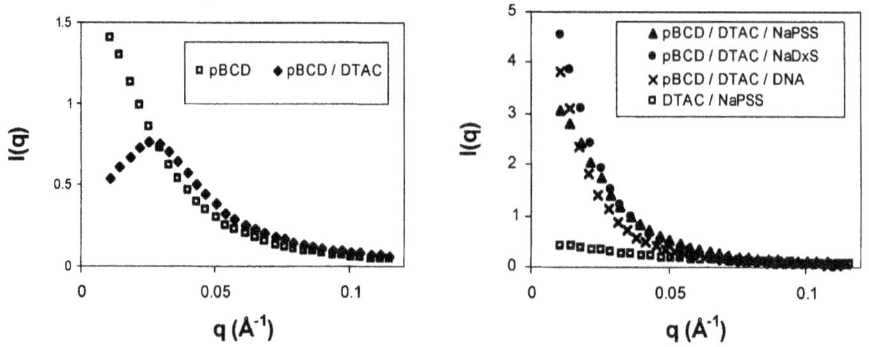

Fig. 3 Small-angle neutron scattering results for **a** a p-β-CD solution and a p-β-CD/DTAC binary mixture and **b** DTAC/sodium polystyrene sulfonate (*NaPSS*) binary mixtures and p-β-CD/DTAC/polyanion ternary mixtures. The β-CD concentration is equal to 14×10^{-3} mol l^{-1} and the molar ratios with the other compounds are the same as in Fig. 2

influence of the DTAC concentration was studied on the solubility of p-β-CD/DTAC/NaDxS mixtures where the polymers concentrations were held constant at a given 1/1 stoichiometry: the concentrations of β-CD and sulfate were14 × 10^{-3} mol l^{-1}. Clear solutions were obtained until a DTAC concentration of 7×10^{-3} mol l^{-1}, where a precipitate was observed. Previous studies [6] have shown that diffuse aggregates are obtained at low DTAC concentration (below 1.5×10^{-3} mol l^{-1}), whereas aggregates of increasing compactness are formed at higher DTAC concentrations. When the density of connectors reaches the stoichiometry a half positive charge (DTAC) per negative charge (NaDxS), the ternary complexes are so compact that they start to be insoluble. Similar experiments have been performed on p-β-CD/DTAC/NaPSS aqueous mixtures and the same qualitative behavior is observed. However, the solubility domain is increased as the precipitation occurs when the DTAC-to-sulfonate concentration ratio is close to 1, i.e. for a DTAC concentration twice as large as in NaDxS ternary complexes. These solubility differences of the ternary systems follow qualitatively the same solubility differences as the DTAC/polyanion binary systems. They can be related to the differences of backbone architectures of the polyanions (NaDxS with its branched structure is more rigid than NaPSS which is linear) and to their chemical structures (the NaDxS backbone is hydrophilic, whereas the NaPSS backbone is hydrophobic and may participate in DTAC micellar aggregates or in inclusion complexes with CD cavities).

Conclusion

Soluble ternary complexes were prepared by mixing in aqueous solutions a β-CD polymer, a cationic surfactant (DTAC) and a polyanion. Two different interaction mechanisms act together to link the three components: electrostatic interactions are involved between the positive charges of the surfactant and the negative charges of the anionic polymer and inclusion complex interactions link the surfactant to the neutral β-CD polymer. Viscosity and small-angle neutron scattering experiments prove that the formation of ternary complexes does not depend on the chemical nature nor on the rigidity of the three polyanions studied (DNA, NaDxS and NaPSS). The solubility limits of the ternary complexes are reached for highly compact aggregates when the concentrations of negative charges and positive charges are of the same order. The NaPSS ternary systems are much more soluble than the NaDxS ones. Future work will focus on the comparison of the structural properties of NaPSS and NaDxS ternary systems.

References

1. Pollard H, Remy JS, Loussouarn G, Demolombe S, Behr JP, Escande D (1998) J Biol Chem 273:7507
2. Rackstraw BJ, Martin AL, Stolnik S, Roberts CJ, Garnett MC, Davies MC, Tendler SJB (2001) Langmuir 17:3185
3. Rädler JO, Koltover I, Salditt T, Safinya CR (1997) Science 281:78
4. Renard E, Barnathan G, Deratani A, Sébille B (1997) Macromol Symp 122:229
5. Wang L, Bloomfield VA (1991) Macromolecules 24:5791
6. Galant C, Amiel C, Wintgens V, Sebille B, Auvray L (2002) Langmuir 18:9687

Progr Colloid Polym Sci (2004) 126: 47–50
DOI 10.1007/b93986

F. Bordi
C. Cametti
D. Gaudiuo
T. Gili
S. Sennato
A. Di Biasio

Polyelectrolyte coupling to charged lipid monolayers and to cationic liposomes

F. Bordi · C. Cametti · D. Gaudiuo
T. Gili · S. Sennato
Dipartimento di Fisica,
Universita' di Roma "La Sapienza",
Piazzale A. Moro 5, 00185 Rome, Italy

F. Bordi · C. Cametti (✉) · D. Gaudiuo
T. Gili · S. Sennato · A. Di Biasio
Istituto Nazionale per la Fisica della
Materia, Unita' di Roma 1, Rome Italy
e-mail: cesare.cametti@roma1.infn.it

A. Di Biasio
Dipartimento di Matematica e Fisica,
Universita' di Camerino, Camerino, Italy

Abstract We have investigated molecular interactions between a charged monolayer of different lipid compositions with a polyelectrolyte at the air–water interface by means of Langmuir isotherm measurements and between charged liposomes in an aqueous suspension by means of light scattering measurements, respectively. The condensing effect of the ionic character of the subphase the stability of the mixed monolayers at the appropriate lipid composition and the aggregation processes in the liposome suspension induced by an oppositely charged polyelectrolyte are briefly discussed.

Keywords Liposome · Lipid monolayers · Liposome aggregates

Introduction

Monolayers of charged lipids coupled to oppositely charged polyelectrolytes at the air–water interface represent a well-defined model system to study the dynamical properties of biological membranes since it has been shown that a variety of important phenomena occurring in bilayer vesicles can be successfully investigated in monolayers at an air–water interface [1, 2].

A very sensitive means of measuring the structural arrangement of various lipid components in biomembranes is to study the surface pressure as a function of the lipid molecular area (π–A curves), from which the thermodynamic properties of the monolayer, such as miscibility, stability and surface activity of lipids, can be deduced [3, 4]. Moreover, the versatility of lipid monolayers at the air–water interface is useful for constructing phase diagrams of amphiphile-charged polymer systems. We have undertaken a systematic investigation of interactions between different lipids which, attracting and/or packing with each other in different ways, could naturally separate into domains (rafts) within the membranes. Rather than expect that lipids remain homogeneously distributed within the membrane, one should expect that domains form spontaneously [5]. Organization of membranes into domains is biologically interesting, as membrane domains can strongly affect membrane functions by concentrating interacting species in particular regions of the surface or by excluding diffusing molecules from some regions [6]. The aim of this investigation is to evaluate the persistence of these domains and the physicochemical conditions by which lipid domains are created, stabilized or disrupted. In this study, we have investigated the stability of lipid domains by means of surface tension measurements to determine their basic thermodynamic properties (miscibility and stability in a mixed monolayer, molecular surface area). We used planar lipid monolayers to study the domain formation of dioleoyltrimethylammoniumpropane (DOTAP), a cationic lipid, in a background of dipalmitoylphosphatidylcholine (DPPC), a zwitterionic phospholipid, and the phase separation for different lipid compositions has been explored in the presence of an aqueous subphase containing a simple uni-univalent electrolyte (NaCl) or a charged polymer (polyacrylate sodium salt, NaPAA) of different molecular weight (5.1 and 225 kD). Furthermore, we studied, by means of dynamic light scattering techniques, the aggregation of a

DOTAP liposome suspension induced by the same polyelectrolyte, evidencing the simultaneous presence, in a steady-state regime, of two different structured aggregates of different average size (in the range of 500 nm and 1–2 μm, respectively). We argue that our phenomenological finding is generic to a large class of polyelectrolytes (including DNA). The different effects observed in planar systems (charged lipid monolayers at the air–water interface) and in spherical surfaces (charged liposomes in aqueous suspension) are briefly discussed. This investigation is mainly devoted to the basic phenomena of complexation of natural and synthetic polyelectrolytes with oppositely charged surfaces (in particular with a liposomal surface) in the light of possible implications of biophysical interest.

Experimental

Materials

DOTAP and DPPC were commercially obtained from Sigma Chem. Co. and were used without further purification. NaPAA of molecular weight 5.1 and 225 kD was obtained from PolyScience.

Liposome preparation

An appropriate amount of DOTAP or a DOTAP–DPPC mixture was dissolved in a methanol–chloroform solution (1:1 v/v) and was placed in a glass vessel. The organic solvent was removed under vacuum by rotary evaporation. The lipid film obtained was resuspended in water at a temperature of 43 °C, well above the phase-transition temperature, for 2 h. The resulting aqueous lipid mixture was sonicated at room temperature for 1 h in pulsed-power mode until the solution appeared optically transparent in white light. The solution consisted of stable liposomal vesicles of about 90-nm diameter with a polydispersity of about 0.20.

Langmuir monolayers and surface tension measurements

In the surface tension measurements, 20–25 ml chloroform solutions containing DOTAP or DPPC, both at a concentration of 1 mg/ml, or a mixture of the two lipids at varying compositions, were deposited onto the aqueous subphase. The measurements were carried out by means of a Wilhelmy plate attached by a thread to a force transduction using a computer-controlled Langmuir-film balance (KSV trough, MINITROUGH). After the deposition, the solvent was allowed to evaporate for 5 min before the beginning of compression (20 mm/min), while, when polyelectrolytes were present, this waiting time was increased to 30 min to allow the possible adsorption of the polymer chains. All the measurements were performed at 25.0 ± 0.2 °C. The results are mean values of at least three different measurements. We investigated the pressure–area isotherms, recording the surface pressure during the monolayer compression, of DPPC, DOTAP and mixed DOTAP–DPPC films at the air–water interface in the presence of an ultrapure water subphase, of simple monovalent electrolyte (NaCl at different concentrations between 10^{-5} and 1 mol/l) and finally in the presence of a negatively charged polyelectrolyte (NaPAA) of different molecular weights, 5.1 and 225 kD, at different concentrations between 10^{-5} and 10^{-2} monomol/l. The appropriate content of polyelectrolyte solution was added to the water subphase before spreading the lipid solution.

Dynamic light scattering measurements

Owing to Brownian motion of particles moving in the solution, scattered light fluctuates in intensity. The homodyne intensity–intensity correlation function, $g^{(2)}(q,t)$, was measured at a scattering angle of 90° with a Brookhaven (BI-9000AT) digital correlator using a laboratory-built setup equipped with a 10-mW He–Ne laser. Here, q is the amplitude of the scattering vector given by $q=(4\pi n/\lambda)\sin(\theta/2)$, where n is the refractive index of the medium and λ is the wavelength of the excitation light in the vacuum. For a Gaussian distribution of the intensity profile of the scattered light, $g^{(2)}(q,t)$ is related to the electric field correlation function, $g^{(1)}(q,t)$, by

$$g^{(2)}(q,t) = A\left\{1 + b\left[g^{(1)}(q,t)\right]^2\right\}, \qquad (1)$$

where A and b are the experimental baseline and a constant depending on the optical system. In the present work, we analyzed the autocorrelation functions by using the program CONTIN, which employs the constrained regularization method to calculate the intensity-weighted distribution of the hydrodynamic radii. More detailed information about the size distribution of polydisperse systems by means of a Laplace inversion of the correlation function can be found elsewhere [7, 8].

Results and discussion

The mean molecular areas (at the collapse pressure) for DOTAP, DOTAP/DPPC (0.6:0.4 w/w) and DPPC monolayers as a function of the ionic composition of the aqueous subphase are shown in Fig. 1. As can be seen (Fig. 1a), in the presence of a simple uni-univalent electrolyte solution (NaCl, in the concentration range from 10^{-5} to 1 mol/l), there is a marked condensing effect in the charged pure and mixed monolayers between 0.05 and 0.1 mol/l. The charged polyions present (Fig. 1b) shift this condensing effect towards lower concentrations and, moreover, a molecular weight dependence is evidenced. For both subphases investigated, the mean molecular area, to a first approximation, does not change appreciably in the case of DPPC monolayers, whose uncharged (zwitterionic) structure prevents or reduces electrostatic interactions with aqueous counterions. The thermodynamic stability of the mixed DOTAP–DPPC monolayers was evaluated from the excess free energy of mixing, ΔG_{ex}, calculated according to the expression

$$\Delta G_{ex} = \int_0^{\pi} (A_{12} - A_1X_1 - A_2X_2)\,d\pi, \qquad (2)$$

where A_1 and A_2 are the molecular areas of the monolayer of component 1 and component 2 and A_{12} is the measured value of the mixed monolayer at a given pressure, π, X_i ($i=1, 2$) are the mole fractions of the two components. ΔG_{ex} is shown in Fig. 2 as a

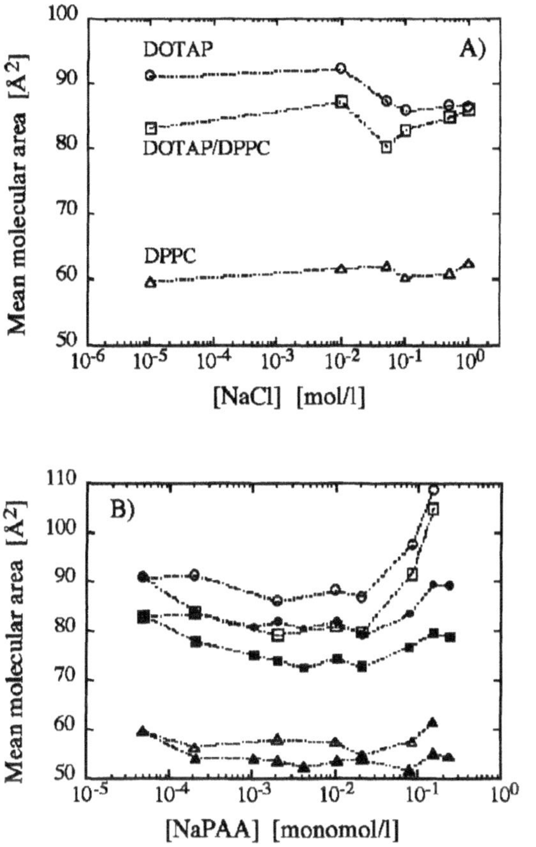

Fig. 1A, B The mean molecular area of dipalmitoylphosphatidylcholine (*DPPC*), dioleoyltrimethylammoniumpropane (*DOTAP*) and mixed DOTAP–DPPC monolayers as a function of the ionic strength of the subphase. **A** In the presence of NaCl electrolyte concentration in the range 10^{-5} to 1 mol/l. **B** In the presence of polyacrylate sodium salt (*NaPAA*) of two different molecular weights: $M_w = 5.1$ kD (*open symbols*); $M_w = 225$ kD (*closed symbols*). DOTAP (*circles*), DOTAP–DPPC (0.6:0.4 w/w) (*squares*), DPPC (*triangles*). The *dotted lines* are drawn for visual purposes only

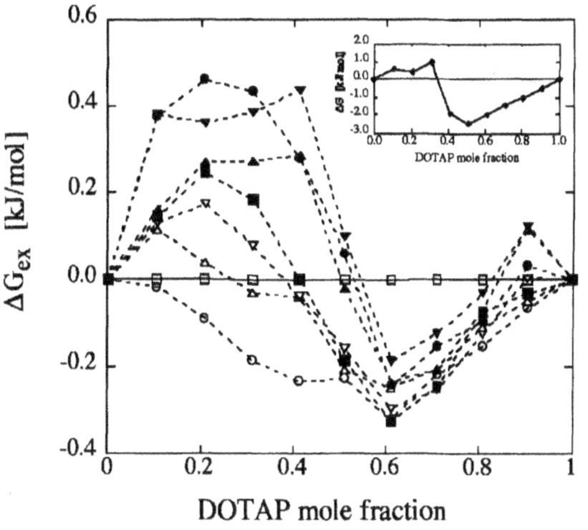

Fig. 2 The excess energy, ΔG_{ex}, of mixed DOTAP–DPPC monolayers as a function of composition at different surfaces pressures: 5 mN/m (*open squares*); 10 mN/m (*open circles*); 15 mN/m (*open up triangles*); 20 mN/m (*open down triangles*); 25 mN/m (*closed squares*); 30 mN/m (*closed circles*); 35 mN/m (*closed up triangles*); 40 mN/m (*closed down triangles*). The *inset* shows the ΔG_{ex} values calculated at the collapse pressure (varying from 45 to 70 mN/m, according to the monolayer composition). The *dotted lines* are drawn to guide the eye only. The errors in ΔG_{ex} derived from the surface pressure–molecular area integration are quoted to within a few percent

function of DOTAP–DPPC composition at different pressures, from 5 mN/m to the collapse pressure. At all the pressures investigated, there is a minimum at a DOTAP-to-DPPC mole fraction of 0.6, suggesting, at this concentration, a structure with a higher thermodynamic stability compared with that of the pure components. Deviation from ideality evidences a different molecular organization of the two lipids, resulting in the formation of cationic lipid domains in a zwitterionic lipid matrix.

The influence of the geometry (planar or spherical interface) in the external leaflet of the bilayer in a liposome structure or in the monolayer at an interface was investigated by means of dynamic light scattering measurements, evaluating the mean average size of the complexes resulting from liposome interactions, in the presence of NaPAA solution. The problem of the correspondence between a monolayer and a bilayer and under what conditions a monolayer at the air–water interface may be equivalent to one leaflet of bilayer liposomes in water has been extensively considered by Feng [1, 2]. The kinetics of the complex formation is characterized by the presence of two different structures, the first of typical size lying in the range 100–500 nm and the second with size continuously increasing with time from some hundred nanometers to some microns. A typical example of this time evolution is shown in Fig. 3 for DOTAP liposomes in the presence of NaPAA polyelectrolyte of molecular weight $M_w = 255$ kD. The different structures are characterized by the lipid-to-polyelectrolyte charge ratio, X. As can be seen, at $X < 1$, the aggregates split into two well-separated distributions, corresponding to the stable and the time-increasing structure, respectively. As far as the first component is concerned, the aggregation mechanism can be described within the usual scheme of polymer-coated colloidal particle stability, resulting in aggregates whose stability is governed by bridging, depletion flocculation or depletion stabilization, according to the polymer concentration regime. A more detailed analysis of this phenomenology is discussed elsewhere [9]. The second, time-dependent, component results in the formation of clusters obeying fractal morphology. The time evolution of the second component is shown in Fig. 4 for a DOTAP liposome suspension at $X = 0.2$. As can be seen, there is an increase in the average size, accompanied by a broadening of the size distribution. This evolution of the average

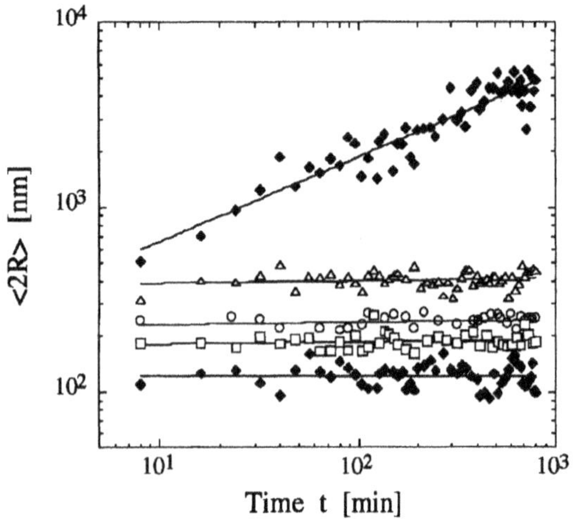

Fig. 3 The time evolution of the average diameter, $<2R>$, of DOTAP–NaPAA complexes, for different values of the charge ratio, X: $X = 8.1$ (*squares*); $X = 3.9$ (*circles*); $X = 1.9$ (*triangles*); $X = 0.2$ (*diamonds*). The polymer molecular weight is 225 kD. The bimodal distribution appears at $X = 0.2$. Both peaks, indicated by the same *diamond symbols*, of the two distributions are shown

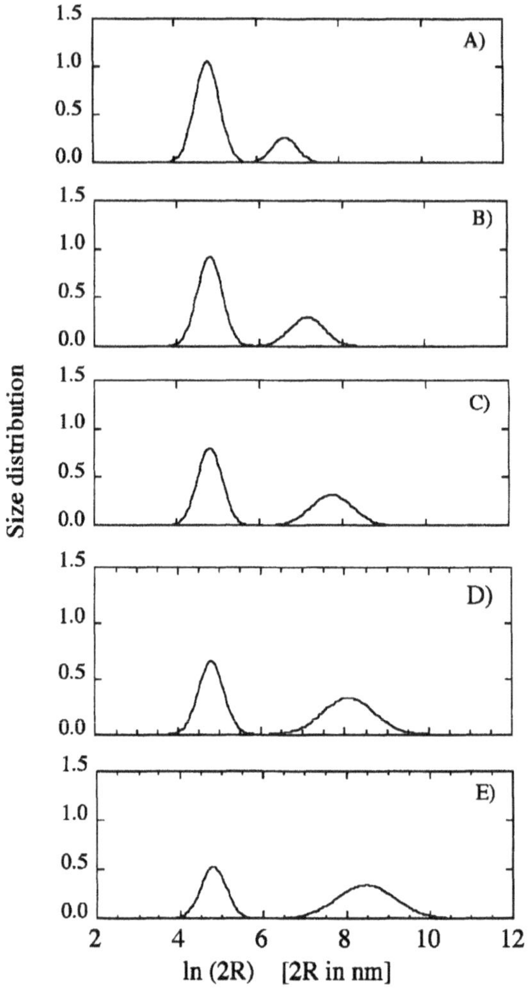

Fig. 4 A–E The size distribution of the DOTAP–NaPAA complexes at different times from the beginning of the aggregation process. **A** $t = 30$ min, **B** $t = 80$ min, **C** $t = 200$ min, **D** $t = 400$ min, **E** $t = 600$ min. $X_i = 0.2$

hydrodynamic radius, R_H, of the cluster is described by the scaling law:

$$R_H(t) = R_H(0)\left(\frac{t}{t_0}\right)^{z/D}, \quad (3)$$

where D is the fractal dimension of the cluster and z is the kinetic exponent. In all the systems investigated, when this second structure appears, both in the pure and in the mixed liposome systems, the exponent z/D assumes values very close to that predicted by the diffusion-limited cluster aggregation regime. It is noteworthy that the formation and the time evolution of these complexes differ from the usual behavior observed in salt-induced aggregation of charge-stabilized colloidal particles, where, at long time, all the aggregation regimes yield a cluster of infinite size. In this case, in contrast, there is the simultaneous presence of two populations, the growth of the large component being reminiscent of that observed in usual colloidal systems. These characteristics can be relevant for constructing self-assembling structures of technological utility.

References

1. Feng S (1999) Langmuir 15:998–1010
2. Feng S, Zhu G, Low MY (2000) Langmuir 16:7401–7410
3. Gong K, Feng S, Go ML, Soew PH (2002) Colloids Surf 207:113–125
4. Bordi F, Cametti C, De Luca F, Gili T, Gaudino D, Sennato S (2003) Colloids Surf B 29:149–157
5. Edidin M (1977) Curr Opin Struct Biol 7:528–532
6. Jacobson K, Dietrich C (1999) Trends Cell Biol 9:87–91
7. Provencher SW (1982) Comput Phys Comm 27:229–242
8. Hallett FR, Watton J, Krygsman P (1991) Biophys J 59:357–362
9. Bordi F, Cametti C, Gili T, Gaudino D, Sennato S (2003) Bioelectrochemistry 59:99–106

Progr Colloid Polym Sci (2004) 126: 51–54
DOI 10.1007/b93981

R. Pieri
A. Chittofrati
P. Baglioni

Dynamic surface tension and Gibbs elasticity of perfluoropolyether surfactant films

R. Pieri (✉) · A. Chittofrati
Solvay Solexis, v.le Lombardia 20,
20021 Bollate, Milan, Italy
e-mail: riccardo.pieri@solvay.com
Tel.: +39-2-38356205
Fax: +39-2-3835-6355

P. Baglioni
Department of Chemistry, University of Florence, v. della Lastruccia, 50019 Sesto Fiorentino, Florence, Italy

Abstract Axisymmetric drop shape analysis has been applied to provide a first insight into the adsorption kinetics and surface viscoelasticity of perfluoropolyether carboxylic salts at the water–air interface. The two surfactants $Cl(C_3F_6O)_nCF_2COONH_4$, with two or three perfluoroisopropoxy units, *n*, have been examined in aqueous solution at molar concentrations well below their critical micelle concentration (cmc), first to compare, at constant reduced concentration [log*c*–log(cmc)], the rate of surface tension reduction with time up to 2,500 s, typically sufficient to attain equilibrium conditions. Then the surface area of each drop was varied with a sinusoidal profile at a frequency in the range 0.016–0.2 s^{-1}, to obtain the Gibbs elasticity, ε_0, and the relaxation frequency, ω_0. The adsorbed films of both perfluoropolyether surfactants displayed ε_0 decreasing upon concentration increase, while ε_0 increased with the surfactant chain length.

Keywords Axisymmetric drop shape analysis · Perfluoropolyether · Surface elasticity · Dynamic surface tension

Introduction

The dynamics of amphiphile adsorption at the water–air interface plays a crucial role in a wide range of applications where the characteristics of the surface film or the rate of the reduction of surface tension are recognized to have a large impact on the surfactant performance. A rich scientific literature deals with the issue of dynamic surface tension and surface elastic properties of hydrocarbon surfactants, for instance Refs. [1, 2, 3, 4, 5], while very few examples of such studies appear to be available on fluorosurfactants [6, 7] or perfluoropolyether (PFPE) surfactants, that we believe to be particularly interesting for their spontaneous packing features driven by a large tail volume and peculiar cohesive energy among their tails [8]. In particular, no such data are available to date for the PFPE carboxylic surfactant salts that are the subject of this work.

The aim is a first insight at the air–water interface, via axisymmetric drop shape analysis (ADSA), of the kinetics of surface tension reduction and surface elasticity given by two homologous ammonium salts differing in the length of the PFPE tail made of perfluoroisopropoxy units. They are found to display a rather high Gibbs elasticity modulus, ε_0, with respect to the typical range of hydrocarbon-based amphiphiles.

Experimental

Materials

The surfactants studied are two high-purity laboratory samples, belonging to the general PFPE class (Structure 1), $Cl\text{–}(C_3F_6O)_n\text{–}CF_2COO^-NH_4^+$ where the number of perfluoroisopropoxy units, *n*, is 2 or 3. To identify each salt, *n* and the counterion are used, as reported in a previous work [9] where their main features in aqueous solutions at 25 °C were described in detail. At relatively high concentration in water, they appear to form liquid-crystalline phases, primarily of lamellar type as shown by optical microscopy, with results qualitatively similar to the L_α phases identified earlier, via NMR, on a mixture with a narrow molecular weight distribution of the same class of surfactants [10].

In practice, these salts display high capability of spontaneous packing, as depicted for PFPE carboxylates having the same chain structure but perfluoromethoxy end groups. [11, 12, 13]. Some qualitative analogies may be described among the structuring of these PFPE carboxylic salts and the salts of perfluoroalkanoic acids [14] and also double-chain hydrocarbon surfactants [15], although there are structural differences. This is an indicator of strong interchain interactions, quite unexpected for these laterally branched fluorinated chains.

Having assessed the very small dependence of the critical micelle concentration (cmc) upon temperature up to 40 °C, we used the cmc values in Ref. [9] here. Water was of Millipore Milli-Q grade.

The work is focused on aqueous premicellar solutions, at the same reduced concentration [log*c*−log(cmc)] for both salts, to compare first the rate of surface tension reduction, then the response to surface area deformations.

The temperature was kept 28.0 ± 0.2 °C in all measurements, to enable sufficient solubility for the $n = 3$ salt and to account for the Krafft temperature constraints. A T_k well below our working temperature was found by Kallay via conductivity of the $n = 2$ salt solutions, while work is in progress for the $n = 3$ analogue. Also, small-angle neutron scattering results at 28 °C are available for micellar solutions of the $n = 2$ surfactant [16] which confirm the presence of spherical-to-ellipsoidal micelles in the appropriate concentration range and give further evidence of the spontaneous structuring of these PFPE surfactants.

Methods

The tensiometer for ADSA was an ITConcept Tracker. The experimental setup is described elsewhere [1, 2]. In these measurements, once the solution equilibrium was reached, the surface area of the pendant drop was perturbed with a sinusoidal profile, changing the oscillation frequency in the range 0.016–0.2 s^{-1} and with amplitude $\Delta A/A = 0.1$–0.2.

Also, the dynamic surface tension was determined by a Lauda TVT2 drop-volume tensiometer, according to the quasi-static method [17, 18]. We used a 2.5-ml syringe, with a capillary outer radius of 1.37 mm.

Principle of the ADSA technique

ADSA, based on the analysis of the shape of a pendant drop via the Gauss–Laplace equation, provides the dynamic surface tension of both static and oscillating interfaces [1, 2].

After an equilibration time, typically on the 1,000-s scale, the drop is subjected to sinusoidal changes of its volume to monitor the dynamic surface tension response to the surface area fluctuations.

The viscoelastic modulus is defined as

$$\varepsilon = \frac{d\gamma}{d\ln A}, \tag{1}$$

and is a function of the oscillation frequency ω [19]:

$$|\varepsilon| = \left(\varepsilon_r^2 + \varepsilon_i^2\right)^{1/2} = \frac{\varepsilon_0}{\left(1 + 2\sqrt{\frac{\omega_0}{2\omega}} + \frac{\omega_0}{\omega}\right)^{1/2}}, \tag{2}$$

where ε_0 is the Gibbs elasticity and ω_0 is the relaxation frequency.

The surface elasticity can be obtained from the experimental values of ε as a function of ω, by a two-parameter fitting routine.

Results and discussion

Drop-volume measurements were used to roughly estimate a diffusion coefficient for the PFPE surfactant, according to the Ward–Tordai equation [20]:

$$\Gamma(t) = 2c_0\left(\frac{Dt}{\pi}\right)^{1/2} - 2\left(\frac{D}{\pi}\right)^{1/2}\int_0^{t^{1/2}} c(0, t-\tau)\,d\tau^{1/2} \tag{3}$$

and its asymptotic solutions at short and long times [21]. Values of the diffusion coefficient in the range 10^{-6}–10^{-5} $cm^2\ s^{-1}$ were obtained, as fairly reasonably expected in view of the molecular weight and size of these two surfactants.

The rate of surface tension reduction, shown in Figs. 1 and 2 for dilute $n = 2$ and $n = 3$ solutions, is on the same time scale as the literature data [7] for perfluoroalkyl-analogues in comparable conditions.

A typical ADSA experimental profile of the dynamic adsorption behavior of the PFPE surfactant is reported in Fig. 3. The first part of the profile, at $t < 500$ s, confirms the profile seen in drop-volume measurements.

The relationship between the elasticity modulus and the oscillation frequency is shown in Figs. 4 and 5, for the $n = 2$ and $n = 3$ salts, respectively. The fitted curves obtained according to Eq. (2) are also shown in the same figures.

The experimental data used for the fitting are summarized in Tables 1 and 2. The values of Gibbs elasticity, ε_0, decrease upon increasing concentration, while ε_0 increases with the chain length. The latter observation seems to be confirmed by some preliminary results on a mixture of the PFPE surfactant for $n = 3$ and $n = 4$. It is worth noting that the Gibbs modulus for the $n = 3$ salt is significantly higher than the values in the range 10–30 mN m^{-1}, typically reported in the

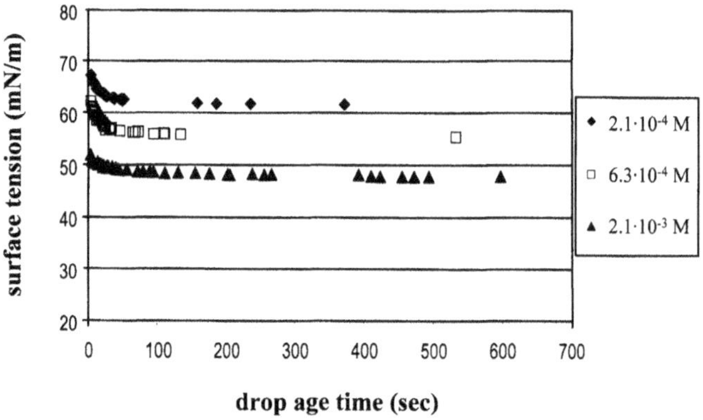

Fig. 1 Surface tension as a function of drop time by TVT2 measurements on $n = 2$ NH_4 solutions

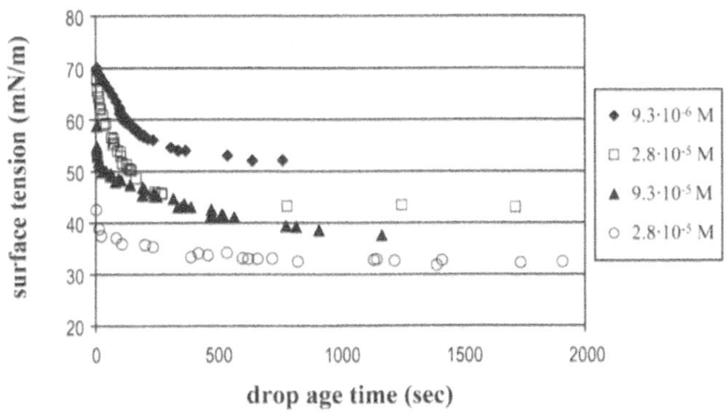

Fig. 2 Surface tension as a function of drop time by TVT2 measurements on $n = 3$ NH_4 solutions

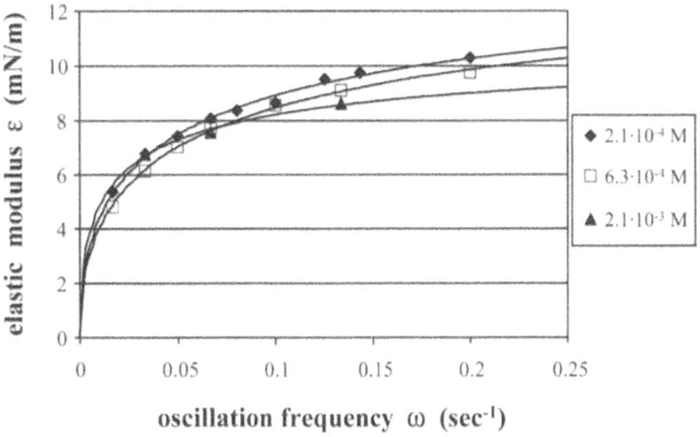

Fig. 4 Elastic modulus as a function of oscillation frequency, for $n = 2$ NH_4 solutions

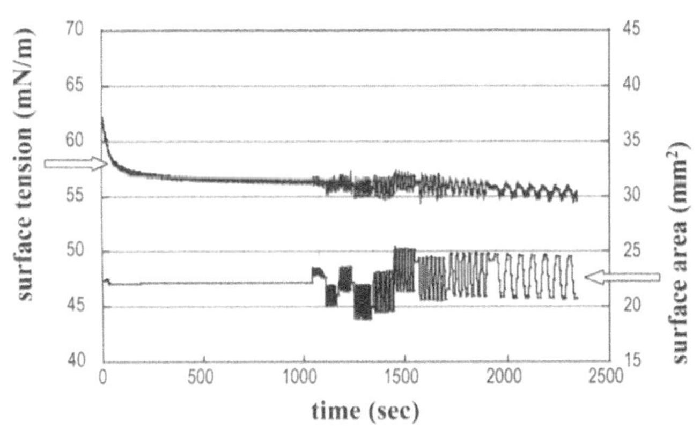

Fig. 3 Dependence of surface tension and surface area on time. Axisymmetric drop shape analysis measurement with area changes on $n = 2$ NH_4 6.3×10^{-4} M solution

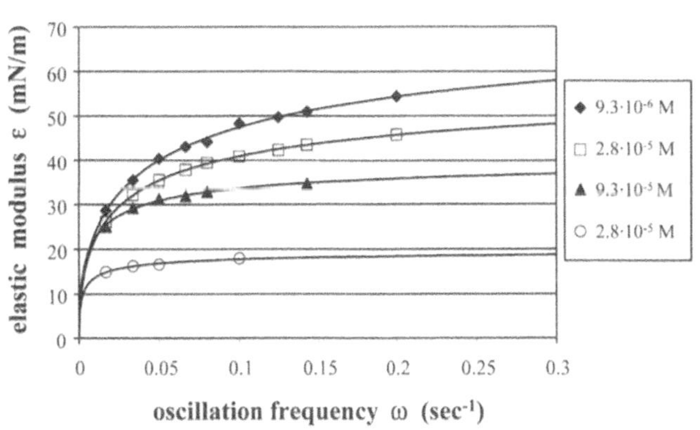

Fig. 5 Elastic modulus as a function of oscillation frequency, for $n = 3$ NH_4 solutions

Table 1 Surface elasticity of aqueous solutions of the ammonium salt of $n = 2$

C (M)	2.1×10^{-4}	6.3×10^{-4}	2.1×10^{-3}
Log$c$–log(cmc) (cmc = 2.1×10^{-2} M)	−2	−1.5	−1
ε_0 (mN m^{-1})	15.3 ± 0.4	15.4 ± 0.3	11.4 ± 0.4
ω_0 (s^{-1})	$7.34 \pm 0.89 \cdot 10^{-2}$	$9.34 \pm 0.71 \cdot 10^{-2}$	$2.40 \pm 0.52 \cdot 10^{-2}$
R^2	0.998	0.999	0.999

Table 2 Surface elasticity of aqueous solutions of the ammonium salt of $n = 3$

c (M)	9.3×10^{-6}	2.8×10^{-5}	9.3×10^{-5}	2.8×10^{-4}
Log$c$–log(cmc) (cmc = 9.3×10^{-4} M)	−2	−1.5	−1	−0.5
ε_0 (mN m^{-1})	79.6 ± 1.6	61.7 ± 0.4	42.0 ± 0.7	20.2 ± 0.5
ω_0 (s^{-1})	$6.64 \pm 0.56 \cdot 10^{-2}$	$3.92 \pm 0.13 \cdot 10^{-2}$	$1.03 \pm 0.12 \cdot 10^{-2}$	$0.35 \pm 0.09 \cdot 10^{-2}$
R^2	0.997	0.999	0.993	0.973

literature for phospholipids and hydrocarbon-based surfactants.

Conclusions

The rate of surface tension reduction, by drop-volume, of premicellar aqueous solutions of two PFPE carboxylic salts is found to be consistent with the few literature data on different fluorosurfactant structures.

Having assessed equilibrium conditions, we were able to use ADSA to provide a preliminary insight into viscoelastic properties of the water–air interface. The Gibbs elasticity modulus decreases with concentration, and increases with chain length. Particularly high values were found with the longer-chain surfactant with respect to hydrocarbon-based surfactants.

References

1. Benjamins J, Cagna A, Lucassen-Reynders EH (1996) Colloids Surf A 114:245
2. Lucassen-Reynders EH, Cagna A, Lucassen J (2001) Colloids Surf A 186:63
3. Susnar SS, Chen P, del Rio OI, Neumann AW (1996) Colloids Surf A 116:181
4. Miller R, Li JB, Bree M, Loglio G, Neumann AW, Möhwald H (1998) Thin Solid Films 327:224
5. Li JB, Kretzschmar G, Miller R, Möhwald H (1999) Colloids Surf A 149:491
6. Hirt DE et al (1990) Colloids Surf 44:101
7. Eastoe J, Paul A, Rankin A, Wat R (2001) Langmuir 17:7873
8. Marchionni G, Ajroldi G, Righetti MC, Pezzin G (1993) Macromolecules 26:1751
9. Chittofrati A, D'Aprile F, Lenti D, Maccone P, Pieri R, Visca M (2001) Prog Colloid Polym Sci (submitted)
10. Caboi F, Chittofrati A, Lazzari P, Monduzzi M (1999) Colloids Surf A 160:47
11. Chittofrati A, Boselli V, Visca M, Friberg SE (1994) J Dispersion Sci Technol 15:711
12. Giannetti E, Chittofrati A, Sanguineti A (1997) Chim Ind 79:22
13. Caboi F, Chittofrati A, Monduzzi M, Moriconi C (1996) Langmuir 12:6022
14. Tiddy GJT (1971) Symp Faraday Soc 5:150
15. Monduzzi M, Knackstedt MA, Ninham BW (1995) J Phys Chem 99:17772
16. Gambi C, Giordano R, Chittofrati A, Pieri R, Baglioni P, Texeira J (2002) Appl Phys A 74:S436
17. Miller R, Hofmann A, Hartmann R, Schano KH, Halbig (1992) J Colloid Interface Sci 60:50
18. Razafindralambo H, Blecker C, Delhaye S, Paquot M (1995) J Colloid Interface Sci 174:373
19. Joos P (1999) Dynamic surface phenomena. VSP, Utrecht
20. Ward AFH, Tordai L (1946) J Chem Phys 14:453
21. Fainerman VB, Makievski AV, Miller R (1994) Colloids Surf A 87:61

Progr Colloid Polym Sci (2004) 126: 55–59
DOI 10.1007/b93982

J. Miñones Jr
P. Dynarowicz-Łątka
O. Conde
E. Iribarnegaray
M. Casas

Interactions of amphotericin B with phospholipids in Langmuir monolayers

J. Miñones Jr · O.Conde
E. Iribarnegaray · M. Casas
Department of Physical Chemistry,
Faculty of Pharmacy,
University of Santiago de Compostela,
15706 Santiago de Compostela, Spain

P. Dynarowicz-Łątka (✉)
Department of General Chemistry,
Faculty of Chemistry,
Jagiellonian University, Ingardena 3,
30-060 Kraków, Poland
e-mail: ucdynaro@cyf-kr.edu.pl
Tel.: +48-12-6336377 ext. 2236;
Fax: +48-12-6340515

Abstract Mixtures of amphotericin B and synthetic phospholipids, differing in the saturation of the phosphatidylcholine acyl chains, such as dipalmitoyl phosphatidylcholine, palmitoyl-oleoyl phosphatidyl choline and dioleoyl phosphatidylcholine, have been studied as floating Langmuir monolayers at the air–water interface. The films were spread at room temperature on aqueous subphases of different pH (3, 6 and 10) and characterized by the surface pressure–area isotherms and compressional modulus–surface pressure plots. The interactions were examined by analyzing the phase transitions appearing on the course of the isotherms. Although no significant difference in the affinity of amphotericin B to saturated/unsaturated phospholipids has been found, the results obtained reveal the influence of the apolar structure of phosphatidylcholine chains on the stoichiometry of amphotericin B–phosphatidylcholine complexes.

Keywords Langmuir monolayers · Mixed Films · Air/water interface · Interactions · Amphotericin B

Introduction

It is well known that amphotericin B (AmB), the polyene antibiotic, has potent antifungal activities [1]. Owing to its poor solubility in water as well as in the majority of solvents used in intravenous drug administration, it is administered clinically as a colloidal suspension in an aqueous medium containing sodium deoxycholate and glucose (Fungizone). However, this formulation has been found to be toxic [1]. It has been reported that AmB encapsulation into liposomes can significantly reduce the drug toxicity without any loss of antimycotic activity [2]. The reason for the lower toxicity of liposomal AmB is still not well understood, even though there seems to be a certain relationship between vehicle-forming phospholipid and AmB toxicity, given that AmB incorporated in saturated phospholipids is not toxic, whereas AmB formulations composed of unsaturated phospholipids are almost as toxic as Fungizone [3]. To elucidate the reason for the reduced toxicity of liposomal AmB as well as to find out how the degree of saturation of phosphatidylcholine acyl chains influence the characteristics of AmB, the Langmuir monolayer technique was employed. Three synthetic phospholipids were chosen for the investigations, having the same phosphatidylcholine headgroup bound to the apolar part composed either by two saturated chains (L-α-phosphatidylcholine dipalmitoyl, DPPC), two unsaturated chains (L-α-phosphatidylcholine dioleoyl, DOPC) or one saturated chain and one unsaturated chain (L-α-phosphatidylcholine β-palmitoyl-γ-oleoyl, POPC). Mixed monolayers of AmB and the particular phospholipid were spread on aqueous subphases of different pH (3, 6 and 10) at room temperature, and the surface pressure–area (π–A) isotherms were recorded upon compression. The interactions were studied and compared by analyzing the phase transition from the compressibility modulus–surface pressure (C_s^{-1}–π) plots.

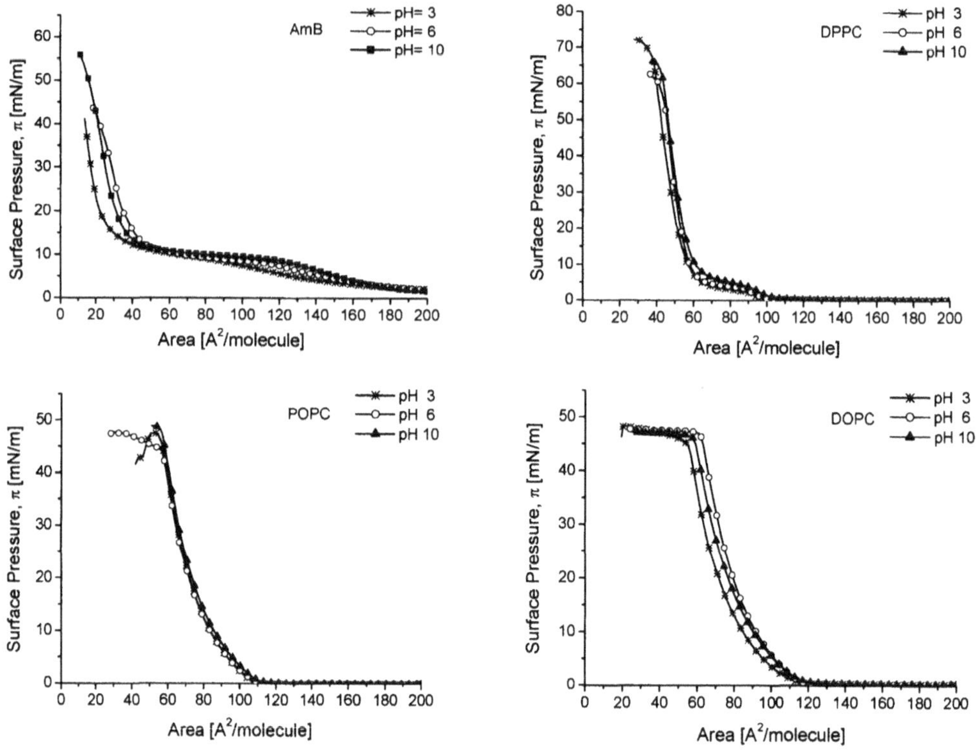

Fig. 1 Surface pressure–area (π–A) isotherms for monolayers of amphotericin B (*AmB*), L-α-phosphatidylcholine dipalmitoyl (*DPPC*), L-α-phosphatidylcholine dioleoyl (*DOPC*) and L-α-phosphatidylcholine β-palmitoyl-γ-oleoyl (*POPC*) spread on water at pH 3, 6 and 10 and at 20 °C

Experimental

DPPC, DOPC and POPC (99% purity) were supplied by Sigma, whereas AmB (purity above 95%) was supplied by Squibb (Bristol-Myers Laboratories). The phospholipids investigated were dissolved in a 4:1 v:v mixture of chloroform (Merck, p.a.) and absolute ethanol (Merck, p.a.). An AmB spreading solution was prepared in a mixture of dimethylformamide and 1 M HCl (3:1 v/v). Mixed solutions were prepared by mixing appropriate volumes of the respective stock solutions. The solutions were stored at 4 °C in a desiccator saturated with the spreading solvent.

A two-compartment Langmuir–Blodgett trough (KSV-5000, Finland), equipped with two symmetrical barriers, was used to record the π–A isotherms. The surface pressure was measured to a precision of ±0.05 mN/m with a platinum Wilhelmy plate. Each monolayer was prepared by dropping aliquots of approximately 200 μl (4.25×10^{16} molecules) with a Microman Gilson microsyringe, precise to ±0.2 μl, on an aqueous subphase prepared with ultrapure water from Milli-Q water purification systems. The pH of the subphase was adjusted with HCl/NaOH. Once spread, the monolayers were left for around 10 min before the compression was initiated, for solvent evaporation. The monolayers were compressed with a barrier speed of 8.5 Å^2/molecule min. The subphase temperature (20 °C) was controlled thermostatically to within 0.5 °C by a circulating water system.

Results

The π–A isotherms of AmB and the phosphatidylcholines (DOPC, POPC and DPPC) spread on aqueous subphases at three different pH (3, 6 and 10) at 20 °C are shown in Fig. 1. AmB (pK 5.5 and 10.0 [4]) is charged positively at pH 3, negatively at pH 10, while at pH 6 it exists as a zwitterion. Over the whole range of pH values investigated, the characteristic of the AmB isotherm is quite similar, and at $\pi \sim 10$ mN/m the plateau transition (A) can clearly be distinguished. This transition has been attributed [5] to the gradual change in orientation of AmB molecules from a horizontal to a vertical position. Further compression, giving rise to the surface pressure beyond the plateau, causes tight molecular packing and finally dissolution of monolayer material into the bulk phase [6].

At 20 °C, both POPC and DOPC exhibit an expanded monolayer, while DPPC shows a typical liquid-expanded–liquid-condensed phase transition (at $\pi = 4$ mN/m) evidenced as a plateau (L) in the course of the isotherm (Fig. 1). The parameters characteristic of phosphatidylcholine isotherms (such as lift-off areas and the collapse pressures) are in good agreement with literature data [7]. As can be seen, no significant differences in the course of the isotherms can be observed for the phosphatidylcholines spread on subphases of different pH.

The surface behavior of mixed AmB–phosphatidylcholine monolayers is also found to be similar at different subphase pH. Therefore the detailed analysis will be discussed for the systems at pH 6, in the case of which both AmB and phosphatidylcholine are electrically neutral.

The isotherms recorded for mixed AmB monolayers with DOPC and POPC are shown in Figs. 2 and 3, respectively. As can be seen, mixed films of X_{AmB} ranging

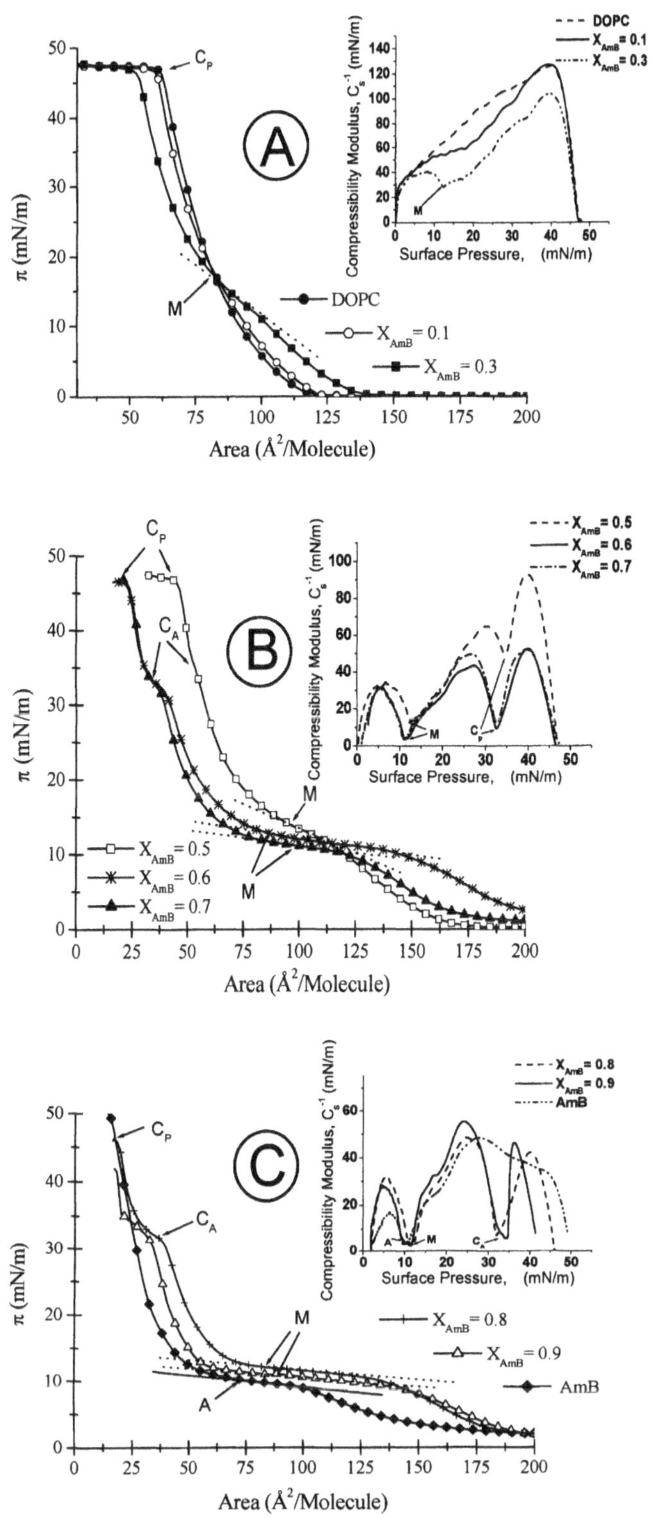

Fig. 2A–C *π–A* isotherms for mixed monolayers of AmB and DOPC spread on water at pH 6, 20 °C. *Insets*: Compressional modulus versus surface pressure dependencies for the respective mixed films

from 0.9 to 0.3 (AmB–DOPC) and from 0.9 to 0.5 (AmB-POPC) exhibit a phase transition (M) at a surface pressures of around 12 mN/m (AmB–DOPC) and 13.5 mN/m

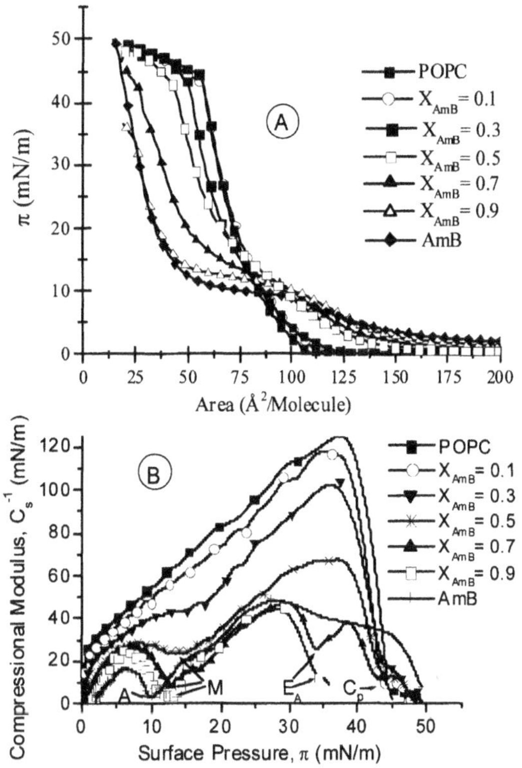

Fig. 3. A *π–A* isotherms for mixed monolayers of AmB and POPC spread on water at pH 6, 20 °C. **B** Compressional modulus versus surface pressure dependencies for the respective mixed films

(AmB–POPC), which appears as a pseudo-plateau region. Its length and the flatness increase as the proportion of AmB in the mixed film is augmented. The longest and the most horizontal plateau is achieved for a mixed monolayer of $X_{AmB} = 0.9$. For both mixtures investigated, the plateau surface pressure for the AmB monolayer ($\pi_A = 10$ mN/m) increases with the addition of the phospholipid, and simultaneously the pseudo-plateau of the mixtures (M) becomes gradually more inclined and its length diminishes. Finally, for both systems at $X_{AmB} < 0.3$ the transition M practically disappears.

Apart from this plateau, for mixed monolayers rich in AmB (X_{AmB} ranging from 0.9–0.5 for AmB–DOPC and from 0.9–0.7 for AmB–POPC) another "kink" is distinguished at higher surface pressures (around 35 mN/m), which can be attributed to the ejection of AmB (E_A) molecules from the surface. This is confirmed with Brewster angle microscopy images (Fig. 4a) showing "ditches" that remain in the film after AmB expulsion. For comparison, the images for pure AmB (Fig. 4b) and DOPC (Fig. 4c) taken at the surface pressure corresponding to the "kink" are also included. On the other hand, for mixed films containing X_{AmB} ranging from 0.1 to 0.8 (for AmB–DOPC) and from 0.1 to 0.7 (for AmB–POPC) another collapse, corresponding to that of pure

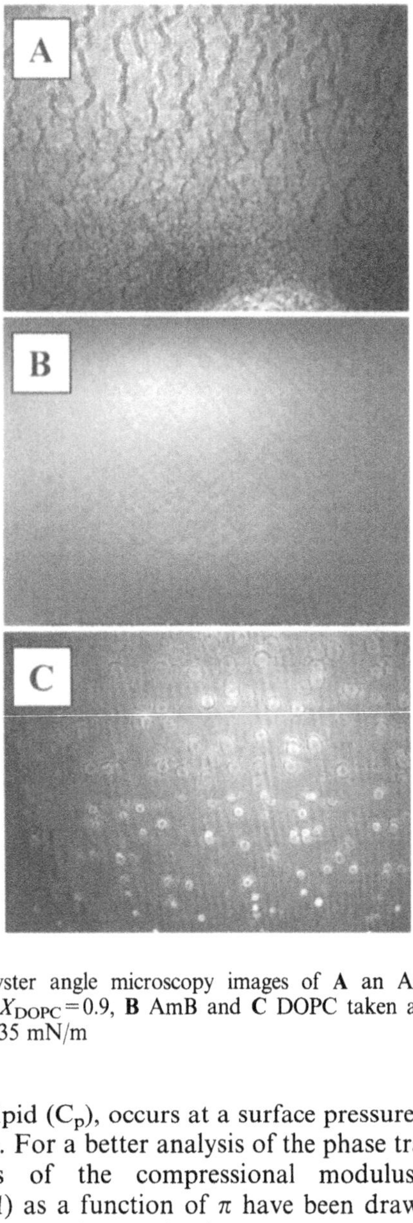

Fig. 4 Brewster angle microscopy images of **A** an AmB–DOPC mixture of $X_{DOPC} = 0.9$, **B** AmB and **C** DOPC taken at a surface pressure of 35 mN/m

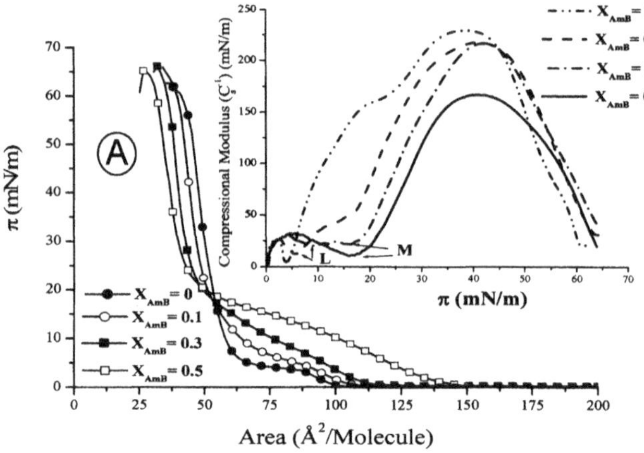

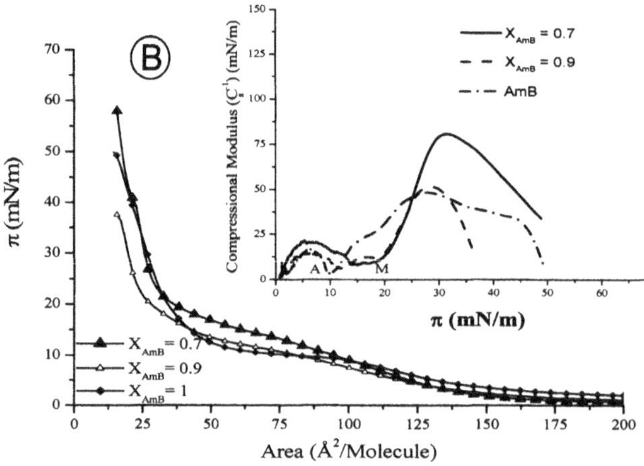

Fig. 5A, B π–A isotherms for mixed monolayers of AmB and DPPC spread on water at pH 6, 20 °C. *Insets*: Compressibility modulus versus surface pressure dependencies for the respective mixed films

phospholipid (C_p), occurs at a surface pressure of about 45 mN/m. For a better analysis of the phase transitions, the plots of the compressional modulus ($C_s^{-1} = -A\,d\pi/dA$) as a function of π have been drawn on the same figures as the π–A isotherms.

For the AmB–DPPC system (Fig. 5) the liquid-expanded–liquid-condensed transition of DPPC ($\pi = 4$ mN/m) gradually disappears when AmB is incorporated into the monolayer ($X_{AmB} = 0.1$–0.3) and its surface pressure shifts to higher values (5.7 and 9 mN/m, respectively). The corresponding $1/C_s$–π plots show that the phase transition of DPPC, seen as a minimum L, shifts to higher pressures for mixtures with increasing amount of AmB ($X_{AmB} = 0.1$–0.3). Simultaneously, another minimum (M) can be distinguished at about 16 mN/m, thus demonstrating the existence of a second phase transition, which is also observed (although less defined) in mixtures of high AmB content ($X_{AmB} > 0.7$) (Fig. 5b), apart from having the transition A characteristic of pure AmB.

In summary, it can be concluded that the progressive addition of a phospholipid into the AmB monolayer causes the disappearance of phase transition A, typical of AmB, and provokes the appearance of a new phase transition, M, at higher surface pressures.

Discussion

The analysis of the plateau region appearing in the π–A isotherms enables us to draw conclusions regarding the interaction between molecules in the mixtures investigated. Since transition M was observed to be shifted to higher surface pressures compared to the transition pressure of pure AmB (A), it can be concluded that the addition of the phospholipid molecules hinders the reorientation of AmB from a horizontal to a vertical position in mixed monolayers [8]. The stronger the interaction between the mixture components, the higher the pressure for transition M. From Figs. 2, 3 and 5 one can see that transition M occurs at the highest pressure for AmB–DPPC (16 mN/m), then for mixtures with

POPC (13.5 mN/m) and, finally, for AmB–DOPC (12 mN/m). Besides the values of the transition surface pressure (π_M), the length and the flatness of pseudo-plateau M in the π–A isotherms can also be useful for quantifying the interactions. For the AmB–DPPC system, plateau M reaches the maximum length at $X_{AmB}=0.5$–0.7 (Fig. 5). Simultaneously, $1/C_s$ attains its minimum value (maximum flatness). Consequently, the strongest interaction seems to occur when the film components are in the proportion of approximately 2:1, which corresponds to $X_{AmB}=0.66$ (a value intermediate between 0.5 and 0.7). Therefore, following our previous studies [8], it can be postulated that the film components form a stable complex, composed of two horizontally oriented AmB molecules and one DPPC molecule in the vertical position.

For the other systems investigated here (AmB–POPC and AmB–DOPC), the interaction between film components seems to be weaker since transition M appears at lower surface pressures (12–13 mN/m) than for the AmB–DPPC system (16 mN/m). Moreover, the stoichiometry of the complexes is seen to be different as well. The fact that the π–A isotherms of these systems show the longest and the most horizontal plateau at $X_{AmB}=0.9$ (Figs. 2, 3) suggests that the strongest interaction corresponds to the complex formation of 9:1 stoichiometry (AmB:DOPC/POPC). It is possible that the presence of unsaturated bond(s) in the hydrophobic tails of the phospholipids enable these molecules to attain the vertical orientation at the air–water interface, thus making the polar groups of a phospholipid less accessible to AmB molecules and, in consequence, reducing the strength of interaction between film-forming components and leading to the formation of complexes of different stoichiometry, rich in AmB molecules, in order to attain the closest possible molecular arrangement with the phospholipid.

It is evident that complexes composed of AmB and either DOPC or POPC, contain a higher proportion of AmB; however, they are less stable than those formed by AmB and DPPC. The lower toxicity of the liposomal formulation of AmB formed by saturated phospholipids, compared to unsaturated ones, can be explained as being due to the AmB–phospholipid interactions, which are stronger for saturated than for unsaturated phospholipids. This leads to the presence of a moderate quantity of AmB in the host cells which is enough to provoke the lysis of fungal cells (fungicidal effect); however, it is not sufficient to damage the host cells.

References

1. Medoff G, Brajtburg J, Kobayashi GS, Bolard J (1983) Annu Rev Pharm Toxicol 23:303
2. Juliano RL, Lopez-Berestein G, Hopfer RL, Mehta R, Mehta K, Mills K (1985) Ann NY Acad Sci 446:390
3. Juliano RL, Grant CWM, Barber KR, Kalp MP (1987) Mol Pharmacol 31:1
4. Newton DW, Kluza RB (1978) Drug Intell Clin Pharm 12:546
5. Saint-Pierre-Chazalet M, Thomas C, Dupeyrat M, Gary-Bobo CM (1988) Biochim Biophys Acta 944:477
6. Rey Gómez-Serranillos I, Miñones J Jr, Seoane R, Conde O, Casas M (1999) Prog Colloid Polym Sci 112:29
7. Mingotaud AF, Mingotaud C, Patterson LK (1993) Handbook of monolayers, vol 1, no 3. Academic, San Diego, CA, pp 796 (DPPC), 862 (DOPC), 892 (POPC)
8. Miñones J Jr, Miñones J, Conde O, Rodriguez-Patino JM, Dynarowicz-£ătka P (2002) Langmuir 18:2817

Progr Colloid Polym Sci (2004) 126: 60–63
DOI 10.1007/b93978

Barbara Gzyl
Maria Paluch

Monolayers of lipids at the water–air interface

B. Gzyl (✉) · M. Paluch
Department of Physical Chemistry and Electrochemistry, Faculty of Chemistry, Jagiellonian University, ul. Ingardena 3, 30-060 Kraków, Poland
e-mail: gzyl@chemia.uj.edu.pl

Abstract Spread monolayers of dipalmitoyl phosphatidylcholine, methyl palmitate and their mixtures were studied. For this purpose, the surface pressure (π)–average molecular area (A) isotherms were recorded using a Langmuir film balance. From the π-A results information on the mutual miscibility of the two components in the two-dimensional state was deduced. Then classical surface chemistry thermodynamic analysis was performed on these isotherms, which involved calculating the excess free energy of mixing, ΔG^{E}_{mix}. The negative values of ΔG^{E}_{mix} indicate that the compounds in the two-dimensional state experience mainly attractive interactions.

Keywords Langmuir monolayer · Dipalmitoyl phosphatidylcholine · Methyl palmitate

Introduction

Biological membranes are very complex structures. Lipids are the major constituents of biological membranes. Membranes are formed of a bilayer of lipid compounds in which the other compounds, such as proteins and sterols, are immersed or bound to the two interfaces [1]. Studies of lipid–lipid interactions in these membranes are important to understand their organisation. The results obtained from studies on isolated membranes are difficult to interpret; therefore, it is necessary to work on well-defined model membranes. Of the several model systems used to study lipid interactions, the monolayer technique has been applied extensively [2, 3, 4, 5]. Experiments with monolayers have the advantage that the arrangement of molecules can be easily controlled by changing the molecular area and the surface pressure of the monolayer. The study of monolayers is a useful means to learn the interfacial distribution of the molecules and the possible interactions among them in the two-dimensional state. Monolayers were first used as membrane models in 1925, in the well-known experiment carried out by Görtel and Grendel [6]. Since then monolayer techniques have been widely used to mimic the molecular organisation of lipids in biomembranes. One particular monolayer compound that has received much attention over the years is the phospholipid dipalmitoyl phosphatidylcholine (DPPC), the principle component of cell membranes [7, 8, 9].

In this paper we report a study of spread monolayers of DPPC and methyl palmitate (MP), and also their mixtures at different molar ratios, at the water–air interface. For this purpose the isotherms of surface pressure,π, versus average molecular area, A, were recorded using a Langmuir film balance. An isotherm is one of the most fundamental characteristics of a monolayer system, and the analysis and interpretation of isotherms provide an easy route to learn about the state of the monolayer. The way mean molecular areas depend on the composition of the mixture can be used to predict possible interactions between molecules in the mixed monolayer.

Experimental

DPPC and MP were supplied by Aldrich Chemie, Germany, and were used without further purification. Chloroform (the spreading solvent) from POCH, Poland, was distilled before use. The water used in all the experiments was distilled four times.

The π–A isotherms were determined with a Langmuir trough (KSV 1000, Finland) at 20 °C. The subphase temperature was controlled with a U3 thermostat. Before each measurement, the water cleanness was checked for possible adsorbed impurities by measuring the surface pressure in the entire area interval; this was always less than 0.1 mN m^{-1}in the whole field. The monolayers were obtained by spreading the solution of lipids in chloroform with a 100-μl microsyringe (Hamilton). After deposition about 20 mm was allowed before compression to permit the solvent to evaporate and the film to spread homogeneously. The compression speed was 6 mm min^{-1}. The compression speed was fixed at that level, because below it no difference in the π–A isotherms was observed. The measurement accuracy was ± 0.01 mN m^{-1} for the surface pressure, ± 0.01 A^2 per molecule for the area and ± 0.05°C for temperature. Each isotherm was recorded at least twice.

Results and discussion

The experimental isotherms corresponding to the monolayers formed by DPPC, MP and their mixtures in different molar fractions, at 20 °C, are presented in Fig. 1.

The surface parameters obtained from the isotherms of pure component are reported in Table 1. A_{lim} is the limiting area per molecule, i.e. the area extrapolated to zero surface pressure under maximum packing conditions, π_{coll} is the collapse pressure determined from the change in the slope of the straight-line part of the isotherm in the high-pressure range and C_s^{-1} is the maximum value of the compressibility modulus defined as

$$C_s^{-1} = -A\left(\frac{\partial \pi}{\partial A}\right)_T . \qquad (1)$$

The values of C_s^{-1} enable the monolayer phases to be defined according to the classification of Davies and Rideal [10].

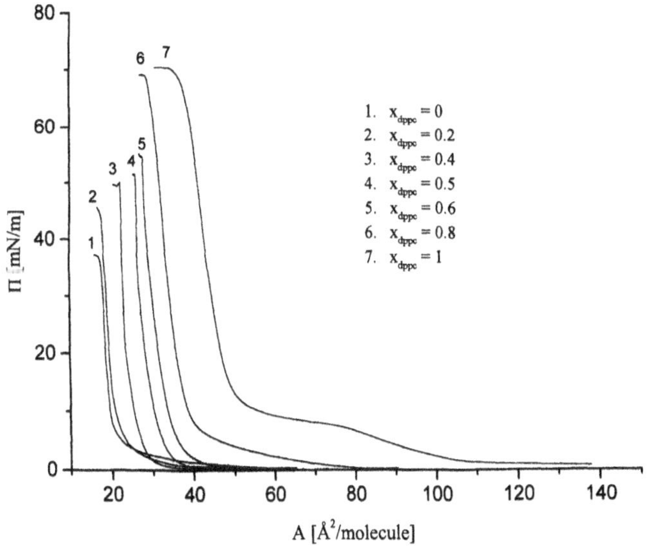

Fig. 1 π–A isotherm of the dipalmitoyl phosphatidylcholine (*DPPC*)–methyl palmitate (*MP*) system on water at 20 °C

Table 1 Limiting area per molecule, A_{lim}, collapse pressure, π_{coll}, and maximum compressibility modulus, $C_{s\,\max}^{-1}$, for dipalmitoyl phosphatidylcholine (*DPPC*) and methyl palmitate (*MP*), at 20 °C

	A_{lim} [Å^2 per molecule]	π_{coll} [mN · m^{-1}]	$C_{s\,\max}^{-1}$ [mN · m^{-1}]
DPPC	48.87	59.31	293.6
MP	19.98	35.09	347.9

The isotherm of the DPPC monolayer is in satisfactory agreement with that previously reported [11]. The characteristic of this isotherm is a plateau representing the region of coexistence of two different phases, the liquid-expanded phase, L_1, and the liquid-condensed phase, L_2. The value of $A_{\text{lim}} = 48.87$ Å^2 per molecule is higher than we could expect for the close-packed double-chain molecules, assuming vertical orientation of alkyl chains (around 2x20 Å^2 ≈40 Å^2). This indicates a tilting of the hydrocarbon chains with respect to the normal to the surface. The value of $C_{s\,\max}^{-1} = 293.6$ mN m^{-1} shows that the monolayer is in the liquid-condensed phase under maximum packing conditions. The collapse pressure is reached at 59.31 mN m^{-1}.

The π–A isotherm for MP monolayers has a steep slope, which indicates that a small decrease in area causes a significant increase in pressure until $\pi_{\text{coll}} = 35.09$ mN m^{-1}. The shape of the curve and very low value of the limiting area, $A_{\text{lim}} = 19.98$ Å^2 per molecule, imply that the MP molecules in the monolayer are arrayed with their hydrocarbon chains almost perpendicular to the interface. This is also confirmed by the value of $C_{s\,\max}^{-1}$, higher than obtained for the DPPC monolayer, indicating more compact packing of hydrophobic chains.

The π–A isotherms of mixed DPPC/MP monolayers, determined for different molar fractions of DPPC (0.2, 0.4, 0.5, 0.6, 0.8) are included between those of DPPC and MP monolayers in a regular sequence of mole fractions. The most striking features of these isotherms are the surface pressures at which the liquid-condensed state appears – they are lower than those observed for monolayers of pure components. This suggests the condensation of mixed monolayers. To validate this suggestion, the maximum compressibility moduli were calculated and compared with those of the pure components (Fig. 2).

For mixtures with similar content of both components $C_{s\,\max}^{-1}$ is much higher than for monolayers of DPPC or MP, 704.8; 849.7 and 708.9 mN m^{-1}, for x_{DPPC} (0.4, 0.5 and 0.6, respectively). Therefore, progressive condensation is observed as the equimolar mixture is approached. Low compressibility is a sign that the molecules are packed tightly and the cohesive forces between them are considerable. The values of the collapse pressures (Fig. 3) are included between those for DPPC and MP monolayers.

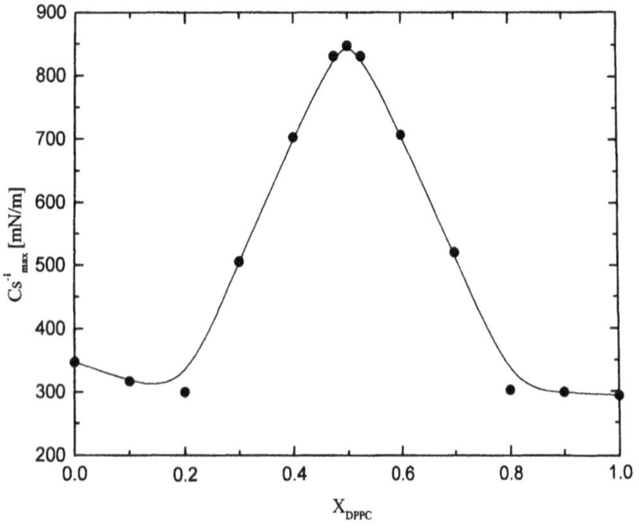

Fig. 2 Maximum compressibility moduli versus molar fractions of DPPC

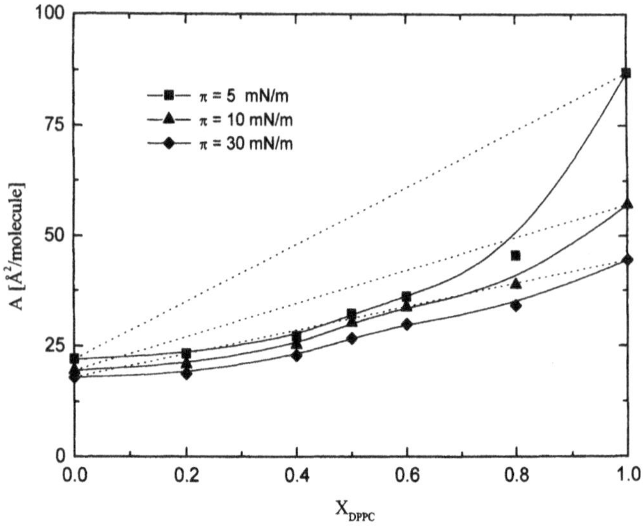

Fig. 4 Surface areas as a function of molar fractions of DPPC, at $\pi = 5$, 10 and 30 mN m^{-1}

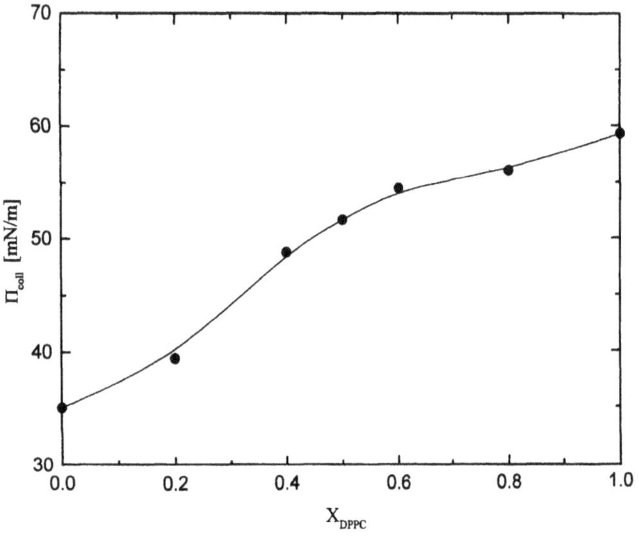

Fig. 3 Collapse pressures versus molar fractions of DPPC

From that figure it is evident that when the DPPC content in the mixture increases, higher collapse pressures are obtained; therefore, the stability of the monolayer against collapse increases. The collapse pressure reflects the strength of attachment of the molecule to the interface.

The behaviour of the mixed monolayers can be analysed by comparing their π–A isotherm with the corresponding theoretical isotherm in which no interactions between the molecules are assumed by the additivity rule:

$$A_{12} = x_1 A_1 + x_2 A_2, \tag{2}$$

where A_{12} corresponds to the mean molecular area of the mixture at the fixed surface pressure, A_1 and A_2 are the molecular areas of pure components 1 and 2 at the same given surface pressure, and x_1 and x_2 are the molar fractions of components 1 and 2, respectively. In the case of ideal miscibility or complete immiscibility of the two components Eq. (2) is valid.

The mean molecular area values against the molar fractions for three different values of pressure (5, 10 and 30 mN m^{-1}) are reported in Fig. 4. From this plot one can easily see the negative deviations from ideal behaviour which are indications of two-dimensional miscibility of the two components at the water–air interface. The negative deviations may be due to the geometrical rearrangement of the two components, which results in a decrease in area, thus causing enhanced dispersive interactions.

The miscibility of these films can also be deduced from their collapse characteristic. According to the phase rule, first applied to the two-dimensional state by Crisp [12], two miscible components have collapse pressures dependent on the mixture composition. Our experimental results (Fig. 3) show that the collapse pressures of the mixtures change with composition. Thus, we can conclude that DPPC and MP give true mixed films.

The final proof of the statement comes from the behaviour of C_s^{-1} as a function of the molar fractions of DPPC. In fact, if the two components are immiscible or if they behave like an ideal mixture, C_s^{-1} follows the same relationship as the surface areas do (see Eq. 2). The positive deviations from additivity illustrated in Fig. 5 indicate the presence of more condensed phases for the mixtures in comparison to the pure components, and therefore the possible attractive interactions.

After having demonstrated the nonideal behaviour of the mixtures, the excess free energy of mixing was calculated following the Goodrich method [13]. Briefly,

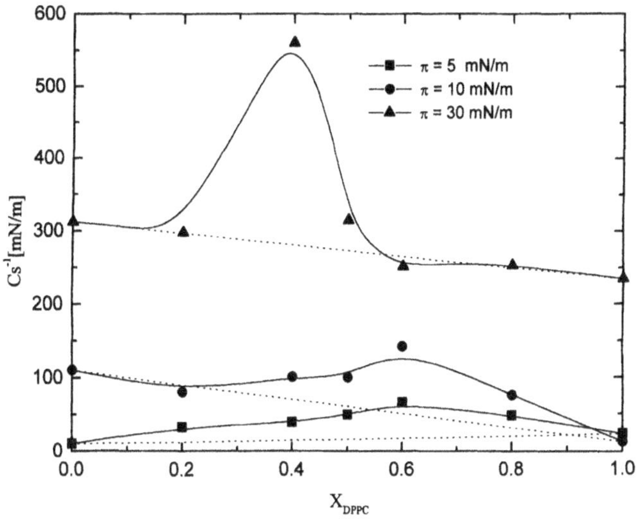

Fig. 5 Compressibility moduli as a function of molar fractions of DPPC, at $\pi = 5$, 10 and 30 mN m^{-1}

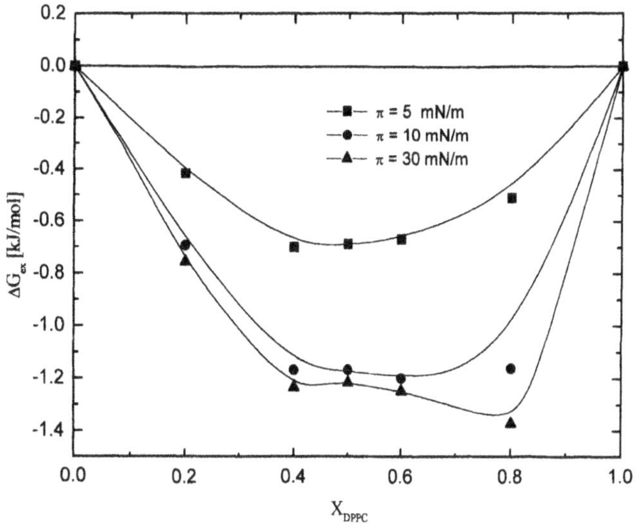

Fig. 6 Excess free energy of mixing of the DPPC–MP system as a function of molar fractions of DPPC, at $\pi = 5$, 10 and 30 mN m^{-1}

this involves integrating under the π–A curves of the mixture and of each of the pure components to evaluate the integrals for the expression

$$\Delta G^{\mathrm{E}}_{\mathrm{mix}} = N_{\mathrm{A}} \int_{\pi_1}^{\pi_2} (A_{12} - x_1 A_1 - x_2 A_2)\mathrm{d}\pi, \quad (3)$$

where π_1 and π_2 are two fixed surface pressures and N_{A} is the Avogadro constant. The upper limit for the integrations may be arbitrarily selected; the lower limit of the integration should be a surface pressure below which the monolayer components may be assumed to mix ideally. The excess free energy of mixing values, $\Delta G^{\mathrm{E}}_{\mathrm{mix}}$, obtained from integration of the π–A isotherm between $\pi \rightarrow 0$ and $\pi = 5$, 10 and 30 mN m^{-1} are reported in Fig. 6.

The negative values of $\Delta G^{\mathrm{E}}_{\mathrm{mix}}$ indicate the greater thermodynamic stability of the mixture with respect to the pure components and the presence of attractive interactions between the components.

Conclusions

Monolayers of DPPC, MP and their mixtures were investigated at the air–water interface at 20 °C. The negative deviations from additivity of the areas and similarly the negative values of the surface free energy of mixing, taken together with the positive deviations of the compressibility modulus and the variation in the collapse pressure as a function of molar fractions of DPPC, clearly indicate the reciprocal miscibility between the components. The miscibility was not ideal but with prevalently attractive energies which could be attributed to hydrophobic interactions among the chains. Since a monolayer is a valid model for cellular membranes, the information obtained on miscibility between lipid components in the monolayer can be important to predict properties of more complex biological aggregates.

References

1. Israelachvili JN (1985) Intermolecular and surface forces. Academic, London, p 263
2. Bringezu F, Brezesinski G, Möhwald H (1998) Chem Phys Lipids 94:251
3. Sánchez-Gonzalez J, Cabrerizo-Vílchez MA, Gálvez-Ruiz MJ (1999) Colloids Surf B 12:123
4. Krasteva N, Vollhardt D (2000) Colloids Surf A 171:49
5. Bringezu F, Brezesinski U (2001) Colloids Surf A 183–185:391
6. Görter, Grendel F (1925) J Exp Med 41:439
7. Thoma M, Möhwald H (1995) Colloids Surf A 95:193
8. Klopfer KJ, Vanderlick TK (1996) J Colloid Interface Sci 182:220
9. Zhao J, Vollhardt D, Brezesinski U, Siegel S, Wu J, Li JB, Miller R (2000) Colloids Surf A 171:175
10. Davies JT, Rideal EK (1963) Interfacial phenomena. Academic, New York, p 265
11. Gzyl B, Paluch M (2001) Prog Colloid Polym Sci 118:22
12. Crisp DJ (1949) Surface chemistry. Butterworth, London, p 17
13. Goodrich FC (1957) Proc 2nd Int Congr Surf Activity 1:85

Progr Colloid Polym Sci (2004) 126: 64–67
DOI 10.1007/b93983

G. Carignano
A. Chittofrati
R. Pieri
M. Visca

Wetting of a poly(tetrafluoroethylene) surface by aqueous solutions of perfluoropolyether carboxylic salts of different chain length

G. Carignano (✉) · A. Chittofrati
R. Pieri · M. Visca
Solvay Solexis CRS (Solvay Group),
v.le Lombardia 20, 20021 Bollate,
Milan, Italy
e-mail: gabriella.carignano@solvay.com
Tel.: +39-2-38352174;
Fax: +39-2-38356355

Abstract In a previous study, the wetting of poly(tetrafluoroethylene) by aqueous solutions of a mixture of perfluoropolyether carboxylic salts was evaluated via dynamic contact angle measurements, assessing surfactant adsorption. In the present work, the two main components of that mixture, differing in chain length only, are used to explore their relative contribution to adsorption. Aqueous solutions of the perfluoropolyether surfactant with the longest chain enable higher adsorption efficiency and superior wetting ability to be achieved than with the shortest-chain term in solutions with identical values of the equilibrium surface tension. The poly(tetrafluoroethylene)–solution contact time and the relative rate of adsorption of the two surfactants are relevant for the contact angle measurements.

Keywords Perfluoropolyether · Contact angle · Surface · Adsorption

Introduction

Fluorinated surfactants attract great interest as wetting agents since they can reduce the water surface tension down to 15–20 mN/m. The wetting properties of aqueous solutions of a well-known mixture of carboxylic perfluoropolyether (PFPE) salts have recently been investigated on low-energy surfaces, such as polyethylene or poly(tetrafluoroethylene) (PTFE) [1]. The surfactant mixture, at appropriate concentration, was shown to wet completely the surface. On PTFE, surface modification by selective surfactant adsorption/desorption was suggested.

The aim of the present work is to focus, via dynamic contact angle (DCA) measurements, on the relative contribution to the adsorption of the two main components of the surfactant mixture, differing in their chain length only. The comparison between solutions at the same surface tension indicates higher adsorption efficiency for the longer-chain component, resulting in lower contact angles.

Experimental

Materials

The surfactants studied are two high-purity laboratory samples, belonging to the general PFPE class [2, 3, 4] (Structure 1) $Cl\text{–}(C_3F_6O)_nCF_2COO^-NH_4^+$, where the number of perfluoroisopropoxy units, n, either 2 or 3, identifies each term. Such surfactants were the main components of the PFPE mixture (PFPE—NH_4) described in Ref. [1], which contained terms with $n=2$–5, in amounts decreasing with the increase in n. The solution behaviour of the same individual terms used here has been described elsewhere [5]. All solutions were prepared with Milli-Q grade water using Millipore apparatus

PTFE films (10 mm × 20 mm × 0.2 mm) were obtained by uniform coating of a glass fibre fabric (Fothergill, UK) with an ALGOFLON PTFE dispersion (Solvay Solexis, I). Their mean roughness was 2 μm. Each film sample was thoroughly cleaned by several washing cycles with methanol, then dried in an oven at 30 °C for 20 h. The washing procedure was assessed to provide an advancing contact angle, θ_A, of water against this PTFE substrate of 122 ± 3° and a receding contact angle, θ_R, of 87 ± 3° in the experimental conditions used throughout this work.

Methods

The value of the equilibrium surface tension for each surfactant solution at 25 ± 0.1 °C was measured by the Du Noüy method, with a Lauda TE1C tensiometer. Each value is the average of at least five measurements, after equilibrium was attained, and is obtained by the pertinent Harkins–Jordan factor [6], with the standard deviation decreasing from 2 to 0.1 mN/m as the concentration increases from 0.01 to 1 g/l.

The DCA was measured with a KSV Sigma 70, upon four subsequent immersion cycles of the substrate into the solution. The immersion–emergence speed range was varied from 0.5 to 20 mm/min, allowing 1 min between two subsequent cycles. The study of the wetting of the PTFE surface was carried out at three different probe speeds (0.5, 6 and 20 mm/min), to explore the consequences of kinetics. The timeframe selected for the DCA measurements was meant to ensure a constant surface tension of the solutions according to the results of the rate of surface tension reduction with the individual $n=2$ and $n=3$ salts described in another article in this issue [7].

The critical surface tension, γ_c, of this PTFE fabric was reported earlier [1], but it is worth recalling that its value by Zisman's method, 20.3 mN/m, is slightly higher than the 18 mN/m often reported for smoother PTFE surfaces [8].

Results and discussion

Surface tension measurements

The surface tension curves of the two individual PFPE surfactant terms, $n=2$ and $n=3$, are compared in Fig. 1 to the curve reported earlier for the PFPE—NH_4 mixture ($n=2$–5) used in our previous wetting work on the same surfactant class and PTFE substrate [1]. As expected, the increase with n of the surfactant (average) molecular weight shifts the curve towards a lower concentration range while displaying a major contribution of the adsorption of the higher n terms, in the PFPE—NH_4 mixture, at the water–air surface at equilibrium. However, the individual components of relatively low n values enable a sufficient reduction in γ for most wetting purposes.

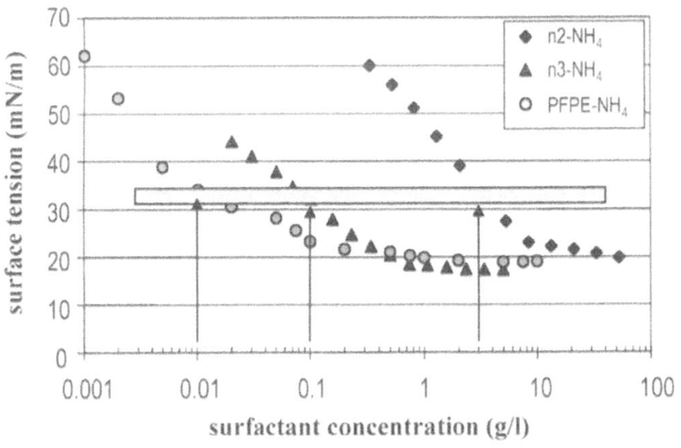

Fig. 1 Equilibrium surface tension measurements of perfluoropolyether (*PFPE*) surfactant ammonium salts

We chose to measure the DCA of the solutions against PTFE substrates keeping constant the surface tension of the different solutions (grey bar in Fig. 1), to gain an insight into the relative surfactant adsorption/desorption well below the critical micelle concentration (cmc) of each surfactant or mixture (arrows in Fig. 1). The equilibrium value of the surface tension of each solution was between 31 and 34 mN/m, as detailed in Table 1, where the individual $n=2$ and $n=3$ terms are compared to their mixture at a molar ratio of $n=2$ to $n=3$ of 2 (labelled $n=2/n=3$—NH_4) and also to the mixture with a much greater molecular weight distribution (PFPE–NH_4) previously reported in Ref. [1].

The experimental error of the data reported in Table 1 was within 2.0 mN/m for the PFPE–NH_4 mixture and within 1.0 mN/m for the other salts.

DCA measurements

The DCA study at the three very different values of the immersion–emergence speed of each PTFE film into each solution was strictly meant to monitor the consequences of adsorption kinetics at almost constant intrinsic wetting capacity of each solution, as driven by the almost constant value of the solution surface tension.

Indeed the adsorption rate at constant probe speed varies with the chain length of the surfactants: as an example, the DCA profiles obtained at a probe speed of 6 mm/min with $n=2$–NH_4 can be compared to those of the PFPE–NH_4 mixture in Figs. 2 and 3, respectively. In all cases, the large thermodynamic hysteresis (θ_A–θ_R) may reasonably be ascribed to the fairly high roughness of the PTFE fabric surface.

The variation of the advancing contact angles θ_{adv} recorded for each surfactant solution of Table 1 at each different probe speed is summarised in Fig. 4. The

Table 1 Surface tension of perfluoropolyether (*PFPE*) ammonium salt solutions at 25 °C

Salt	Concentration (g/l)	Concentration (mol/l)	Surface tension (mN/m)
$N = 2$—NH_4	3.00	6.26×10^{-3}	30.9 ± 0.1
Mixture $n = 2/n = 3$—NH_4 (2/1 molar ratio)	0.17	3.18×10^{-4}	32.4 ± 1.0
$N = 3$-NH_4	0.10	1.55×10^{-4}	30.8 ± 1.0
PFPE—NH_4	0.01	1.62×10^{-5}	33.8 ± 2.0

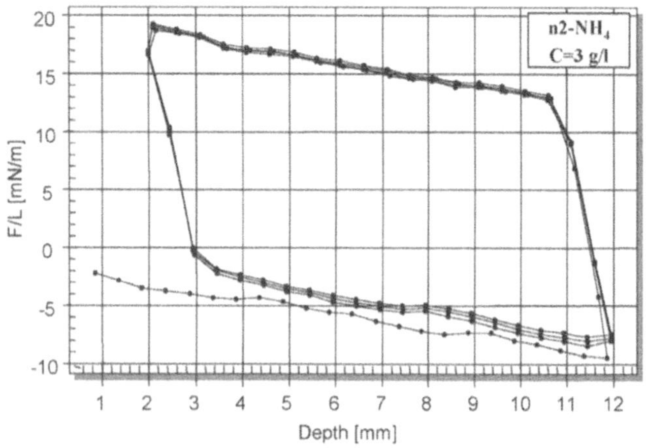

Fig. 2 Dynamic contact angle (*DCA*) profile analysis of 3 g/l $n=2$—NH_4 solution on poly(tetrafluoroethylene) (*PTFE*) at 25 °C and 6 mm/min probe speed

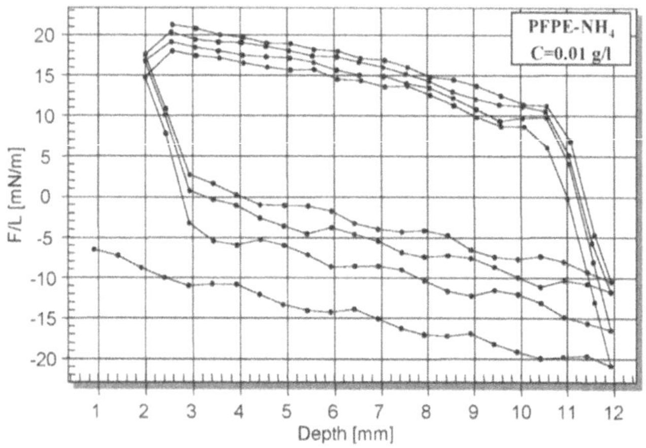

Fig. 3 DCA profile analysis of 0.01 g/l PFPE—NH_4 solution on PTFE at 25 °C and 6 mm/min probe speed

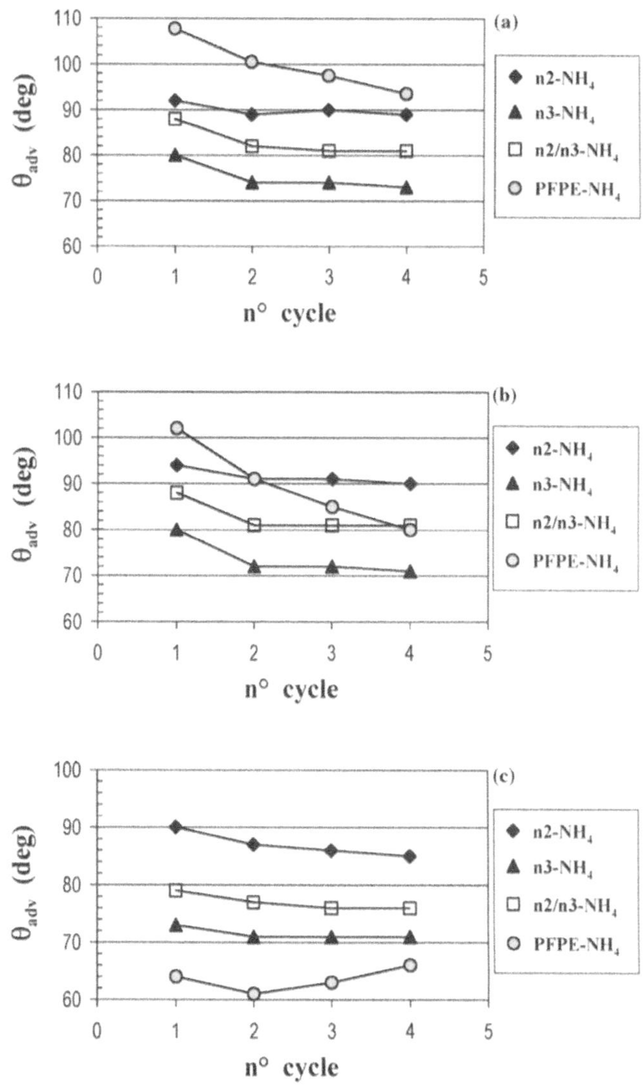

Figs. 4–6 Advancing contact angles of PFPE salt solutions (see Table 1) on PTFE at 25 °C as a function of the cycle number at **a** high (20 mm/min), **b** medium (6 mm/min) and **c** low (0.5 mm/min) probe speeds

dependence of the contact angle of each cycle upon the surfactant chain length agrees with expectations for the $n=2$—NH_4 and $n=3$–NH_4 solutions, where the wetting efficiency markedly increases with the chain length.

At all probe speeds examined, the kinetic hysteresis (θ_A variation among the different cycles) remains small enough to conclude that the adsorption onto PTFE is not diffusion-limited for $n=2$ and $n=3$ surfactants in solution per se, or in their binary mixture.

The decrease in θ_{adv} in the cycles subsequent to the first one suggests increased hydrophilic character of the PTFE, which reasonably accounts for the surfactant adsorption onto the fabric surface. In this respect, the actual surfactant concentration must play a role and it is worth recalling that the molar concentration of $n=2$ is about 40 times higher than the concentration of $n=3$. However, the $n=3$ solution displays much lower θ_{adv} than the $n=2$ solution at comparable surface tension, strongly supporting the adsorption capability onto the surface to increase with n for the individual components.

Such individual contributions should be averaged in the $n=2/n=3$ mixture well below its cmc, accounting for the competition between the actual concentration of each component and its specific affinity towards the substrate. As a consequence, the binary surfactant mixture displays overall wetting behaviour intermediate between that of the individual components.

A more complex situation is found with the PFPE—NH_4 mixture, where the slow adsorption of even longer tail components, namely $n=4$ and $n=5$, appears to be detectable by the lowering of θ_{adv} at a slow immersion rate only, as apparent by the comparison of the pertinent curves in Figs. 4, 5 and 6. The actual

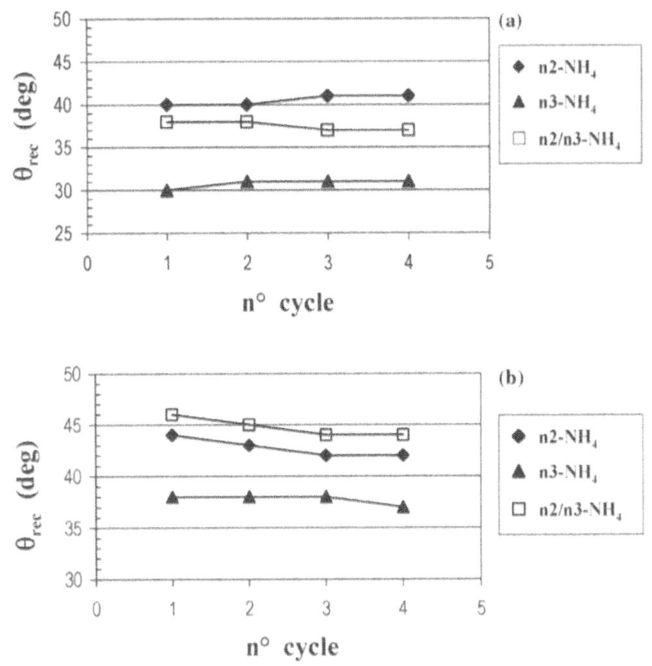

Figs. 7, 8 Receding contact angles of PFPE salt solutions (see Table 1) on PTFE at 25 °C as a function of the cycle number at **a** high (20 mm/min) and **b** low (0.5 mm/min) probe speeds

concentrations of these $n=4$ and $n=5$ components are well below those of the $n=2$ and $n=3$ analogues, but they appear able to compete with the adsorption of shorter analogues when sufficient contact time is allowed between surface and solution, to compensate their much slower diffusion.

Indeed, at a high probe speed, 20 or 6 mm/min, the contact time between the solid sample and the solution does not allow the $n=4$ and $n=5$ components to reach the surface, as pointed out by the large kinetic hysteresis. Much longer contact times, as reached with the lowest probe speed, allow diffusion limits to be exceeded in such case too.

The variations of the receding contact angles θ_{rec} measured for the $n=2$, $n=3$ and $n=2/n=3$ surfactant solutions of Table 1 at the lower and higher probe speed are summarised in Figs. 7, 8. All the θ_{rec} values depend on the probe speed; however, in both cases the $n=3$ solution displays lower θ_{rec} compared to $n=2$; moreover the $n=2/n=3$ mixture gives higher or lower values than those of $n=2$, depending on the probe speed.

Owing to the rather weak PTFE–surfactant interaction, characteristic of the fluorinated materials, a rearrangement of the composition and concentration of the adsorbed film in the receding experiment may occur because of the surface mobility of the adsorbed molecules. This could result in a modification of the local concentration in the receding solution film, with desorption and dewetting possibly occurring. However, further work is needed to better understand the mechanisms impacting on θ_{rec}.

Although a quantitative picture of adsorption energy is lacking, no doubt is left of the capability of these surfactants to increase the surface energy of PTFE by tuned adsorption.

Conclusions

The wetting features of aqueous solutions of fluorinated surfactants on a PTFE surface were studied at almost constant surface tension. PFPE ammonium salts differing in the number of perfluoroisopropoxy units in the chain were used either pure or in mixtures. Evidence of adsorption efficiency increasing with chain length was found when the immersion speed was tuned to overcome individual diffusion constraints.

The lower contact angles in the $n=3$—NH_4 salt solution, with respect to the $n=2$—NH_4 solution, reflect a higher efficiency for the adsorption of the longer-chain term, in agreement with previous results for the adsorption of the same surfactants onto PTFE latex particles [9]; this is further support for the expected strong dependence of adsorption efficiency upon the chain length.

The lowest contact angles were obtained with the PFPE-NH4 polydisperse mixture at a low probe speed, as a consequence of the superior adsorption efficiency of the longer chain with $n > 3$.

References

1. Pieri R, Carignano G, Chittofrati A, D'Aprile F, Visca M (2001) Prog Colloid Polym Sci (submitted)
2. Marchionni G, Srinivasan P(1992) Synthesis, properties and commercial applications, paper 14. Communication at UMIST Manchester, 6–8 January
3. Tonelli C, Di Meo A, Fontana S, Russo A (2002) J Fluorine Chem 118:107–121
4. Tonelli C, Di Meo A, Barchiesi E (2002) J Polym Sci Part A Polym Chem 40:4266–4280
5. Chittofrati A, D'Aprile F, Pieri R, Lenti D, Maccone P, Visca M (2001) Prog Colloid Polym Sci (submitted)
6. Harkins WD, Jordan HF (1930) J Am Chem Soc 52:1751
7. Pieri R, Chittofrati A, Baglioni P (2004) Prog Colloid Polym Sci
8. Wu S (1982) In: Polymer and interface adhesion. Dekker, New York, p 170
9. Lenti D, D'Aprile F, Chittofrati A, Visca M (1999) Communication at XIII ECIS conference, Dublin

Progr Colloid Polym Sci (2004) 126:68–73
DOI 10.1007/b93977

Daniel J.V.A. dos Santos
José A.N.F. Gomes

Chain length effect on the structure of alkyltrimethylammonium chloride monolayers between two immiscible liquids

D.J.V.A. dos Santos (✉)
José A.N.F. Gomes
CEQUP/Departamento de Química da Faculdade de Ciências da Universidade do Porto, R. Campo, Alegre 687, 4069-007 Porto, Portugal
e-mail: dasantos@fc.up.pt

Abstract The main purpose of this article is to understand at the microscopic level the possible effect of the chain length on the properties of an adsorbed monolayer between two immiscible liquids. We report a molecular dynamics simulation study of an alkyltrimethylammonium chloride monolayer adsorbed in the water/isooctane interface. The surfactants studied have 8, 12 and 16 carbon atoms in the chain and are studied at a constant area of 45 Å^2 per surfactant. It was found that the chain widths of the number density profile distributions increase with the chain length and the tilt angle distribution does not change with the chain length. There is also no correlation between the percentage of trans conformations in an alkyltrimethylammonium chain and the length of the chain. The probability of finding a given number of trans conformations per chain is also similar in the three different surfactants. An increase of four methylene groups in the central part of the chain just displaces this distribution by about two trans conformations. The mean-square displacement of the terminal methyl group of the alkyltrimethylammonium chains is also very similar in the different surfactant types. Our results support the conclusions of Bell et al. [(1998) *J. Phys. Chem. B* 102:218] concerning the structures of the alkyltrimethylammonium monolayers (they are very similar and independent of the chain length).

Keywords Self-assembly · Monolayers · Molecular dynamics

Introduction

Molecular monolayers and thin films have been extensively studied and play an important economic and scientific role. They are used extensively, for example, in the mining and oil industry for extraction and separation, or in the household industry for fabric softeners [1]. Clear evidence of the extensive experimental work in this field of chemistry is the appearance in the literature of several reviews [1, 2, 3, 4, 5, 6].

The knowledge of the relationship between the surfactant molecular structure and the macromolecular interfacial structure is important to understand and predict the performance of the surfactant. Yet, only recently the molecular behaviour of the monolayers could be obtained thanks to the advent of new experimental techniques, like vibrational sum-frequency-generation spectroscopy [7, 8], neutron reflection [9, 10, 11, 12, 13, 14], or scanning tunnelling microscopy [15, 16], that can provide information on interfacial phenomena at a molecular level. With this new information, the data obtained by computer simulation can now be more thoroughly compared and validated.

Extensive theoretical [17, 18, 19] and experimental work (see the reviews already cited and the references therein) has been performed for Langmuir and Langmuir–Blodgett interfaces but only recently a small number of simulation studies have been performed on monolayers between two immiscible liquids [20, 21, 22, 23].

Monolayers between two immiscible liquids are suitable for studies by electrochemistry techniques, namely cyclic voltammetry. They are easily built, mechanically stable, and the potential drop across the interface can be accurately controlled externally [24, 25, 26, 27]. The molecular behaviour of a monolayer between two immiscible liquids is very difficult to study even by the newest experimental techniques. Nevertheless, Conboy et al. [28] have applied vibrational sum-frequency spectroscopy to a series of monolayers adsorbed at the deuterated water and carbon tetrachloride interface and Lu and coworkers [29, 30] have studied a monolayer of monododecyl pentaethylene glycol at the solution/air interface with and without an oil layer of dodecane and dodecyltrimethylammonium ($C_{12}TA$) bromide and hexadecyltrimethylammonium ($C_{16}TA$) bromide monolayers mixed with dodecane at the water/air interface by neutron reflectometry.

Lyttle et al. [31] have studied by neutron reflection a series of alkyltrimethylammonium (C_nTA) bromide monolayers with different chain lengths. They found that the monolayer thickness is constant and independent of the length of the carbon chain. For this to occur, the density of the monolayer and the mean tilt of the surfactant chains must increase as the chain length increases. On the other hand, Bell et al. [7] studied this same series of monolayers by sum-frequency spectroscopy and ellipsometry and found that neither the mean chain tilt nor the density varies significantly with the chain length.

Molecular dynamics can provide insightful information and help to clarify this issues. On the other hand, there is some information at the molecular level that despite the advances in the experimental techniques is still not yet accessible, but can be obtained by this type of computer simulation.

Simulation details

For the water potential we used the simple point-charge model [32]. For the $C_{16}TA$ we used the charges calculated by Böcker et al. [33] with the van der Waals and ligand forces parameterisation of the CHARMM22 [34] force field except for the dihedral angle, which was replaced by the Ryckaert–Bellemans [35, 36] description.

To obtain the charges for the other C_nTA ions, C_8TA and $C_{12}TA$, the Gaussian98 [37] ab initio computer program was used with the 6-31G* basis set in a way similar to that used by Böker at al. [33]. The charges on the sites of the united atom model were obtained by a Mulliken population analysis and they were found to be independent of the surfactant carbon tail length.

For isooctane (ISO) we used the same parameters as for the C_nTA carbon chains with no partial charges. For the chloride ion we used the parameters developed by Smith and Dang [38]. The water angle and all bond lengths were constrained using the SHAKE [39] algorithm with a tolerance of 10^{-5} Å. To obtain the Lennard-Jones parameters between sites of unlike atoms the Lorentz–Berthelot [34, 40] mixing rule was used (as required by the CHARMM22 potential).

All simulations were performed with a modified version of the DL_POLY [41] molecular dynamics package in the *NAT* ensemble (300 K and an area of 45 $Å^2$ per head group) using a Nosé–Hoover [42, 43] thermostat in the implementation of Melchionna et al. [44].

Periodic boundary conditions were applied in the monolayer surface (*x*- and *y*-axes) but not for the longitudinal axis, in which the system interfaces with vacuum. In all simulations the velocity Verlet algorithm was used for the integration of the equations of motion with a time step of 2 fs. A 10-Å cutoff was used for the short-range interactions and a 12-Å one for the long-range interactions, with a smoothing function [45] applied between 11 and 12 Å. A multiple time step was also used for interactions greater than 11.5 Å with an update frequency of eight time steps. A system check in the microcanonical ensemble was performed to verify the validity of the parameters and good energy conservation was found.

The system was prepared by constructing a square lattice monolayer with 100 $C_{16}TA$ ions with 45 Å per head group. This corresponds to a cross-section of about 67×67 $Å^2$. All the dihedral angles of the $C_{16}TA$ tails were set in a trans conformation and parallel to the *z*-axis. Chloride ions were added at a distance of about 3–4 Å from the $C_{16}TA$ head positions. A rapid equilibration was performed with the positions of the nitrogen atoms frozen while the tails were randomised – changing the orientation and the dihedral conformations. In all the following simulations the nitrogen atom positions were free to move according to Newtonian dynamics.

A water box with 4,500 molecules and a box containing 753 isooctane molecules, with the same cross-section of about 50×50 Å, were equilibrated for 150 ps in the *NpT* ensemble with periodic boundary conditions and simulation parameters similar to those described for the $C_{16}TA$ chloride lattice.

The water molecules split across the boundaries were reunited and a system corresponding to a Langmuir interface was constructed by joining together the $C_{16}TA$ chloride layer and the water box with a gap of 3–4 Å between them. A new equilibration of 300 ps was performed in the *NAT* ensemble with periodic boundary conditions in all axes except the longitudinal *z*-axis. The molecules of the isooctane box which were split across the boundaries were also joined together. Finally, the monolayer between water/isooctane was constructed by joining the oil box to the Langmuir interface using a gap of 3–4 Å between the $C_{16}TA$ tails and the oil molecules. This system has a length of about 100 Å along the major *z*-axis. The molecules interface with a vacuum at both ends of the longitudinal *z*-axis and have periodic boundary conditions in the monolayer plane (*x*-axis and *y*-axis).

This system was equilibrated for, at least, 1,200 ps. The other two monolayer systems with surfactants containing eight and 12 carbon atoms were constructed by subtracting from the first equilibrated system the excess carbon atoms in the surfactant tails. Both systems were equilibrated for at least 700 ps. In this way we obtained three systems with monolayers adsorbed between water and isooctane that only differ in the number of atoms in the carbon chains (8, 12 and 16).

After this equilibration, a production run of 1,000 ps on the three systems was started. During this production run the trajectories of the particles present in the systems were written to three different files every 32 steps and were analysed later.

Results

The simulated system containing the $C_{16}TA$ monolayer is represented by the average density profile in Fig. 1. The

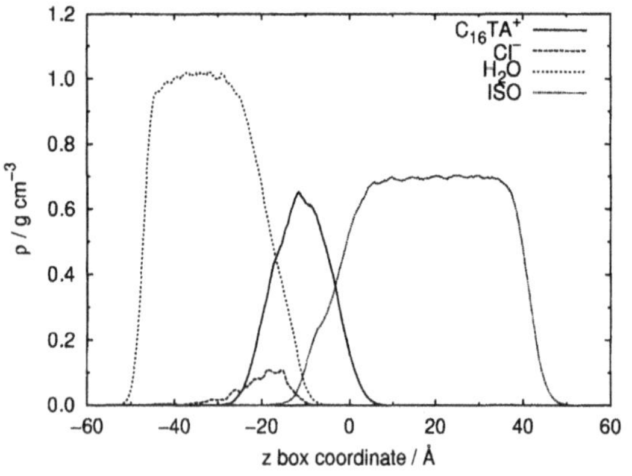

Fig. 1 Average density profiles for the water, chloride ions isooctane and hexadecyltrimethylammonium ($C_{16}TA$) ions as a function of the box coordinate

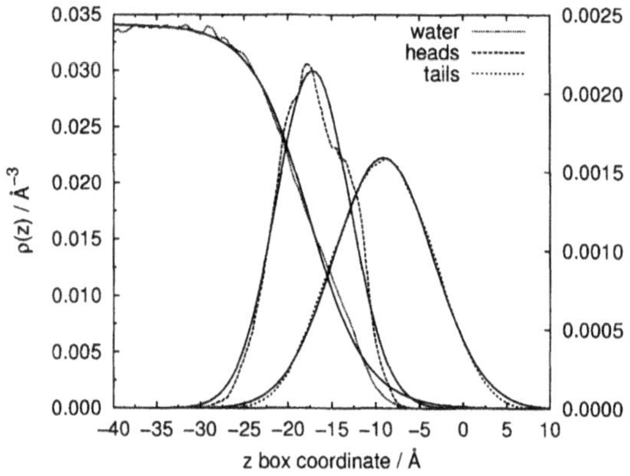

Fig. 2 Profiles for the water and $C_{16}TA$ (heads and tails) number densities at the water/isooctane interface. The fits are also displayed. The water scale is on the left-hand side

other two systems are also very similar and for all we could see the density profiles match the bulk densities of the liquids (water 0.997 and isooctane 0.6980 g cm^{-3}) [46].

The simulation with the C_8TA monolayer is intended to test our results a step further since the carbon tail is very short. We performed a similar simulation with the C_8TA surfactant adsorbed at the water/2-heptanone interface and found that the monolayer was not stable. In the physically short simulation time, the surfactant started to dissolve into the 2-heptanone lamellae and also started to form a double layer. The stability of this monolayer when adsorbed in the water/isooctane interface is due to the fact that isooctane is much more hydrophobic than 2-heptanone. The water solubility in 2-heptanone is rather high, for an oil.

In the density profile for the three systems we could see a diffuse electrical double layer formed by a negatively charged diffuse barrier of chloride ions and a positively charged diffuse barrier of C_nTA heads ($n=8$, 12 and 16). The penetration of the water molecules stops at the point where there are no more heads to be solvated. The diffuse layer of chloride ions is located mainly near the C_nTA heads, but some of the ions diffuse throughout the water slab.

The number density profiles were calculated for the water and C_nTA heads and tails, for the three different surfactants. As an example, in Fig. 2 we present $C_{16}TA$ and water number density profiles. The fits to the water profiles were done using the same hyperbolic tangent function that is commonly used to fit experimental data [6]:

$$\rho = \frac{\rho_0}{2}\left[1 \pm \tanh\left(\frac{z - z_0}{\zeta}\right)\right] \tag{1}$$

where ζ is the width of the distribution and z_0 is the centre.

For the C_nTA heads and tails, a Gaussian function was used:

$$\rho = \rho_0 \exp\left[-4\left(\frac{(z - z_0)^2}{\sigma^2}\right)\right], \tag{2}$$

where σ is the width of the distribution and z_0 is also the centre. The parameters obtained by fitting the simulation results are presented in Table 1.

Although the experimental results presented in Table 1 were obtained for a series of C_nTA monolayers where the counterion was bromide and not chloride, and at the water/air interface instead of the water/isooctane interface, the fitting parameters agree quite well. Aromatic anions can cause a major change in the structure of monolayers [8], but in our case, since the nature of the counterions is similar, this effect is absent and the experimental and simulation results can be directly

Table 1 Widths ζ and σ of the water and alkyltrimethylammonium (C_nTA) head and tail profiles as defined by Eqs. (1) and (2). δ_{ij} are the separations between the distribution centres (chains, c, heads, h, water, s). The experimental values were taken from Refs. [6, 31]

Surfactant	σ_h/Å	σ_c/Å	ζ/Å	δ_{ch}/Å	δ_{hs}/Å	δ_{cs}/Å
C_8TA – this work	11.80 ± 0.09	11.67 ± 0.03	6.45 ± 0.06	4.90 ± 0.04	0.22 ± 0.07	5.12 ± 0.05
$C_{12}TA$ – this work	12.0 ± 0.1	13.42 ± 0.03	6.59 ± 0.07	6.70 ± 0.05	0.22 ± 0.08	6.93 ± 0.05
$C_{12}TA$ – exp	12.5 ± 3	16.0 ± 1	5.5 ± 0.5	6.5 ± 1	1 ± 1	7.0 ± 1
$C_{16}TA$ – this work	12.0 ± 0.1	15.96 ± 0.04	6.50 ± 0.06	8.07 ± 0.05	0.44 ± 0.07	8.51 ± 0.05
$C_{16}TA$ – exp	14 ± 3	16.5 ± 1	6.5 ± 1	8.5 ± 1	2 ± 1.5	9.0 ± 1

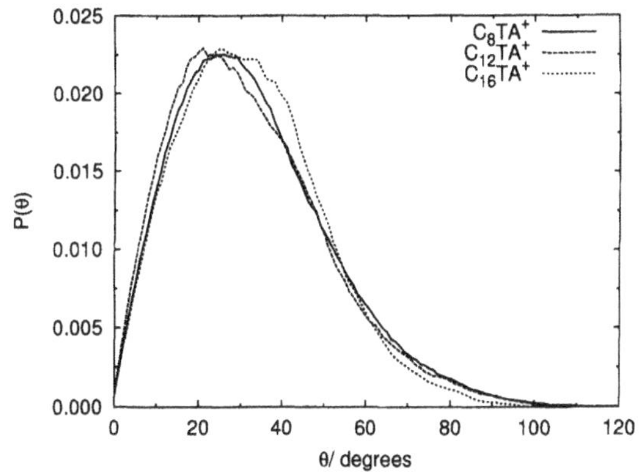

Fig. 3 Tilt angle distributions for the alkyltrimethylammonium (C_nTA) ions adsorbed at the water/ISO interface

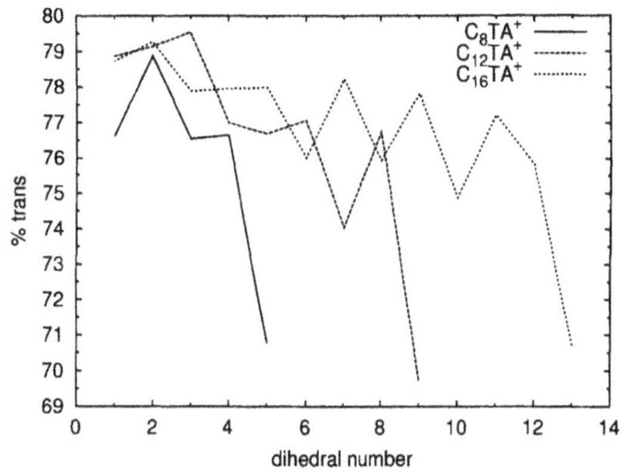

Fig. 4 Percentage of trans conformations in the C_nTA hydrocarbon chains as a function of the dihedral number. On the left-hand side we found the water lamellae and on the right-hand side the ISO lamellae

compared. The presence of the isooctane oil layer instead of air can change some properties, namely the average mean tilt angle (Eq. 3). The tilt angle distribution can be different if a monolayer is adsorbed at the water/air instead of at the water/oil interface, but cannot modify a consistent change in behaviour with the chain length at the same interface.

The major difference between the fitting parameters obtained by simulation and those obtained by neutron reflection refer to the values for σ of the chains. The experimental values for σ_c let Penfold and coworkers state that the thickness of the monolayer is independent of the chain length (about 16 Å for the surfactant series). On the other hand, our results agree with the conclusion of Bell et al. [7] since there is an increase in the width of the chain distribution profile as the chain length increases.

The tilt angle, θ, is defined as

$$\theta = \arccos\left(\mathbf{Z}\cdot\frac{\mathbf{R}_{1N}}{|\mathbf{R}_{1N}|}\right), \qquad (3)$$

where $\mathbf{R}_{1N}$ is the vector between the nitrogen atom and the terminal methyl group in the same molecule, $\mathbf{R}_1$–$\mathbf{R}_N$, and $\mathbf{Z}$ is the unit vector of the z-axis. This angle was calculated over all the C_nTA ions and a probability distribution was obtained for each surfactant type.

As can be seen in Fig. 3 the tilt angle distribution does not change when changing the chain length. For the three distributions, the maximum probability is located between 20 and 30° and the average angle is about $33 \pm 18°$. We can conclude that for a surfactant monolayer at the same coverage density, the tilt angle distribution and consequently the mean tilt do not change when the length of the chain changes. Bell et al. also found that neither the density nor the mean tilt angle varies significantly with the chain length.

The percentage of trans conformations as a function of the dihedral number in the chains of the C_nTA series is presented in Fig. 4. Similar behaviour occurs for this C_nTA series: when one goes through the C_nTA chain from the water into the oil phase, the number of trans conformations decreases and the terminal dihedral has the greater percentage of gauche conformations. The percentage of trans conformations for the terminal dihedral of the chain is almost the same for the three different surfactants. This also occurs with the penultimate and with the dihedral angles that are closer to the surfactant heads. This behaviour means that when we increase the chain length by introducing methylene groups in the middle of the chain, the average conformation remains unchanged and we are just introducing dihedral angles with similar conformations in the central part of the chain. This also agrees with the findings of Bell et al. [7].

The independence of the conformation of the chains upon the chain length is also clearly seen when we calculate the probability of finding a given number of trans conformations per chain and discover that they are similar (Fig. 5). An increase of four methylene groups in the central part of the chain just displaces this distribution by about two trans conformations.

Several order parameters can be obtained by the following tensor [47]:

$$S_{ij} = 1/2\langle 3\cos\theta_i \cos\theta_j - \delta_{ij}\rangle, \qquad (4)$$

where θ_i is the angle between the ith molecular axis and the monolayer normal – z-axis. The order parameter is a measure of the deviation from the monolayer normal. For each CH_2 unit we can define the z-axis as the vector from the C_{n-1} to the C_{n+1} unit; the y-axis is the vector in the plane of these two units and is perpendicular to the z-axis; the x-axis is perpendicular to the y-axis and the z-axis.

The chain order parameter is usually defined as

$$S_{zz} = 1/2\langle 3\cos\theta_i \cos\theta_j - 1\rangle \qquad (5)$$

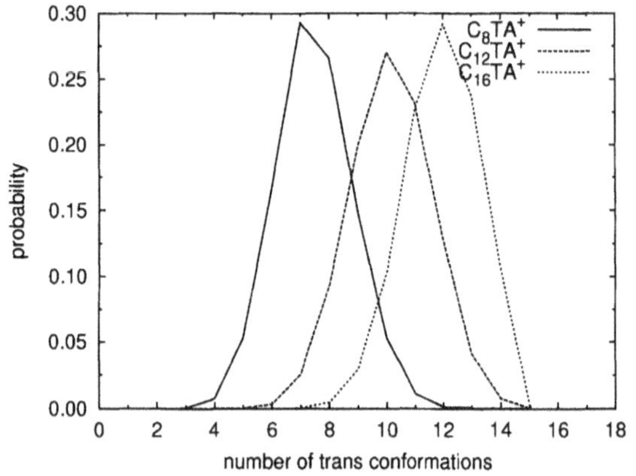

Fig. 5 Probability of finding a given number of trans conformations for each C_nTA surfactant type

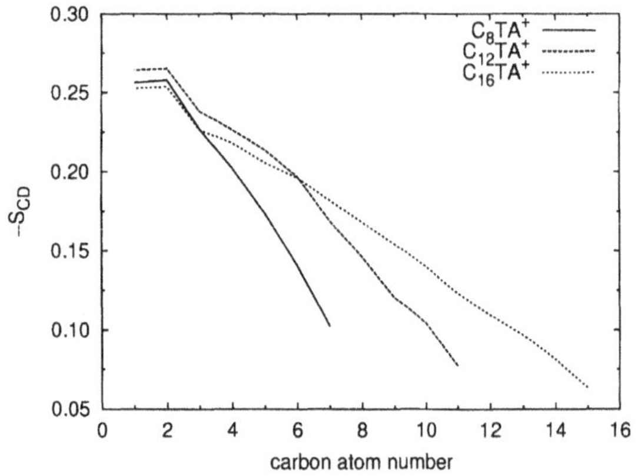

Fig. 6 Average deuterium order parameter as a function of methyl number for each C_nTA surfactant type. The water slab is located on the left-hand side and the oil slab on the right-hand side

and is related to the local order of the long molecular axis. The experimentally observed deuterium NMR order parameter can be obtained using Eq. (6):

$$S_{CD} = 2/3S_{xx} + 1/3S_{yy}. \qquad (6)$$

The values for the chain order parameter and for the deuterium order parameter (Fig. 6) were calculated and found to be in good agreement with the experimental values for other surfactant types [22, 23, 48].

For all three different surfactants, the randomness in the orientation increases from the head to the tail and similar behaviour as described for the percentage of trans conformations in the C_nTA hydrocarbon chains occurs. The addition of methylene groups modifies the slope of the curve but maintains the same characteristics: the order parameters for the carbon atoms near the heads and at the end of the tails remain the same, the only difference remains at the central part of the chains.

The bond order parameter was calculated using the bond vector as the molecular axis and is a measure of the average inclination of the bond axes with respect to the monolayer normal. The differences between the three surfactant types and the evolution of the bond order parameter as one goes from the head to the tail of a given surfactant type is similar to the description already made for the other order parameters. However, the seesaw-like behaviour is more pronounced for the surfactants with the longest chains.

The in-plane mean-square displacement for the terminal methyl group of the C_nTA chains was calculated and was found to be very similar between the different surfactant types, but a positive correlation exists with the chain length. This was to be expected since for the longer chains a trans-to-cis conformation conversion in the central part of the chain makes a large movement of the terminal methyl group.

We can use the Einstein relation [40] for two dimensions and valid for long times to calculate the diffusion coefficient parallel to the interface:

$$D = \lim_{t\to\infty} \frac{1}{4t} \left\langle |r_i(t) - r_i(0)|^2 \right\rangle, \qquad (7)$$

where $r_i(t)$ is the particle position at time t. Using this relation to calculate the diffusion constant, D, for the terminal methyl group, we find values of 6.80×10^{-6}, 5.99×10^{-6} and 5.59×10^{-6} cm^2 s^{-1}, for $C_{16}TA$, $C_{12}TA$ and C_8TA, respectively. These values are in fair agreement with those obtained by Tarek et al. [49] for the lateral diffusion of the tetradecyltrimethylammonium mass centre.

Summary and conclusions

In the present study we reported on a molecular dynamics simulation study of a C_nTA chloride monolayer (with $n = 8$, 12 and 16) between the water/isooctane interface. Both the collective and the individual properties of the C_nTA ions were studied and a comparison was made in order to try to understand the influence of the chain length on the static and dynamic microscopic properties of the monolayer. Our results are in good agreement with available experimental and simulation data.

A diffuse electrical double layer was found, formed on one side by a diffuse negative layer of chloride ions and on the other by a diffuse positively charged layer formed by the C_nTA heads.

When analysing the number density profile distributions, we could find that when varying the surfactant chain length the head-group widths remain unchanged; however, the chain widths increase with the chain length. The tilt angle distribution does not change with the chain length; the maximum probability is located between 20 and 30° and the average angle is about $33 \pm 18°$.

The average number of trans conformations per chain exhibits similar behaviour for this C_nTA series. When the chain length increases, the average number of trans conformations remains unchanged and we are just introducing dihedral angles with similar conformations in the central part of the chain. The probability of finding a given number of trans conformations per chain is also similar in the three different surfactants. An increase of four methylene groups in the central part of the chain just displaces this distribution by about two trans conformations.

The mean-square displacement of the terminal methyl group of the C_nTA chains is very similar among the different surfactant types. Nevertheless, when the chain length increases, the diffusion coefficient also increases because a trans-to-cis conformation conversion in the central part of the chain leads to a large movement of the terminal methyl group.

The data that best characterise the monolayer and surfactant behaviour, namely the width of the number density profiles for the chains, the distance between the distribution centres of the heads and tails, the tilt angle distribution and the percentage of trans conformations, lead us to suggest, in harmony with the findings of Bell et al. [7], that the structures of the C_nTA monolayer studied are very similar and are independent of the chain length.

Acknowledgements D.J.V.A.d.S. thanks the Programa PRAXIS XXI for a doctoral scholarship (BD/15576/98). It is also a pleasure to thank Natália Cordeiro for stimulating discussions. Finally, I am deeply indebted to Elsa Henriques for the manipulation of the Quanta computer program.

References

1. Swalen JD, Allara DL, Andrade JD, Chandross EA, Garoff S, Israelachvili J, McCarthy TJ, Murray R, Pease RF, Rabolt JF, Wynne KJ, Yu H (1987) Langmuir 3:932
2. Shinoda K, Lindman B (1987) Langmuir 3:135
3. Whitesides GM, Laibinis P (1990) Langmuir 6:87
4. Gelbart WM, Ben-Shaul A (1996) J Phys Chem 100:13169
5. Miranda PB, Shen YR (1999) J Phys Chem B 103:3292
6. Lu JR, Thomas RK, Penfold J (2000) Adv Colloid Interface Sci 84:143
7. Bell GR, Manning-Benson S, Bain CD (1998) J Phys Chem B 102:218
8. Bell GR, Bain CD, Li ZX, Thomas RK, Duffy DC, Penfold J (1997) J Am Chem Soc 119:10227
9. Simister EA, Lee EM, Thomas RK, Penfold J (1992) J. Phys Chem 96:1373.
10. Lu JR, Hromadova M, Simister BA, Thomas RK, Penfold J (1994) J Phys Chem 98:11519
11. Lu JR, Li ZX, Smallwood J, Thomas RK, Penfold J (1995) J Phys Chem 99:8233
12. Lu JR, Li ZX, Thomas RK, Penfold J (1996) J Chem Soc Faraday Trans 92:403
13. Lu JR, Thomas RK (1998) J Chem Soc Faraday Trans 94:995
14. Penfold J (2002) Curr Opin Colloid Interface Sci 7:139
15. Wu P, Zeng Q, Xu S, Wang C, Yin S, Bai C-L (2002) CHEMPHYSCHEM 2:750
16. Xu S-L, Wang C, Zeng Q-D, Wu P, Wang Z-G, Yan H-K, Bai C-L (2002) Langmuir 18:657
17. da Gama MMT, Gubbins KE (1986) Mol Phys 59:227
18. Vincze A, Horvai G, Leermakers FAM (1995) Electrochim Acta 40:2875
19. Rieu JP, Vallade M (1996) J Chem Phys 104:729
20. Schweighofer K, Essmann U, Berkowitz M (1997) J Phys Chem B 101:3793
21. Schweighofer K, Essmann U, Berkowitz M (1997) J Phys Chem B 101:10775
22. Dominguez H, Smondyrev AM, Berkowitz ML (1999) J Phys Chem B 103:9582.
23. Dominguez H, Berkowitz ML (2000) J Phys Chem B 104:5302
24. Wandlowski T, Marecek V, Samec Z (1988) J Electroanal Chem 242:277
25. Kakiuchi T, Nakanishi M, Senda M (1989) Bull Chem Soc Jpn 62:403
26. Kakiuchi T, Kondo T, Kotani M, Senda M (1992) Langmuir 8:169
27. Cunnnane VJ, Schiffrin DJ, Fleischmann M, Geblewicz G, Williams D (1988) J Electroanal Chem 243:455
28. Conboy JC, Messner MC, Richmond GL (1997) J Phys Chem 101:6724
29. Lu JR, Thomas KK, Binks BP, Fletcher PDI, Penfold J (1995) J Phys Chem 99:4113
30. Lu JR, Li ZX, Thomas RK, Binks BP, Crichton P, Fletcher PDI, McNab JR, Penfold J (1998) J Phys Chem B 102:5785
31. Lyttle PJ, Lu JR, Su TJ, Thomas RK, Penfold J (1995) Langmuir 11:1001
32. Berendsen HJ, Postma JPM, van Gunsteren WF, Hermans J (1981) In: Pullman B (ed) Intermolecular forces. Reidel: Dordrecht, pp 33
33. Böcker J, Schlenkrich M, Bopp P, Brickmann J (1992) J Phys Chem 96:9915
34. MacKerell RAD Jr et al (1998) J Phys Chem B 102:3586
35. Ryckaert. J-P, Bellemans A (1975) Chem Phys Lett 30:123
36. Ryckaert J-P, Bellemans A (1978) Faraday Discuss Chem Soc 66:95
37. Frisch MJ et al (2001) Gaussian 98, revision A.11.2. Gaussian, Pittsburgh, PA
38. Smith PB, Dang LX (1994) J Chem Phys 100:3757
39. Ryckaert J-P, Ciccotti C, Berendsen HJC (1977) J Comput Phys 23:327
40. Allen MP, Tildesley DJ (1990) Computer simulation of liquids. Oxford University Press, Oxford
41. Forrester PW, Smith W (1998) DL_POLY, version 2.11. CCLRC Daresbury Laboratory, Daresbury, UK
42. Nosé S (1991) In: Meyer M, Pontikis V (eds) Computer simulation in materials science. Kluwer, Dordrecht, pp 21
43. Hoover WG (1985) Phys Rev A 31:1695
44. Melchionna S, Ciccotti G, Holian BL (1993) Mol Phys 78:533
45. Brooks CL III, Pettitt BM, Karplus M (1985) J Chem Phys 83:5897
46. Weast RC (1990) Handbook of chemistry and physics. CRC, Boca Raton, FL
47. Egberts B, Berendsen HJC (1988) J Chem Phys 89:3718
48. Sheng Q, Shulten K, Pidgeon C (1995) J Phys Chem 99:11018
49. Tarek M, Tobias DJ, Klein ML (1995) J Phys Chem 99:1393

Progr Colloid Polym Sci (2004) 126: 74–78
DOI 10.1007/b93980

Patrycja Dynarowicz-Łatka
Katarzyna Kita
Piotr Milart
Demétrio A. Silva Filho

Langmuir films from *n*-alkyl substituted 5′-phenyl-*m*-terphenyl carboxylic acids

P. Dynarowicz-Łatka (✉) · K. Kita
P. Milart
Department of General Chemistry,
Faculty of Chemistry,
Jagiellonian University,
Ingardena 3, 30-060 Kraków, Poland
e-mail: ucdynaro@cyf-kr.edu.pl
Tel.: +48-12-6336377 ext. 2236
Fax: +48-12-6340515

D.A.S. Filho
Universidade Estadual de Campinas,
Instituto de Física Gleb Wataghin,
Campinas, SP, Brazil

Abstract We report studies on the influence of the introduction of an *n*-alkyl group in the 4′ position of an aromatic carboxylic acid (namely 5′-phenyl-*m*-terphenyl-4-carboxylic acid, PTCA). For this purpose, 4′-methyl and 4′-*n*-propyl PTCA were synthesized and investigated at the air/water interface. The monolayers were characterized by measuring the surface pressure and electric surface potential upon film compression. The dependence of the size of the *n*-alkyl substituent on the plateau transition and on the dipole moment is studied and compared with the basic PTCA.

Keywords Langmuir monolayers · Aromatic carboxylic acids · Air/water interface

Introduction

It is well known that the chemical structure of the hydrophobic part of a molecule as well as the nature of the polar head group dictate the monolayer characteristics of classical film-forming compounds. Although many studies have focused on the influence of the chemical structure on monolayer properties from aliphatic amphiphiles, only little work has been devoted to the surface characterization of purely arenic analogues. Our previous studies on film-forming properties of molecules with a solely aromatic, uncondensed hydrophobic moiety consisting of nonrigid polyphenyl system [1, 2, 3] have proved that the compound possessing a symmetrical triphenylbenzene ring (5′-phenyl-*m*-terphenyl core) and the carboxylic group attached to the distal 4 position, abbreviated as PTCA (Scheme 1, compound A), and its purely aromatic 4′ derivatives ($R = -C_6H_5$; $-C_6H_4-C_6H_5$) are capable of forming stable floating monolayers at the air/water interface on their own, without the necessity of mixing with any builder material, despite the fact that these molecules do not have any alkyl chains, which are generally assumed to be necessary for spreading. Interestingly, their monolayer characteristics differ significantly from those of model long-chain aliphatic ampiphiles, of which stearic acid is the most common example. Namely, the surface pressure–area, $\pi-A$, isotherms of PTCA and its 4′ derivatives exhibit a broad plateau with about a twofold decrease in area, separating two regions of liquid-condensed character [2, 3]. Upon detailed analysis, on the basis of surface pressure and electric surface potential measurements complemented with quantitative Brewster angle microscopy experiments, this plateau transition was ascribed to a gradual inclination of film molecules upon compression, followed by multilayer formation [4]. Our former studies have demonstrated that the introduction of one [2] or two [3] benzene rings in the 4′ position of PTCA improves significantly the monolayer stability. This is logical since the basic, unsubstituted PTCA, which consists of four phenyl rings, lies just on the borderline regarding the capability of monolayer formation, and the introduction of the additional hydrophobic substituent brings better film-forming properties.

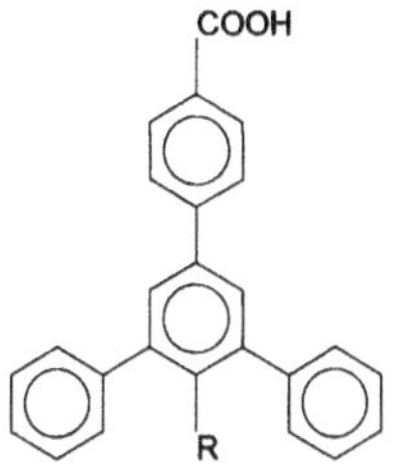

Compound	R
A (PTCA)	-H
B (methyl-PTCA)	$-CH_3$
C (propyl-PTCA)	$-C_3H_7$

Scheme 1 Chemical structures of the compounds investigated

In this work we have furthered our studies in researching to what extent the presence of an *n*-aliphatic substituent of the 5′-phenyl-*m*-terphenyl core modifies the monolayer properties. For this purpose, the methyl and *n*-propyl groups were introduced in the 4′ position of PTCA (Scheme 1, compounds B and C, respectively) and their influence was examined with the Langmuir techniques, by recording the surface pressure (π) and the electric surface potential, ΔV, upon monolayer compression.

Experimental

Synthesis

The synthesis of 5′-phenyl-1,1′:3′,1″-terphenyl-4-carboxylic acid (PTCA), 5′-phenyl-4′-methyl-1,1′:3′,1″-terphenyl-4-carboxylic acid (methyl-PTCA) and 4-(4-carboxyphenyl)-2,6-diphenylpyrylium perchlorate (Scheme 2, compound 1) was described earlier [6].

5′-Phenyl-4′-propyl-1,1′:3′,1″-terphenyl-4-carboxylic acid

A mixture of pyrylium salt (Scheme 2, compound 1) (1.60 g, 3.5 mmol), anhydrous sodium valerate (1.24 g, 10.0 mmol) and valeric anhydride (10 cm^3) was stirred and heated at 170 °C for 2 h. The mixture was cooled to room temperature and the solvent was removed under reduced pressure (around 10 mm Hg). The oily dark-brown residue was boiled with 30 cm^3 25% aqueous NaOH solution for 1 h. The solution was cooled with ice to 5 °C and acidified with 1:1 hydrochloric acid. The separated brown oil was isolated from the almost colourless aqueous layer and treated with 10 cm^3 acetic acid. The mixture was refluxed for few minutes and, after cooling, a light-brown precipitate was separated. The crude carboxylic acid was recrystallized from a small amount of acetic acid and then sublimed in a vacuum (230–240 °C, 1 mm Hg). This procedure gives a colourless product. Yield: 0.23 g (16.8%). M.p. 275–276 °C. Anal. Calcd for $C_{28}H_{24}O_2$: C, 85.67; H, 6.18. Found: C, 85.70; H, 6.13. IR (KBr) ν (cm^{-1}): 3,454, 2,960, 2,865, 2,668, 254 (OH + CH); 1,683 (C = O); 1,606 (Ar rings); 1,422, 1,284. ^1H NMR (dimethyl sulphoxide-d_6, DMSO-d_6, δ) (ppm): 12.90 (s, 1H, broad, –COOH); 7.98 and 7.84 (2d, 4H, AA′BB′, $J_{ortho} = 8.4$ Hz, protons of the ring with –COOH group); 7.51–7.36 (m, 12H, other aromatic protons); 2.54 (t, 2H, $J = 8.0$ Hz, benzylic protons); 1.10–1.03 (m, 2H, $-CH_2-$); 0.42 (t, 3H, $J = 7.3$ Hz, $-CH_3$). The protons in the propyl group are shielded by the anisotropic field of both neighbouring phenyl rings. As a result, the signals of $-CH_2-CH_3$ protons are strongly shifted upfield. The same effect was observed for 2,4,6-triphenyl-4-propylpyridinium tetrafluoroborate [7].

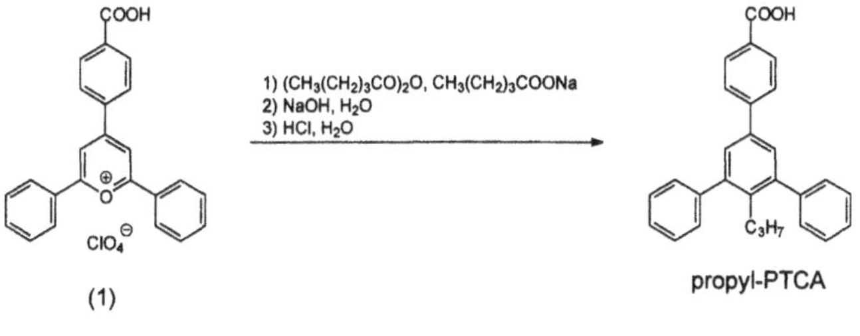

Scheme 2 Synthetic path leading to compound C (5′-phenyl-4′-propyl-1,1′:3′,1″-terphenyl-4-carboxylic acid, *n-propyl-PTCA*)

Langmuir monolayers

Ultrapure water was used as a subphase for the Langmuir monolayers, which were prepared by spreading an aliquot (around 150 μl) of a chloroform (Merck, p.a.) solution (concentration around 0.5 mg ml^{-1}) onto the air/water interface. Purified water was supplied by a Milli-RO/Milli-Q system (Millipore; resistivity 18 MΩ cm, pH 5.6). π–A isotherms were measured in a computer-controlled Langmuir–Blodgett film balance (KSV Instruments, Helsinki, Finland, model 5000). The surface pressure of the floating monolayer was measured to an accuracy of 0.1 mN m^{-1} using a Wilhelmy plate (made of chromatography paper, ashless Whatman Chr 1) connected to an electrobalance. Simultaneously, the surface potential was monitored using a vibrating plate located around 2 mm above the water surface. The reference electrode, made from platinum foil, was placed in the water subphase. The surface potential measurements were reproducible to ± 10 mV.

After spreading, the solutions were left for 5 min to ensure complete solvent evaporation before compression was started. The monolayers were usually compressed with a barrier speed of 25 mm min^{-1} (equivalent to a compression rate of 7.5×10^{17} Å^2 min^{-1}) at 20 °C unless otherwise specified. The spreading isotherms reported here are the averages of at least three experiments.

Characterization methods

Melting points were determined with a Mel-Temp II apparatus in open capillaries and are uncorrected. Elemental analyses were carried out with a EuroVector EA 3000 analyser. IR spectra were recorded using a Bruker Equinox 55 spectrometer and KBr pellets. ^1H NMR spectra were taken at 500.13 MHz with a Bruker AMX 500 spectrometer using DMSO-d_6 as a solvent and tetramethylsilane as an internal standard.

Results and discussion

The influence of the introduction of the alkyl substituent into the apolar core of the parent compound, PTCA, is illustrated in Fig. 1. The general characteristic of the π–A isotherm for both unsubstituted PTCA and its derivatives remains unchanged, i.e. all the compounds under investigation exhibit a broad plateau, which spans over a

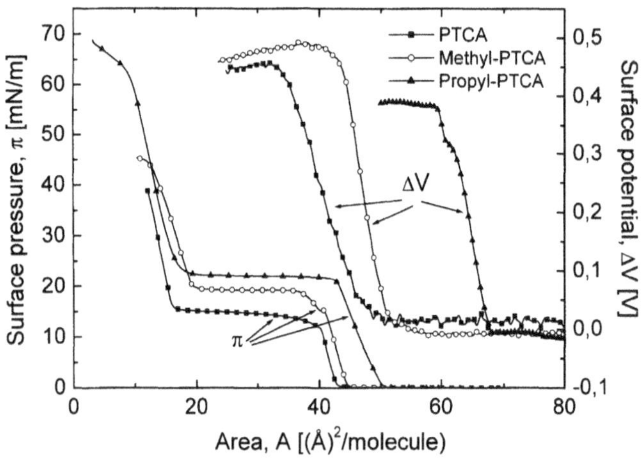

Fig. 1 Surface pressure (π)–area (A) and surface potential (ΔV)–A isotherms of the compounds investigated spread on a pure water subphase. Compression speed 7.5×10^{17} Å^2 min^{-1}, 20 °C

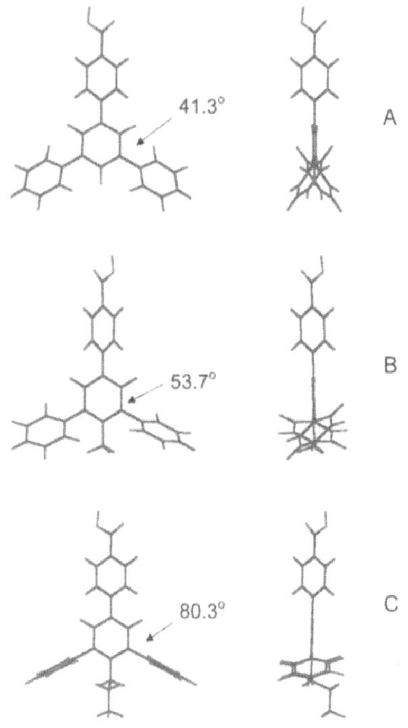

Scheme 3 Projections of the molecules investigated along the normal to the surface

region corresponding to an about twofold decrease in the area per molecule and separates two liquid-condensed regions. The onset for the surface-pressure rise for the 4′-methyl derivative is only shifted by around 2 Å^2 towards larger areas compared to PTCA. In turn, the pressure lift-off area for 5′-phenyl-4′-propyl-1,1′:3′,1″-terphenyl-4-carboxylic acid (4′-propyl-PTCA) is increased by 5 Å^2 with respect to the methyl derivative. These little differences in lift-off areas are not surprising since the values for the onset of the surface-pressure rise can be considered as representing the areas of projection of a molecule at the interface. The presence of a small alkyl group, which is screened by the two neighbouring phenyl rings may slightly push two phenyl groups out, thus increasing the molecular area; however, it does not contribute significantly to this value, as evidenced by very similar experimental data of lift-off areas for all the compounds investigated. Although the total contribution to the molecular area caused by the substituent in the 4′ position is only small, since the aromatic rings in PTCAs are free to rotate along the single bond connecting the rings, it can be supposed that the alkyl group, which is introduced between two side phenyl rings, may induce the conformational changes in PTCA.

To confirm this hypothesis, we performed semiempirical computations with the Austin model 1 (AM1) [8] parameterization for the free molecules in a vacuum. The calculations were performed with the quantum chemistry program HyperChem [9]. Our calculations revealed that the dihedral angle between the planes of the phenyl rings in PTCA (around 41°) (Scheme 3, compound A) increases with the introduction of the methyl or the propyl group in the 5′-phenyl-*m*-terphenyl core, leading to a more pronounced out-of-plane configuration of the molecule. At the same time, the angle between the central benzene ring and the ring bearing the COOH group remains practically the same (40.9° for PTCA, 39.7° for methyl-PTCA and 39.6° for propyl-PTCA). As shown in Scheme 3, the side rings are twisted with respect to the central one by nearly 54 and 80° for methyl-PTCA (compound B) and propyl-PTCA (compound C), respectively. These conformational changes caused by the introduction of an alkyl substituent in the 4′ position of the 5′-phenyl-*m*-terphenyl core explain the observed differences in the mean molecular area for the onset of the surface-pressure rise for PTCA and its derivatives.

Another important observation from the surface-pressure measurements is that the plateau transition for 4′-alkyl PTCAs is attained at higher surface pressures compared to PTCA. The bigger the alkyl substituent, the higher the value of the transition pressure observed.

Similarly to surface pressure, the electric surface potential isotherms (see Fig. 1) also show a similar trend for all the compounds investigated, i.e. the potential increases steadily upon compression, reaching a maximum value (0.45; 0.47 and 0.38 V for PTCA, methyl-PTCA and propyl-PTCA, respectively) in the preplateau region, and upon further compression ΔV remains nearly constant.

For the quantitative, comparative analysis it is useful to calculate the effective dipole moment of film-forming molecules upon compression. The effective dipole moment, $\mu_{\perp}$, which is the normal (to the interface) component of the dipole moment of the molecule at the surface, was calculated from surface-potential data, applying the Helmholtz (parallel-plate condenser) model [10]:

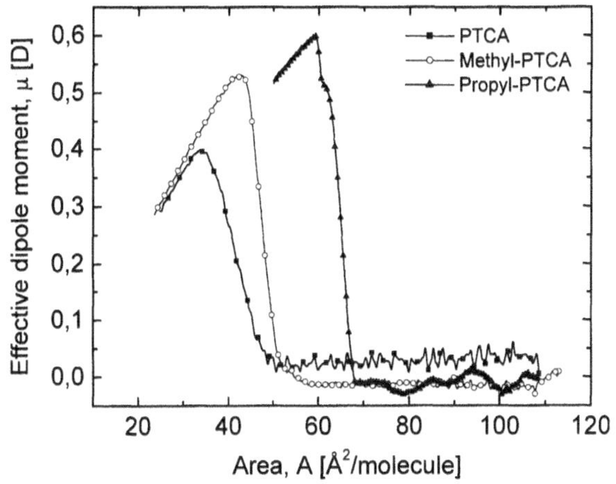

Fig. 2 Effective dipole moment ($\mu_\perp$)–A isotherms of the monolayers investigated on water. Compression speed 7.5 × 10^{17} Å^2 min^{-1}, 20 °C

$$\mu_\perp = \Delta VA\varepsilon_r\varepsilon_0, \qquad (1)$$

where ε_r and ε_0 are the dielectric constants of the monolayer and the permittivity of free space, respectively. In calculating $\mu_\perp$, the value for ε_r was taken as 1. The dependence of the effective dipole moment on area is shown in Fig. 2. The trend of the effective dipole moment isotherm resembles that of the electric surface potential, i.e. the effective dipole moment increases along the monolayer compression, reaching the maximum value at the same area as ΔV_{max} is attained, and afterwards decreases rapidly. As analysed elsewhere [5], the maximum value of the effective dipole corresponds to the perpendicular orientation of film molecules, which occurs at the termination of the preplateau region. The experimental results show the difference in $\mu_\perp^{max}$ is 0.39; 0.53 and 0.60 D for PTCA, methyl-PTCA and propyl-PTCA, respectively, spread on a water subphase. A similar trend can be observed for monolayers formed on 10^{-3} M HCl aqueous solution (0.53; 0.63 and 0.78 D, respectively) where the film molecules are not ionized.

According to the Demchak–Fort model [11], the effective dipole moment can be expressed as

$$\mu_\perp = \mu_1/\varepsilon_1 + \mu_2/\varepsilon_2 + \mu_3/\varepsilon_3. \qquad (2)$$

In Eq. (2), μ_1 is the normal (to the water surface) component of the dipole moment originating from reorientation of water molecules in the subphase due to the presence of the monolayer, μ_2 is the normal dipole moment of the head group, and μ_3 is the normal dipole moment of the hydrophobic part of the molecule. Each of the dipoles is anchored in a medium with effective dielectric constant ε_i. The double-layer contribution [12] is not considered since we will analyse only nonionized monolayers.

Assuming the same vertical orientation of all the molecules investigated at the area corresponding to the maximum values of either the surface potential or the effective dipole moment, and the same contribution from the water molecules (μ_1/ε_1) as well as from the head (–COOH) group (μ_2/ε_2), the difference in $\mu_\perp^{max}$ must be due to different resulting contribution from the hydrophobic part of the molecules (μ_3/ε_3). Certainly, a further assumption is that the head group orientation is the same for all the compounds investigated, which seems to be quite reasonable at the termination of the preplateau region. Under these conditions, the contributions from the groups comprising the hydrophobic part of the molecules can be calculated by comparing the experimental results of the dipole moments. As evidenced from Table 1 (column 2) the introduction of the methyl group to the 5′-phenyl-*m*-terphenyl core causes $\mu_\perp^{max}$ to be increased by approximately 0.1 D. This contribution can, in fact, be expected and is consistent with calculated values of dipole moments (Table 1, column 3). The calculations were performed as follows. First, the ground-state geometries of PTCA, methyl-PTCA and propyl-PTCA, using the semiempirical method AM1 [8] and the Hartree–Fock/3-21G* [13] ab initio method, both implemented in the personal computer GAMESS version of the GAMESS (US) quantum chemistry package [14], were optimized. No geometry or symmetry constraints were used during the optimization procedure. A tight convergence limit of 10^{-6} hartree bohr^{-1} was imposed on the largest component of the gradient during all optimizations. The geometry optimization was followed by a frequency calculation in order to assure that a minimum, rather than a saddle point, was found. The dipole moments were calculated after each geometry optimization. The molecule was then oriented as displayed in Table 1, column 1, and the components of the total calculated dipole moment: perpendicular, along the z-axis, μ_z, and parallel, along the $x-y$ plane, μ_{x-y}, were determined (Table 1, column 3). This table also contains the values of the ratios $|\mu_z|/|\mu_{total}|$ and $|\mu_{x-y}|/|\mu_{total}|$. The ratios are guidelines to compare AM1 and HF/3-21G*, since it cannot be expected that these two approaches lead to identical quantitative values. Instead, they allows one to follow the trend.

For the PTCA derivatives investigated, both AM1 and HF/3-21G* indicate that the introduction of the methyl group in the polyphenyl core causes the increase of both the total dipole moment (μ_{total}) and the the perpendicular component of the dipole moment ($|\mu_z|$)of the molecule. This corroborates with the experimental results of the effective dipole moments, showing an increase of around 0.1 D in the value of $\mu_\perp^{max}$ for methyl-PTCA compared to the basic, unsubstituted compound. Further increase in the length of the alkyl chain increases the experimental value of the effective dipole moment by around 0.08 D on average. This is consistent with earlier reports on the small increase in surface potentials/effective dipole moments with the length of the alkyl chain in fatty acid monolayers [15].

Table 1 Comparison of experimental and calculated dipole moment of 5′-phenyl-*m*-terphenyl-4-carboxylic acid and its 4′-alkyl derivatives. μ_Z is the component of the dipole moment normal to the interface and μ_{x-y} is the component of the dipole moment parallel to the interface

Comparison of experimental and calculated dipole moment of PTCA and its 4'-alkyl derivatives

Compound		**Experimental Effective Dipole Moment*, [D]**	**Calculated Dipoles Moments in Vacuum, [D]**					
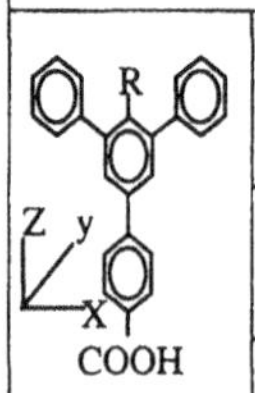	Radical	$\mu_\perp^{max}$	Hamiltonian	μ_z	μ_{x-y}	μ_{total}	$\|\mu_z\|/\|\mu_{total}\|$	$\|\mu_{x-y}\|/\|\mu_{total}\|$
	R=H	0.530 ±0.010	AM1	-2.675	1.127	2.903	0.921	0.388
			HF/3-21G*	-1.957	1.317	2.359	0.830	0.558
	R= CH_3	0.630 ±0.005	AM1	-2.998	1.119	3.200	0.937	0.349
			HF/3-21G*	-2.440	1.321	2.774	0.879	0.476
	R= C_3H_7	0.780 ±0.010	AM1	-3.051	1.047	3.235	0.943	0.332
			HF/3-21G*	-2.361	1.310	2.701	0.874	0.485

Legend

μ_z – Component of the dipole moment normal to the interface
μ_{x-y} – Component of the dipole moment parallel to the interface

*) For monolayers spread on 10^{-3}M HCl aq. solution, calculated using Helmholtz equation at the area corresponding to the maximum value of surface potential.

Also, AM1 calculations show a similar, increasing trend; however, a slight decrease in both μ_{total} and $|\mu_Z|$ is observed with HF/3-21G* computation. It is worth mentioning here that the molecules investigated are rather flexible and even little changes in the geometrical parameters, which can be associated with changes in energy, can modify the values of the dipole moments and can thus be responsible for the observed discrepancies in the dipole moment values calculated with AM1 and HF/3-21G*.

References

1. Czapkiewicz J, Dynarowicz-Łatka P, Janicka G, Milart P (1998) Colloids Surf A 135:149
2. Dynarowicz-Łatka P, Dhanabalan A, Oliveira ON Jr (1999) J Phys Chem B 103:5992
3. Dynarowicz-Łatka P, Dhanabalan A, Oliveira ON Jr (2000) Langmuir 16:4245
4. Ferreira M, Dynarowicz-Łatka P, Miñones J Jr, Caetano W, Kita K, Schadke M, Lösche M, Oliveira ON Jr (2002) J Phys Chem B (in press)
5. Dynarowicz-Łatka P, Dhanabalan A, Cavalli A, Oliveira ON Jr (2000) J Phys Chem B 104:1701
6. Czapkiewicz J, Dynarowicz P, Milart P (1996) Langmuir 12:4966
7. Katritzky AR, Liso G, Lunt E, Patel RC, Thind SS, Zia A (1980) J Chem Soc Perkin Trans I:849
8. Dewar MJS, Zoebish EG, Healy JJP, Stewart J (1985) J Am Chem Soc 107:3902
9. Hypercube (1998) HyperChem Professional, release 5.1. A molecular visualization and simulation software package. Hypercube, Gainesville, FL
10. Davies JT, Rideal K (1963) Interfacial phenomena. Academic, New York
11. Demchak RJ, Fort TJ Jr (1974) J Colloid Interface Sci 46:191
12. Davies JT (1956) J Colloid Sci 11:377
13. Pietro W, Francl M, Hehre W, Defrees D, Pople J, Binkley J (1982) J Am Chem Soc 104:5039
14. Schmidt M, Baldridge K, Boatz J, Elbert S, Gordon M, Jensen J, Koseki S, Matsunaga N, Nguyen K, Su S, Windus T, Dupuis M, Montgomery J (1993) J Comput Chem 14:1347
15. Kuchhal YK, Katti SS, Biswas AB (1974) 49:48

Progr Colloid Polym Sci (2004) 126: 79–82
DOI 10.1007/b93939

K. Debbabi
F. Guittard
S. Géribaldi

Oxosulfonylisocyanate precursors of monopodal nonionic fluorinated surfactants

K. Debbabi · F. Guittard (✉)
S. Géribaldi
Chimie des Matériaux Organiques et Métalliques, Faculté des Sciences, Université de Nice-Sophia Antipolis, Parc Valrose, 06108
Nice Cedex 2, France
e-mail: guittard@unice.fr,
Fax: +33-4-92076156

Abstract Monodisperse nonionic fluorinated surfactants with oligooxyethyl monomethylated ethers as the hydrophilic head and a perfluoroalkyl tail as the hydrophobic moiety linked together through an *N*-sulfonylcarbamate function were synthesized in two steps with good yields. The evaluation of their behaviour at the air–water interface has been investigated at 20 °C by measurement of surface tension, γ_s, versus concentration (Wilhelmy method). From these measurements, the critical micelle concentration (cmc), the maximum surface excess concentration and the minimum area per molecule were calculated. This allows us to show clearly the variation of the cmc (from 1.14×10^{-4} to 7.3×10^{-5} mol l^{-1} for lower γ_s (from 23.5 to 17.3 mM m^{-1}). These compounds exhibit noteworthy properties as fluorinated surfactants. The relationships between these physicochemical parameters and the structure of fluorinated oligooxyethylated surfactants are discussed.

Keywords Nonionic surfactants · Fluorinated · Oligooxyethylated · Surface tension

Introduction

Nonionic oligooxyethylated surfactants are of considerable research interest because of their applications in many fields, such as detergents, solubilizing agents, emulsifying agents, dispersing agents and potential biomaterials for biomedical applications [1, 2, 3, 4, 5, 6, 7]. Furthermore, the introduction of a highly fluorinated tail in a structure confers, compared to its hydrocarbon homologue, a number of specific physicochemical properties, such as higher chemical and thermal stability, lower surface tension at extremely low concentration, higher vapour pressure and high density. The interfacial properties of surfactants provide the basis for various applications and are usually characterized by the critical micelle concentration (cmc), surface tension at the air–water interface, surface area per molecule of surfactant and surface excess concentration. Surface tension as a function of surfactant concentration was the method used to calculate these parameters. So from these facts, we have dedicated this work to the preparation of fluorinated oligooxyethylated surfactants using highly fluorinated oxosulfonylisocyanates as starting materials. The surface-active properties of the oligooxyethylated surfactants were determined from surface tension measurements. The effects of hydrophobic and hydrophilic chain lengths on the properties were examined and were correlated to the molecular structures of the fluorinated oligooxyethylated surfactants.

Experimental

Methods

The surface tensions of aqueous surfactant solutions were measured at 20 °C with a Prolabo tensimat model N3 using the Wilhelmy plate method. The tensiometer was calibrated with tare before use. The platinum plate was cleaned by flaming, while the glassware was cleaned with ultrapure water. The surfactant solutions were prepared by using bidistilled water of low conductivity (purified by a Buchi water purification system to remove colloidal, inorganic and organic impurities). The surfactant solution (0.01 wt%) was freshly prepared as a stock solution and then diluted to the desired concentration for each measurement. The surface tension was measured three times for each concentration and an average error less than 0.1 mN m^{-1} was obtained routinely. IR spectra were recorded at room temperature with a PerkinElmer Paragon 1000 Fourier transform IR spectrometer. The materials were ground with KBr and pressed into pellets. ^{1}H NMR chemical shifts measured in deuterated solvents are given in parts per million from tetramethylsilane (TMS) as an internal standard and ^{19}F NMR ($CFCl_3$ as internal reference) spectra were recorded with a Bruker AC 200 MHz spectrometer, using $(CD_3)_2CO$ as solvent (5–10 wt%). The mass spectra were run on MS-MS apparatus Finnigan MAT TSQ 7000 in atmospheric pressure chemical ionization mode. Distillations were performed with a Kugelrohr from Aldrich with an Alcatel 2004 A pump.

Materials

Chlorosulfonylisocyanate (99%) and oligooxyethylmonomethylated (greater than 97%), were purchased from Aldrich and were used as received. 1*H*,1*H*-Perfluoroheptan-1-ol, 1*H*,1*H*-perfluorooctan-1-ol and 1*H*,1*H*-perfluorononan-1-ol, purchased from Interchim, were used as received. Unless specified the solvents were of unpurified reagent grade. The nonionic oligooxyethylated surfactants were prepared using the following method.

In the first step 15 mmol chlorosulfonylisocyanate dissolved in 30 ml dried chlorobenzene was added dropwise at 20 °C to a solution of 15 mmol perfluoroalkylmethanol in 40 ml dried chlorobenzene. The reaction solution was stirred at 120 °C for 24 h and then concentrated to a light brown viscous residue by vacuum distillation. The residue was distilled using Kugelrohr apparatus to give the desired pure colourless oxosulfonylisocyanate. In the second step 2 mmol oligoethoxymonomethylated ether was added dropwise to 20 ml diethyl ether solution containing 2 mmol oxosulfonylisocyanate (freshly prepared). The reaction solution was stirred at room temperature for 2 h and then concentrated to a white solid or a colourles gel residue by vacuum distillation. The residue was washed with petroleum ether to remove traces of impurties for solids and purified by column chromatography using 60:40 ether/cyclohexane as eluent. The purity was higher than 99.2% as determined from high-performance liquid chromatography measurements.

MS: $F_{6\text{-}2}$ $(M\text{–H})=574.1$, $F_{7\text{-}2}$ $(M\text{–H})=624$, $F_{8\text{-}2}$ $(M\text{–H})=674.1$, $F_{7\text{-}1}$ $(M\text{–H})=580.1$, $F_{7\text{-}3}$ $(M\text{–H})=668.1$.

^{1}H NMR [$(CD_3)_2CO$/TMS, δ]: 5.1(t, 2H, CH_2–R_F, $^{3}J_{H\text{–}F}$ = 13.4 Hz); $q=1$ 3.3 (s, 3H, CH_3), 3.6 (m, 2H, CH_2), 4.3 (m, 2H, CH_2); $q=2$ 3.3 (s, 3H, CH_3),3.5 (m, 2H, CH_2),3.6 (m, 2H, CH_2),3.7 (m, 2H, CH_2), 4.4(m, 2H, CH_2); $q=3$ 3.3 (s, 3H, CH_3), 3.6 (m, 10H, 5×CH_2), 4.3 (m, 2H, CH_2).

^{19}F NMR [$(CD_3)_2CO/CFCl_3$, δ): $p=6$ -81 (CF_3), -113 ($CF_{2\alpha}$), -121 ($CF_{2\beta}$), -122 ($CF_{2\gamma}$), -123 ($CF_{2\delta}$), -126 ($CF_{2\omega}$); $p=7$ -81 (CF_3), -119 ($CF_{2\alpha}$), -121 (2×CF_2), -122 (2×CF_2), -126 ($CF_{2\omega}$); $p=8$ -81 (CF_3), -119 ($CF_{2\alpha}$), -121 (3×CF_2), -122 (2×CF_2), -125($CF_{2\omega}$).

IR, ν(cm^{-1}): 3,241 (ν_{NH}), 1,745 (ν_{CO}), 1,366 ($^{a}\nu_{SO2}$), 1,134 ($^{s}\nu_{SO2}$), 1,200–1,100 (ν_{CF}).

Results and discussion

Synthesis

The synthesis of nonionic surfactants often requires the use of ethylene oxide as starting material, which leads to a polydisperse ethyloxy moiety [8, 9]. However, some applications, such as microemulsion development, require the use of oligoethoxylated nonionic surfactants with a number of ethoxy units perfectly well defined and monodisperse in their formulation [10, 11]. The nonionic fluorinated surfactants synthesized in this work are symbolized by $F_{p\text{-}q}$, where p and q represent the number of difluoromethylene units in the fluorinated tail and the number of ethoxy motifs per oligooxyethylated tail, respectively. The presence of the hydrocarbon spacer between the perfluorinated alkyl group and the oxyethyl hydrophilic moiety could contribute to greatly reducing, although not eliminating, the potential toxicity of these synthetic fluorinated surfactants [12]. The oligooxyethylated surfactants can be obtained using the synthetic pathway described in Fig. 1.

The 1-*F*-alkylmethylalcohols (where the *F*-alkyl moiety is *F*-hexyl, *F*-heptyl or *F*-octyl) react in a first step with chlorosulfonylisocyanate, leading to the formation of *N*-chlorosulfonylcarbamates, 1, which, in a second step and at high temperature, undergo a rearrangement to give the corresponding oxosulfonylisocyanates, 2. It is important to note that this transformation can occur only when the *F*-alkyl tail is close to the oxosulfonyl group. This behaviour is similar to that observed in the preparation of aroxylsulfonylisocyanate [13]. The oxosulfonylisocyanates so obtained react with monomethylated oligooxyethylene to give the corresponding oligooxyethylated nonionic surfactants ($F_{p\text{-}q}$).

The yields as well as the melting points of the nonionic fluorinated surfactants synthesized are collected in Table 1.

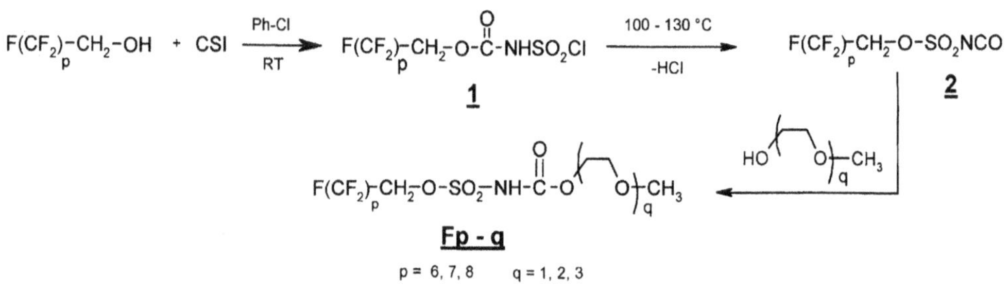

Fig. 1 Synthetic pathway for the synthesis of fluorinated surfactants $F_{p\text{-}q}$

Evaluation of the interfacial properties

The oligooxyethylated surfactants are readily soluble in distilled water at 0.01% in weight in concentration. This low solubility is attributed to their relatively high hydrophobic-to-hydrophilic ratio. In methanol the $F_{p\text{-}q}$ surfactants are soluble, while in octane none of the surfactants are soluble. The enhancement of the length of the fluorinated tail decreases their solubility in water slightly as expected [14].

The surface tensions of the aqueous surfactant solutions obtained for $F_{p\text{-}q}$ surfactants are low and have the main properties of fluorinated surfactants (Table 2). The results obtained show a high surface activity at the air–water interface. In fact, the surface tension values range from 23.5 mN m^{-1} in the case of $F_{6\text{-}2}$ to 17.3 mN m^{-1} in the case of $F_{8\text{-}2}$. For the same number of oxyethylene groups (q), the increase of the length of fluorinated tail leads to a decrease of surface tension, γ^{cmc}. On the other hand, for a fixed value of the fluorinated tail, $p=7$, the surface tension at the cmc increases with increasing number of oxyethylene motifs, as generally expected [15, 16]. This can be put in correlation to previous studies from nonionic fluorosurfactants.

The solution behaviour of surfactants is the result of their amphiphilic nature. As shown in Fig. 2, the surface tension decreases linearly with increasing logarithm of surfactant concentration and then levels off. The inflection point observed for all the surfactants corresponds to the cmc. The cmc results for the oligoethoxylated surfactants are summarized in Table 2.

The increase of the fluorinated chain length leads to a decrease in the cmc values as generally found [16, 17].

Table 1 Yields and melting points of nonionic surfactants $F_{p\text{-}q}$

Compound	p	q	Steps		Melting point (°C)
			Yield[a]	Yield[b]	
$F_{6\text{-}2}$	6	2	70	97	77
$F_{7\text{-}2}$	7	2	57	92	66
$F_{8\text{-}2}$	8	2	67	95	74
$F_{7\text{-}1}$	7	1	57	96	62
$F_{7\text{-}3}$	7	3	57	95	Oil

[a]Yield of oxosulfonylisocyanates distilled (compound 2 in Fig. 1)
[b]Yield of purified and isolated $F_{p\text{-}q}$ compounds

The cmc results are consistent with the hydrophobic domain being the predominant driving force for micellization in aqueous solutions. The examination of the data obtained (Table 2) also shows that the cmc decreases slightly with increasing number of ethoxylated groups. Plots of the logarithm of the cmc for the oligooxyethylated surfactants as a function of the number of fluorinated carbons and ethoxylated groups are shown in Fig. 3. The logarithm of the cmc decreases linearly on increasing either the number of ethoxylated motifs or the number of fluorinated carbons.

The maximum surface excess per nonionic surfactant at the air–water interface was calculated from the slope of the linear portion of the curves obtained by plotting the surface tension against log$c$ and the Gibbs equation [18]. The data recorded indicate that both fluorination of the hydrophobic part and reducing the number of ethoxylated groups lead to a decrease of the surface excess. Under conditions of surface saturation, the Gibbs equation can be used to determine the maximum extent of surfactant adsorption on the surface, from which the surface area per molecule of surfactant was calculated. The results obtained show an increase in the trend of the surface area with the increase in the length of the fluorinated tail and a decrease in the surface area with the

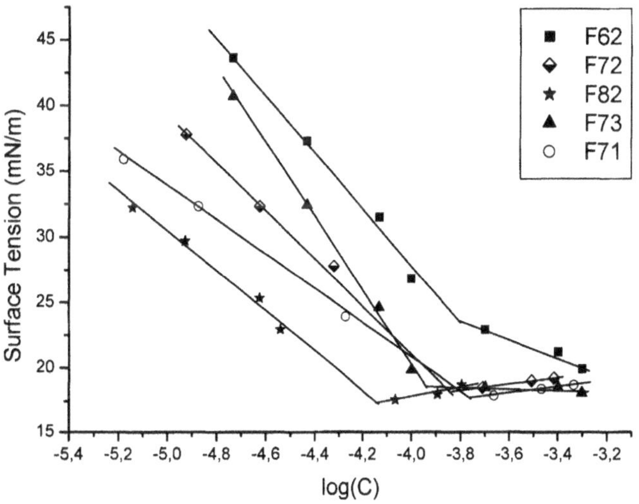

Fig. 2 Surface tension against logarithm of molar concentration for oligooxyethylated surfactants $F_{p\text{-}q}$

Table 2 Values of the critical micellar concentration (*cmc*), the surface tension above the cmc, γ_s^{cmc}, the surface excess, Γ, and the surface area, a, per head group of sulfonamide surfactants

Surfactant	10^4cmc (mol dm^{-3})	γ_s^{cmc} (mN m^{-1})	$-[d\gamma/d\log c]_T$ (mN m^{-1})	$10^6\ \Gamma$ (mol m^{-2})	$10^2\ a$ (nm^2/molecule)
$F_{p\text{-}q}$					
6-2	1.57	23.5	22	3.85	43
7-2	1.47	18.1	16.5	2.9	57
8-2	0.73	17.3	15	2.62	63
7-1	1.72	17.7	13.3	2.33	71
7-3	1.14	18.6	27.8	4.88	34

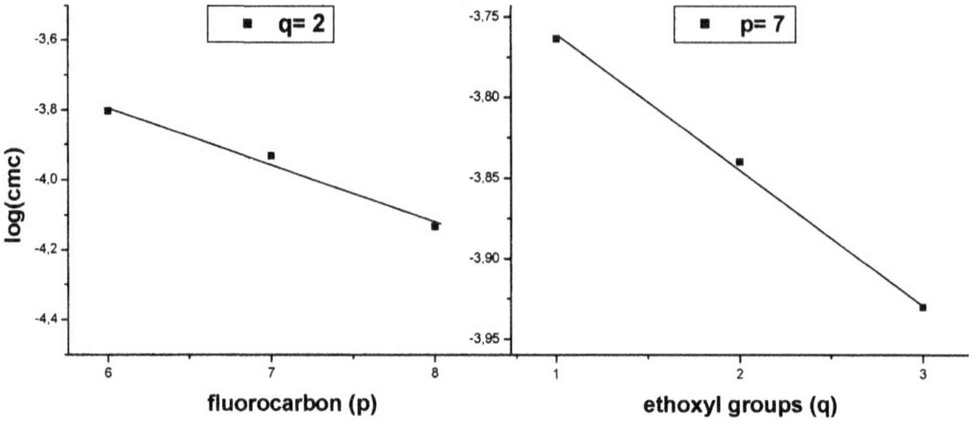

Fig. 3 Logarithm of the critical micellar concentration (*cmc*) versus p and q

increase in the number of oxyethylated groups. The area occupied by a surfactant molecule provides valuable information on the orientation of these molecules at the interface. The small area covered by the adsorbed fluorinated oligooxyethylated surfactants shows very tight packing at the air–water interface.

Conclusion

In this work, we have shown that a series of structurally well defined fluorinated oligooxyethylmonomethylated nonionic surfactants can be obtained by a very efficient pathway with high yields. Results for these monodisperse surfactants containing a perfluoroalkyl tail show that they exhibit high surface activity at the air–water interface and that they decrease efficiently the surface tension of aqueous solutions at extremely low concentration. In fact, the minimum surface tension, 17.3 mN m^{-1}, was obtained with compound $F_{8\text{-}2}$. The variation of the cmc follows the general rule with a decrease with the increase in both the length of the fluorinated tail and/or the number of oxyethyl units. The micellization and adsorption processes are more sensitive to oligooxyethylene units in the surfactant molecule and are also more dependent on the length of the fluoroalkyl chain. This oligooxyethylated series exhibits very interesting physicochemical parameters for use as stabilizing agents for perfluorocarbon emulsion. The synthetic approach described in this work provides opportunities for the synthesis and examination of surface-active properties and adsorption behaviour of new surfactants with more complex structures, such as bipodal and polyhydroxylated structures.

Acknowledgement K.D. thanks CROUS for a grant.

References

1. Schick MJ (1967) Nonionic surfactants. Dekker, New York
2. Schick MJ (1987) Nonionic surfactants: physical chemistry. Dekker, New York
3. Sagitani HJ (1988) J Dispersion Sci Technol 9:1115
4. Lee JH, Kopecek J, Andrade JD (1989) J Biomed Mater Res 23:351
5. Krafft MP, Riess JG (1998) Biochimie 80:489
6. Riess JG (2001) Chem Rev 101:2797
7. Riess JG (2002) Tetrahedron 58:4113
8. Meissner E, Myszkowski J, Szymanowski J, Barkoum R (1993) J Chem Tech Biotechnol 57:49
9. Meissner E, Myszkowski J, Szymanowski J, Barkoum R (1993) J Chem Tech Biotechnol 58:271
10. (a) Ravey JC, Stebe MJ (1994) Colloids Surf A; (b) Choi SM, Ku CY, Chen SH (1999) J Phys Chem B 103:5347
11. LoNostro P, Choi SM, Ku CY, Chen SH (1999) J Phys Chem B 103:5347
12. Yiv HS (1990) US Patent 4,975,468
13. Debbabi K, Beji M, Baklouti A, Guittard F, Geribaldi S (2002) Phosphorus Sulfur Silicon 177:933
14. Mukerjee P, Mysels KL (1994) In: Fluorinated surfactants. Surfactant science series, vol 50. Dekker, New York, p 236
15. Porter MR (1994) Hand book of surfactants. Blackie, London
16. Mureau N, Trabelsi H, Guittard F, Geribaldi S (2000) J Colloid Interface Sci 229:440–444
17. Selve C, Ravey JC, Steb MT, El Moudjahid C, Moumni EM, Delpuech JJ (1991) Tetrahedron 47:411
18. Kissa E (2001) In: Fluorinated surfactants and repellents, 2nd edn. Surfactant science series, vol 97. Dekker, New York

Progr Colloid Polym Sci (2004) 126: 83–85
DOI 10.1007/b94025

Ana Čop
Nikola Kallay

Surface potential of ice in aqueous electrolyte solutions

A. Čop · N. Kallay (✉)
Laboratory of Physical Chemistry,
Department of Chemistry,
Faculty of Science, University of Zagreb,
Marulicev trg 19, P.O. Box 163,
10 001 Zagreb, Croatia
e-mail: nkallay@chem.pmf.hr
Tel.: +385-1-4895502;
Fax: +385-1-4895510

Abstract The surface potential at the ice–water interface was measured with an ice electrode as a function of pH. Equilibration in the basic region was found to be slow and no stable potential was obtained. In the pH region below 8 the equilibration was fast, yielding stable and accurate values of the surface potential. The effect of Na^+, K^+ and Mg^{2+} ions was examined. It was found that these ions do not have a different influence on the surface potential in the entire pH region.

Keywords Ice · Surface potential · Electrical interfacial layer

Charging of the ice–water interface

It is well established that solid surfaces immersed in aqueous liquid media undergo several interfacial ionic reactions resulting in surface charge. There are numerous reports on different types of interfaces, such as silver halides and metal oxides. It is somewhat surprising that one of the important interfaces in nature, namely the ice–water interface was somehow neglected. Generally, an ice surface in a liquid aqueous environment can be charged by nonequilibrium processes, but also owing to equilibration of bulk ionic species with the surface. Fast freezing results in selective incorporation of ions from the bulk phase into the solid ice phase so that relatively high electrostatic potentials can be obtained [1, 2, 3, 4]. Equilibrium processes, such as interactions of ionic species at the interface, result in lower but still significant potentials. Electrokinetic measurements showed that ice is positively charged below pH 3–4, and negatively charged above this value [5, 6]. Adsorption measurements showed that cations are bound to the negatively charged surface in the region above pH 6, and that the number of cations adsorbed increases with pH [7]. Measurements with an ice electrode showed that the electrostatic potential at the ice surface decreases with pH [8]. In the pH region below 8 the approximate linearity of E(pH) was observed with a slope lower than the Nernstian. In the region above pH 8, a substantial deviation from linearity was observed. All these measurements suggested that the ice–water interface undergoes processes similar to metal oxide interfaces, which is not surprising because OH groups at the ice surface may bind and release protons, resulting in a positive or a negative charge, respectively [9].

There are numerous reports on the examination of equilibrium at metal oxide aqueous electrolyte interfaces [10, 11, 12, 13]. For that purpose several experimental techniques were commonly employed, for example, electrokinetics, potentiometry, adsorption measurements, and spectroscopy. Quantitative interpretation of the equilibrium is based on the surface complexation model (SCM) considering specific surface reactions. Electrostatic effects on the surface equilibrium are taken into account on the basis of the assumed structure of the electrical interfacial layer (EIL) that extends from the surface towards the bulk of the liquid phase. There are several variations of both the SCM and the EIL model, and the proper choice of model cannot be easily made on the basis of experimental data. However, despite this shortcoming the general behavior of metal oxide aqueous interfaces is fairly well understood.

According to the SCM amphoteric surface sites (≡OH) on ice undergo charging by protonation and deprotonation, with extents depending on the pH of the aqueous solution:

$$\equiv \mathrm{OH} + \mathrm{H}^+ \rightleftharpoons \equiv \mathrm{OH_2}^+;$$

$$K_\mathrm{p} = \exp(\varphi_0 F/RT)\frac{\Gamma(\equiv \mathrm{OH_2}^+)}{a(\mathrm{H}^+)\Gamma(\equiv \mathrm{OH})}, \quad (1)$$

$$\equiv \mathrm{OH} \rightleftharpoons \equiv \mathrm{O}^- + \mathrm{H}^+;$$

$$K_\mathrm{d} = \exp(-\varphi_0 F/RT)\frac{a(\mathrm{H}^+)\Gamma(\equiv \mathrm{O}^-)}{\Gamma(\equiv \mathrm{OH})}, \quad (2)$$

where Γ denotes the equilibrium surface concentration, φ_0 the potential in the 0-plane affecting the state of the charged surface species, a is the activity in the bulk of the solution, K_p and K_d are thermodynamic surface equilibrium constants of protonation and deprotonation, respectively, while F, R and T have their usual meaning.

Counterions (anions A^- and cations C^+) associate with oppositely charged surface groups according to

$$\equiv \mathrm{OH_2}^+ + \mathrm{A}^- \rightleftharpoons \equiv \mathrm{OH_2}^+ \cdot \mathrm{A}^-;$$

$$K_\mathrm{A} = \exp\left(-\varphi_\beta F/RT\right)\frac{\Gamma\left(\equiv \mathrm{OH_2}^+ \cdot \mathrm{A}^-\right)}{a\left(\mathrm{A}^-\right)\Gamma\left(\equiv \mathrm{OH_2}^+\right)} \quad (3)$$

and

$$\equiv \mathrm{O}^- + \mathrm{C}^+ \rightleftharpoons \equiv \mathrm{O}^- \cdot \mathrm{C}^+;$$

$$K_\mathrm{C} = \exp\left(\varphi_\beta F/RT\right)\frac{\Gamma\left(\equiv \mathrm{O}^- \cdot \mathrm{C}^+\right)}{a\left(\mathrm{C}^+\right)\Gamma\left(\equiv \mathrm{O}^-\right)}. \quad (4)$$

The potential in the β-plane (φ_β) to which associated counterions are exposed is usually lower in magnitude than φ_0. K_A and K_C are thermodynamic equilibrium constants for association of anions and cations, respectively.

In order to quantitatively describe the interfacial equilibrium, with parameters such as equilibrium constants of surface reactions, one would need to know the surface concentrations of the adsorbed species. Since there is no possibility to prepare a colloidal ice–water dispersion with sufficiently high and constant surface area, a method based on the interpretation of surface potential and electrokinetic data was developed [9]. The experimental data for the ice–water interface are shown schematically in Fig. 1.

In this work we examine equilibration of the interfacial processes and the effect of different cations on the surface potential.

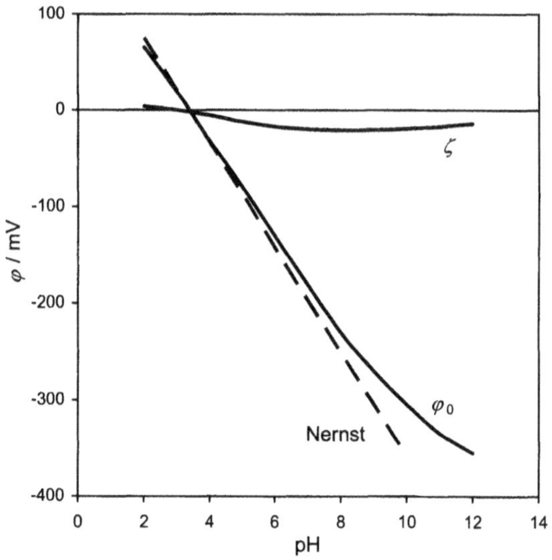

Fig. 1 Schematic representation of the potentials in the electrical interfacial layer for an ice–water system, according to Refs. [8, 9]

Experimental

An ice electrode consists of a body with a platinum wire at the bottom on which the ice layer was formed from pure water by internal cooling with the circulating liquid having a temperature of −7 °C. After formation of the ice layer, the electrode was immersed in the measuring solution. The systems were kept under an argon atmosphere and thermostated, so that the temperature of the measuring solution was 0.02 °C. The pH of the solution was varied by addition of HNO_3, NaOH or KOH. The pH was measured with a glass electrode (Metrohm, model 6.0222.100) in combination with an Ag|AgCl|Cl⁻(aq) reference electrode. The glass electrode was calibrated by three standard thermostated (0.02 °C) buffer solutions (Merck). During the potentiometric experiment (using a Titrino, Metrohm) the temperature regime was kept stable so that the amount of ice on the ice electrode remained constant. The measurements resulted in potentials of the ice electrode as a function of the pH of the solution. Considering the electric circuit, in which several potential drops are present, one may conclude that the only potential drop that depends on pH is the one which was developed at the ice–water interface. Other potential drops are constant and do not depend on the pH of the solution since these interfaces are not exposed to the influence of the H^+ ions from the solution. Accordingly, the measured ice electrode potential is the sum of a constant value and the pH-dependent potential drop at the ice–water interface. This potential drop, is the difference in the electrostatic potentials between the bulk of the solution and the solid ice and corresponds to the surface potential (φ_0). In order to determine its value from the measured ice electrode potential one needs to subtract a constant value. For that purpose it is enough to locate its zero value, which was taken to be at the electrokinetic isoelectric point ($\mathrm{pH_{iep}}$ 3.4) [13].

Results and discussion

In order to examine the rate of equilibration of the ice–water interface the time dependency of the ice electrode potential was measured. Two typical results are represented in Fig. 2. It shows the change of the potential

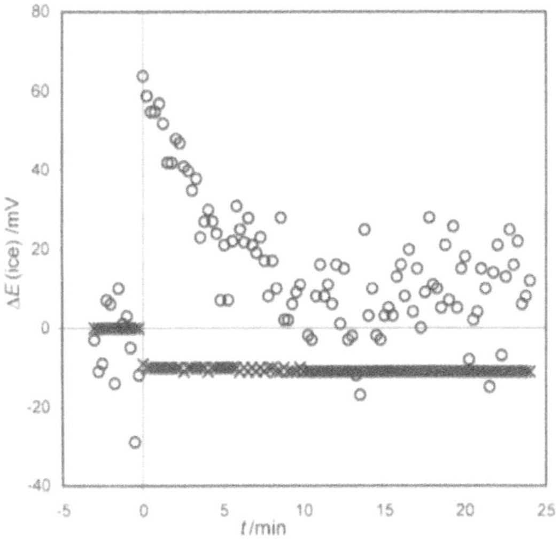

Fig. 2 Stability of the potential of an ice electrode immersed in aqueous $NaNO_3$ solution at 273.2 K. Titrant (HNO_3 or NaOH) was added at $t=0$. The data points for $t<0$ correspond to initial condition. The results in the acidic region (pH_{in} 3.4, pH_{fin} 3.6, *crosses*) and in the basic region (pH_{in} 11.5, pH_{fin} 11.3, *circles*) are represented as the change of the potential of the ice electrode with respect to the initial conditions

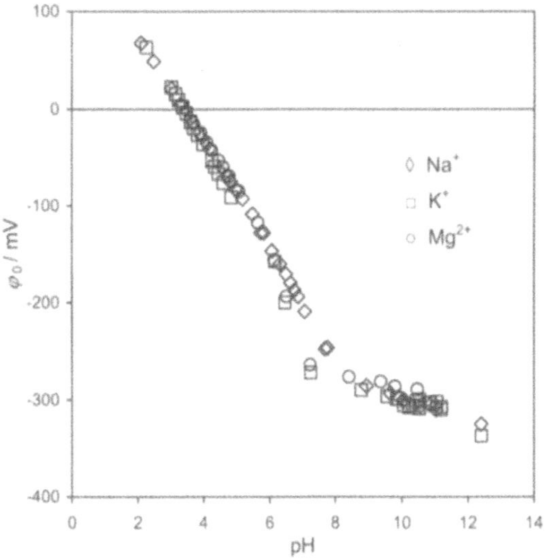

Fig. 3 Surface potential of ice as a function of pH at 273.2 K in the presence of $NaNO_3$ ($c=1\times10^{-5}$ mol dm^{-3}) (*diamonds*), KNO_3 ($c=1\times10^{-5}$ mol dm^{-3}) (*squares*) and $Mg(NO_3)_2$ ($c=1\times10^{-5}$ mol dm^{-3})+KNO_3 ($c=1\times10^{-5}$ mol dm^{-3}) (*circles*). The pH values were adjusted by addition of HNO_3 and NaOH (in the case of Na^+ ion) or KOH (in the case of K^+ and Mg^{2+} ions)

after addition of a base to the system in the acidic region at pH 3.4 and of an acid in the basic region at pH 11.5. The final pH values were 3.6 and 11.3, respectively. In the acidic region the potential of the ice electrode decreased rapidly from 495 to 484 mV, reaching a final stable value after 30 s. The reversibility in the acidic region was additionally confirmed by reversed titration, yielding the same values and stability of the potential. However, in the basic region the noise was found to be within the range of 20 mV. This finding indicated that equilibration of the ice surface in the basic region was not reached, which may be due to slow incorporation of sodium ions into the soft, negatively charged layer at the interface.

To investigate the specificity of counterions the surface potential of ice was measured in the presence of sodium, potassium and magnesium ions. The first two ions are of the same charge, but different in size, while the third one is doubly charged, so more significant binding affinity might be expected. The results are presented in Fig. 3.

It is obvious that no significant difference in φ_0 for these three cations was observed. Even at the relatively high concentration of Mg^{2+} ions the dependency of the surface potential on pH is practically the same as for Na^+ and K^+ ions.

In conclusion one may say that the ice–water interface is charged owing to interactions of surface sites with H^+ ions from the liquid phase and that ice is negatively charged in the region above pH 3.4, which means that in natural conditions the ice surface is always negative. Negatively charged sites at the ice surface bind positively charged counterions, which reduce the electrostatic effects at the surface. Therefore, in addition to nonequilibrium processes, ionic interactions at the ice–water interface result in electrostatic charge and could also be responsible for some natural phenomena (e.g. thunderstorms).

References

1. Workman EJ, Reynolds SE (1950) Phys Rev 78:254
2. Parreira HC, Eydt AJ (1965) Nature 208:33
3. Murphy EJ (1970) J Colloid Interface Sci 32:1
4. Bronshteyn VL, Chernov AA (1991) J Cryst Growth 112:129
5. Romanov VP, Nechaev EA, Zvonareva GV (1975) Kolloidn Zh 38:1009
6. Drzymala J, Sadowski Z, Holysz L, Chibowski E (1999) J Colloid Interface Sci 220:229
7. Nechaev EA, Ivanov IA (1974) Kolloidn Zh 36:583
8. Kallay N, Čakara D (2000) J. Colloid Interface Sci 232:81
9. Kallay N, Čop A, Holysz L, Chibowski E (2003) J Colloid Interface Sci 259:89
10. Kosmulski M, Sprycha R, Szczypa J (2000) In: Kallay N (ed) Interfacial dynamics. Dekker, New York, pp 163–223
11. Kallay N, Kovačević D, Čop A (2000) In: Kallay N (ed) Interfacial dynamics. Dekker, New York, pp 249–271
12. Kovačević D, Čop A, Kallay N (2002) In: Delgado AV (ed) Interfacial electrokinetics and electrophoresis. Dekker, New York, pp 99–121
13. Kallay N, Žalac S (2000) J Colloid Interface Sci 230:1

Progr Colloid Polym Sci (2004) 126: 86–89
DOI 10.1007/b94012

A. F. C. Campos
T. A. P. G. Cotta
J. Depeyrot
F. A. Tourinho

Speciation diagrams as a useful tool to improve nanoparticle surface–ligand complexation for biocompatible magnetic colloids

A. F. C. Campos · T. A. P. G. Cotta
F. A. Tourinho (✉)
Complex Fluids Group,
Instituto de Química,
Universidade de Brasília,
CP 04478, 70919-970
Brasília (DF), Brazil
e-mail: touriinho@fis.unb.br
Tel.: +55-61-3072163,
Fax: +55-61-2734149

J. Depeyrot
Complex Fluids Group,
Instituto de Física,
Universidade de Brasília,
CP 04455, 70919-970
Brasília (DF), Brazil

Abstract Biomedical applications of magnetic carriers are based on the specific interaction between the particle surface and ligands like *meso*-2,3-dimercaptosuccinic acid. Such cross-linking agents are poly-functional Brönsted acids leading in aqueous media to several species the amounts of which vary strongly as a function of pH. Moreover, electrical double-layered magnetic colloids behave as a mixture of strong and weak diprotic acids and the analysis of the equilibria involved allows us to determine the quantitative pH dependence of the superficial density of the charge. Then, the behavior of coated magnetic nanoparticles by complexation of superficial metal ions should be controlled by the pH. In this context, we explore, from a theoretical point of view, the particle surface–ligand complexation in order to improve the synthesis of biocompatible magnetic fluids. The results obtained allow a theoretical prediction of the optimal pH for ligand–surface complexation and the colloidal stability in a physiological medium.

Keywords Biocompatible magnetic colloids · Succimer · Speciation diagrams

Introduction

Currently, magnetic carrier technology is widely used for biological and biomedical purposes such as cell sorting, drug vectoring, contrast agents in magnetic resonance imaging and oncological treatment. Many applications of these materials are based on specific receptor–effector complexes where the latter is coupled with coated iron oxide particles. The receptor is located on the outer monolayer of the membrane and the effector is bonded to the particle surface through nontoxic chelating agents [1]. The coating of the particle is ensured by superficial metal sites which tend to form covalent bonds with electron donor groups from the complexing agent [2]. In the case of *meso*-2,3-dimercaptosuccinic acid (DMSA), it has been reported that the adsorption of the succimer is accompanied by oxidation of thiol groups [3]. Moreover, it has been evidenced very recently [4, 5] that magnetic fluids behave as a mixture of a strong acid (bulk dispersion) and a weak diprotic one (nanoparticle surface). In this model, the charge of the particle surface, responsible for the colloidal stability of electrical double-layered magnetic fluids (EDL-MFs), is obtained through the reaction in aqueous solution of metal ions and the following hydrolysis reactions:

$$\equiv MOH_2^+ + H_2O \underset{}{\overset{pK_1}{\rightleftharpoons}} \equiv MOH + H_3O^+_{(aq)}, \quad (1)$$

$$\equiv MOH + H_2O \underset{}{\overset{pK_2}{\rightleftharpoons}} \equiv MO^- + H_3O^+_{(aq)}. \quad (2)$$

Such equilibria lead to three kinds of superficial sites where most of them are MOH_2^+ in strong acidic media, MO^- in strong basic media and MOH, the intermediate amphoteric sites (where M is a superficial metal). pK_1 and pK_2 are the thermodynamical constants of each respective chemical equilibrium. Then, the surface charge density, σ_0, results from a proton-transfer mechanism between the particle surface and the bulk dispersion. The complexing agents are also polyfunctional Brönsted acids, leading to different species in aqueous media, including amphoteric ones, whose amounts vary strongly as a function of pH. In this context, the speciation diagram of complexing agents has to be taken into account in the nanoparticle surface complexation mechanism of spinel ferrite based MF.

The aim of the present work is, in a first attempt, to explore the speciation in the particle surface–succimer complexation in order to improve the synthesis of biocompatible MFs. In this way, using our formalism of Brönsted complex equilibria involving amphoteric species, we first obtain the quantitative pH dependence of the superficial density of charge in the case of EDL-MFs based on cobalt ferrite nanoparticles [4, 5]. Then, the speciation diagram for charged sites of the particle surface is coupled with that of DMSA in order to analyze the mechanism of surface–DMSA complexation. Finally, we propose a surface reaction which leads to the maximum complexation rate.

Experimental

Sample synthesis and characterization

The investigation of the EDL-MF sample was carried out using the usual procedure [6]. $CoFe_2O_4$ nanoparticles were prepared through hydrothermal coprecipitation of aqueous solutions of $Co(NO_3)_2$—$FeCl_3$ in an alkaline medium. Then, the particles were conveniently peptized in an acidic medium by adjustment of the ionic strength, resulting in stable sols of high quality. The mean nanoparticle size was determined by X-ray diffraction spectra recorded from powder samples, obtained by evaporation of the liquid carrier and was found to be 12.0 nm.

Determination of the pH-dependence of the surface charge density

The analysis of the equilibria in Eqs. (1) and (2) allows us to determine the pH-dependence of the surface charge density[4]:

$$\sigma_0(\mathrm{pH}) = \frac{F}{A} V \left(\frac{10^{-2\mathrm{pH}} - 10^{-(\mathrm{p}K_1+\mathrm{p}K_2)}}{10^{-2\mathrm{pH}} + 10^{-(\mathrm{p}H+\mathrm{p}K_1)} + 10^{-(\mathrm{p}K_1+\mathrm{p}K_2)}} \right) c_\mathrm{T}, \quad (3)$$

where F is the Faraday constant, A the total surface area of particles and V the volume of the dispersion. The total concentration of surface sites, c_T, and the dissociation constants can easily be obtained using simultaneous potentiometric and conductimetric titrations.

Simultaneous potentiometric and conductimetric titrations of 40 ml of the ferrofluid dispersion (volume fraction $\Phi = 1.46\%$ corresponding to 1.7×10^{22} particles/m^3) were performed using a titrant solution of 0.106 mol/l sodium hydroxide. All titrations were repeated twice. The potentiometric readings were done with a Metrohm 713 pH meter using a pH glass double-junction electrode, while the conductivity was measured with a Metrohm 712 conductometer. Both apparatus were gauged using appropriate templates. From the simultaneous titration curves, one can determine, using the Henderson–Hasselbalch equation, the p*K*s corresponding to both Eqs. (1) and (2). Moreover, c_T is calculated using the mass balance. Then, the speciation diagram of surface sites is obtained.

Results and discussion

The results of the potentiometric and conductimetric titrations obtained simultaneously for our MF sample are exhibited in Fig. 1. Using the technique of graphical direction lines [6] the equivalence points were evaluated in order to calculate the concentration of the charged superficial sites. The pH-dependence of the modulus of the surface charge density for the cobalt ferrite EDL-MF sample is shown in Fig. 2. The saturation value is 0.326 C/m, which approximately corresponds to one charge per 0.5 nm^2 or about 905 sites per particle. This value is in excellent agreement with the reported ones [7]. Around pH values about 3.5 and 10.5, σ_0 reaches the maximum and the particle surface is charge-saturated, which agrees with ultrastable magnetic sols obtained experimentally in acidic or basic media. Nevertheless, for extreme values of pH, the high ionic strength does not allow stable ferrofluid dispersions, although the particle surface is charge-saturated.

In the pH range corresponding to a physiological medium, the superficial density of charge goes to zero and the ferrofluid dispersion flocculates, showing therefore that uncoated particles cannot be used for biological purposes.

The thermodynamical constants, associated with the equilibria between the particle surface and the bulk

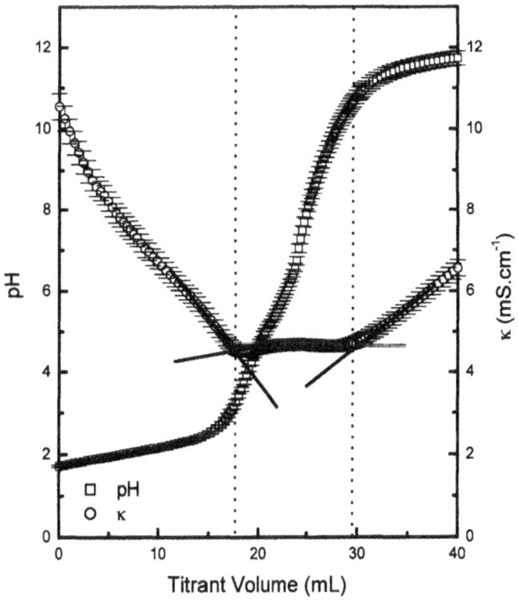

Fig. 1 Simultaneous potentiometric and conductimetric titrations obtained for the magnetic fluid sample

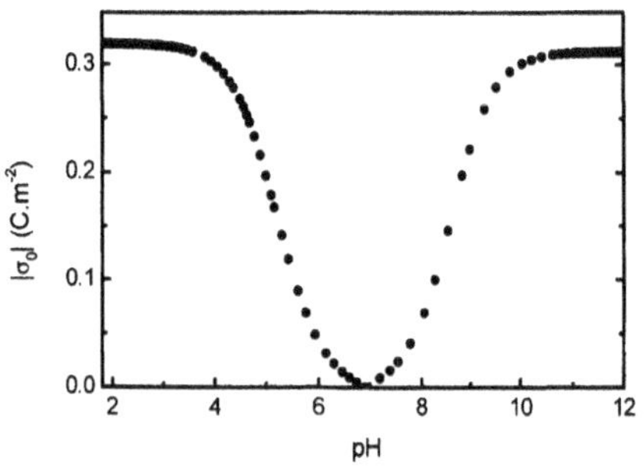

Fig. 2 pH-dependence of the modulus of the superficial density of charge for the electrical double-layered magnetic fluid sample

Table 1 pK_1 and pK_2 values and the modulus of the maximum superficial density of charge for the electrical double-layered magnetic fluid sample

pK_1	pK_2	$\|\sigma_0^{max}\|$ (C/m)
5.2	8.6	0.326

dispersion, determined experimentally, are listed in Table 1. Then, the concentration values of particle surface sites can be calculated and is plotted in Fig. 3 as function of pH. This speciation diagram gives a quantitative description of the particle surface, illustrating the domains of protonation of surface sites. At low (high) pH values in acidic (basic) media (typically pH≤3.5 and pH≥10.5), the particle surface is charge-saturated. For small pH ranges around pK values, Fig. 3 evidences the coexistence of charged and neutral sites.

Close to the neutral region (physiological medium), the molar ratio of charged sites strongly decreases and contrary to the expected result, the cobalt ferrite ferrofluid solution does not present any point of zero charge since the molar ratio of the amphoteric sites does not reach unity. However, an isoelectric point is found and is readily related to the equilibrium constants for the surface-site species.

The pK values corresponding to the equilibria involving the five DMSA species often abbreviated H_4L, H_3L^-, H_2L^{2-}, HL^{3-} and L^{4-} are listed in Table 2. The first and second steps in the dissociation process involve successive loss of protons from the two carboxylic groups; the third and fourth steps involve dissociation of the protonated thiol groups.

The eight curves plotted as a function of pH in Fig. 4 show the five molar ratios for DMSA containing species coupled with the three for nanoparticle superficial sites.

Table 2 Values of equilibrium constants of *meso*-2,3-dimercaptosuccinic acid [8]

pK_{C1}	pK_{C2}	pK_{S1}	pK_{S2}
2.71	3.43	9.65	12.05

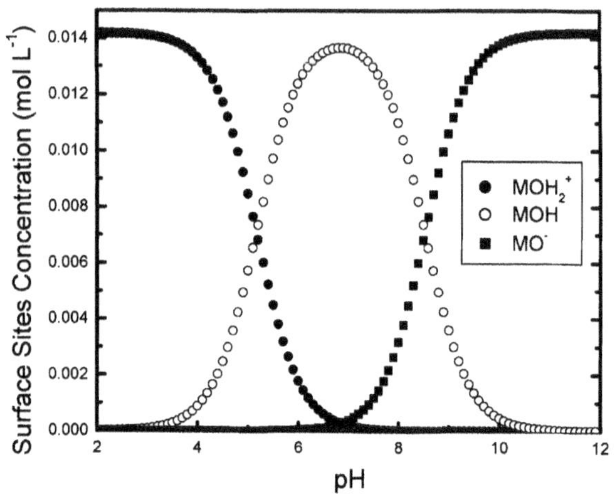

Fig. 3 Speciation diagram of particle surface sites

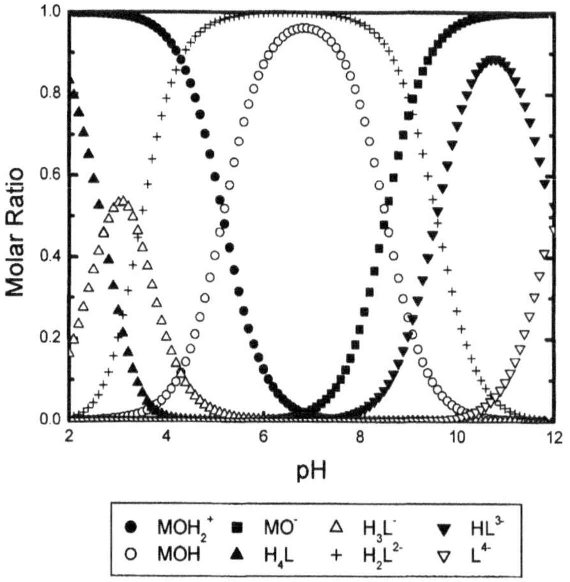

Fig. 4 Coupled speciation diagrams of superficial sites and *meso*-2,3-dimercaptosuccinic acid

Since the first two constants of DMSA are of the same order of magnitude, it suggests that the first two protons dissociate from opposite ends of the molecule. As a consequence of their physical separation, the negative charge created by the first dissociation does not greatly affect the removal of the second proton. Thus, around pH 3 the three first DMSA species coexist. The dissociation of the other two protons, relative to the thiol groups, is more difficult since they are much closer to the negatively charged carboxylate ions created by the initial dissociation.

Moreover, it is well known that carboxylate acid groups form remarkably stable chelates with many metal ions in solution [9]. In this context, the behavior of DMSA-coated magnetic nanoparticles by complexation of superficial metal ions should be controlled by the pH, which also monitors the superficial density of charge. Then, it is therefore necessary to analyze the coupled speciation diagrams of particle superficial sites and DMSA. Based on the free metal ion complexation equilibrium, the most probable surface reaction can be written as

$$\equiv MOH_2^+ + \text{ligand} \rightleftharpoons \equiv M^+ - \text{ligand} + H_2O$$

since the exchange of ligand for OH superficial groups is thermodynamically less favored [9]. In alkaline media, both particles and DMSA ions are negatively charged (see Fig. 4) and no complexation would occur. From these considerations, one can easily conclude that the best pH to coat magnetic nanoparticles by DMSA is around 3. Although the pH range of the maximum molar ratio for $MOH_2{}^+$ superficial sites goes from a strong acid medium to pH 3, H_3L^- and H_2L^{2-} only reach their maximum values for this pH. Once the DMSA is coated to the particle surface, the stability of the magnetic sol in biological buffers is obtained through the H_2L^{2-} species, one end of which is bonded to the surface and the other ensures the charge.

Conclusion

MFs behave as a mixture of a strong acid and a weak diprotic one. The thermodynamical constants of the chemical equilibria involved between the bulk dispersion and the nanoparticle surface are determined experimentally. Then, the pH-dependence of the nanoparticle surface charge density of EDL-MFs based on cobalt nanoparticles is obtained. From the coupled speciation diagrams of both the particle surface and DMSA one can predict the optimal pH for ligand–surface complexation and better understand the colloidal stability in a physiological medium. Our model can be applied for many other cross-linking agents which behave as polyfunctional acids. In a second attempt we will investigate the effect in our theoretical approach of a possible oxidation–reduction process involved in the thiolation of magnetic nanoparticles by a succimer.

Acknowledgements The authors thank the Brazilian agencies CAPES, CNPq and FAP-DF for their financial support.

References

1. Roger J, Pons JN, Massart R, Halbreich A, Bacri JC (1999) Eur Phys A 5:321
2. Fauconnier N, Bee A, Roger J, Pons JN (1996) Prog Colloid Polym Sci 100:212
3. Fauconnier N, Pons JN, Roger J, Bee A (1997) J Colloid Interface Sci 194:427
4. Campos AFC, Tourinho FA, Silva GJ, Lara MCFL, Depeyrot J (2001) Eur Phys J E 6:29
5. Tourinho FA, Campos AFC, Aquino R, Lara MCFL, Depeyrot J (2002) Braz J Phys 32(2B):501
6. Chromiak E, (1995) Analyst 120:149
7. Tourinho FA, Franck R, Massart R (1990) J Mater Sci 25:3249
8. Crisponi G, Cristiani F, Nurchi VM, Pinna R, Estévez MJT (1999) Polyhedron 18:3257
9. Pribil R (1982) Applied complexometry. Pergamon, Oxford

Progr Colloid Polym Sci (2004) 126: 90–92
DOI 10.1007/b94004

Gisèle Volet
Catherine Amiel
Loïc Auvray

Adsorption of amphiphilic copolymers on solid surfaces in different solvents

G. Volet (✉) · C. Amiel
Laboratoire de Recherche sur les Polymères, UMR C 7581 CNRS, 2,
rue Henri Dunant, 94320 Thiais, France
e-mail: volet@glvt-cnrs.fr

L. Auvray
Laboratoire Matériaux Polymères aux Interfaces, Université d'Evry Val d'Essonne, Rue du Père Jarlan, 91025 Evry Cedex, France

Abstract Diblock copolymers of poly(isobutyl vinyl ether-*b*-2-methyl-2-oxazoline) (PIBVE–POXZ) were synthesized by cationic polymerization. Copolymers with molecular weight from 1,000 to 4,000 g mol^{-1} had a composition of 20–50% of the IBVE block. They were adsorbed on porous silica particles from three different solvents: dichloromethane, which is a solvent of both blocks; water and cyclohexane, which are selective solvents. The structures of the adsorbed layers, studied by small-angle neutron scattering, were found to depend strongly on the solvent selectivity. Adsorption from dichloromethane showed adsorbed amounts and thicknesses comparable to homopolymer adsorption. Denser layers were obtained in the case of a selective solvent of one block. In water (solvent of POXZ), the results are coherent with brush structures, PIBVE blocks being collapsed on the surface: the grafting densities were greatly influenced by the copolymer composition and are in good agreement with models of copolymer adsorption from selective solvents. In contrast, adsorption from cyclohexane (solvent of PIBVE) was quite insensitive to copolymer composition. Direct micelle adsorption or bilayer formation could explain this behavior. A variety of interfacial architectures can be obtained by controlling the copolymer composition and the adsorption conditions.

Keywords Adsorption · Amphiphilic copolymer · Small-angle neutron scattering · Poly(isobutyl-vinylether-*b*-2-methyl-2-oxazoline)

Introduction

The great interest in amphiphilic copolymers is that adsorption can occur selectively according to the nature of the solvent. Hydrophilic copolymers from poly(2-methyl-2-oxazoline) (POXZ) were previously studied for adsorption at solid–liquid and liquid–air interfaces [1, 2]. In this work, a diblock copolymer of poly(isobutyl vinyl ether) (PIBVE–POXZ) of various composition [3] was adsorbed on porous silica particles in different solvents. The aim of this work is to show that interfacial architectures can be tailored by controlling both the copolymer composition and the history of adsorption, i.e. the nature of the solvent of adsorption.

Experimental

Adsorption of polymers

The silica particles (CPG, Fluka) used were porous particles with a particle diameter of 10 μm and a pore diameter of 3,000 Å (surface area 6.9 m^2 g^{-1} and pore volume 0.77 cm^3 g^{-1}, given by manufacturer). The polymers were adsorbed over 24 h, from solutions with a polymer concentration of 2 wt%. The ratio of polymer to silica was fixed at 0.2 w/w.

Neutron scattering measurements

Small-angle neutron scattering (SANS) measurements were performed at the Laboratoire Léon Brillouin (ORPHEE reactor, CEN

Saclay) using the PACE spectrometer. The scattering vector length, q, ranged from 10^{-3} to 10^{-1} Å^{-1}. The theory of the SANS applied to adsorbed polymer, the experimental procedures and the analysis of the data have been widely described in the literature [4, 5, 6]. A short theoretical background with the equations used in the present work is given in this section.

By using a suitable isotopic mixture of fully hydrogenated–fully deuterated solvent, it is possible to match the "neutral refractive index" of the silica. Under that condition, the scattering intensity comes only from the adsorbed polymer. The spectra were analysed directly (as they were obtained) by using a step function model for the interfacial density profile, $\Phi(z)$, which is not an a priori assumption for the shape of $\Phi(z)$ but rather provides a useful tool for comparing the data. The scattering intensity of this step is, at contrast matching,

$$I(q) = 2\pi\left(\frac{S}{V}\right)(n_p - n_s)^2 q^{-4} 2(1 - \cos qh)\Phi_s^2, \quad (1)$$

where q is the scattering vector, S/V the specific surface, n_p (or n_s) the scattering length density of the polymer (or solvent), h the thickness of the layer and Φ_s the fraction of the surface occupied by the polymer. A Taylor expansion of Eq. (1) leads to

$$q^2 I(q) = 2\pi\left(\frac{S}{V}\right)(n_p - n_s)^2 \gamma^2 \left(1 - \frac{q^2 h^2}{12}\right) \quad (2)$$

for very small q, where $\gamma = \int_o^h \Phi(z)\mathrm{d}z$ (in angstroms). γ can be easily converted into the amount of polymer per unit area (in milligrams per meter squared). $\Gamma = 0.1\gamma d$, where d is the density of the polymer (in grams per centimeter cubed). The amount of adsorbed polymer, Γ, was determined by extrapolating $q^2I(q)$ at q close to zero and the slope of the linear regime is related to the thickness of the polymer layer, h.

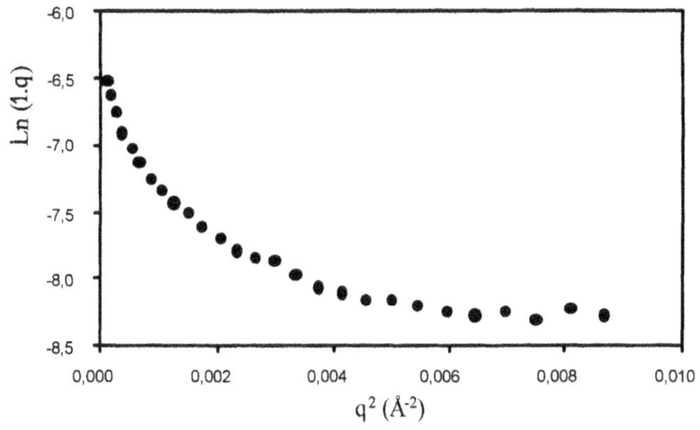

Fig. 1 $\ln(Iq^2)$ versus q^2 for an adsorbed layer of copolymer with 50% of isobutyl vinyl ether in CD_2Cl_2

Table 1 Polymer adsorption in D_2O

Isobutyl vinyl ether (%)	M_w (g mol^{-1})	Γ (mg m^{-2})	h (Å)
0	3,150	1.0	65
20	3,780	1.8	–
27	2,630	4.1	65
30	1,690	5.0	76
37	1,830	5.5	65
46	3,130	7.8	92
50	5,000	7.9	140

Results and discussion

Adsorption in a nonselective solvent

If adsorption occurs in dichloromethane, both blocks of the copolymer are equally soluble. For homopolymer solutions, the amounts adsorbed are 2.8 and 1 mg m^{-2} for POXZ and PIBVE, respectively. For the copolymers, homogeneous adsorbed layers were also obtained as illustrated by the shape of the $\ln(Iq^2)$ versus q^2 curve (Fig. 1) obtained for a copolymer with 50% of IBVE. The comparison of the adsorption of the copolymers made of both blocks at varying composition showed that the amounts adsorbed were in the range 1–1.8 mg m^{-2} and were independent of the IBVE molar ratio. The copolymer adsorbed layers had interfacial architectures comparable to PIBVE homopolymer adsorption, in terms of grafting densities and of layer extensions.

Adsorption in a selective solvent of the POXZ block

In water, stable and very dense adsorbed layers were obtained. The amounts of polymer adsorbed varied with the composition of both blocks (Table 1). The PIBVE block, insoluble in water, anchored the copolymer chain by collapsing on the surface and the soluble POXZ block

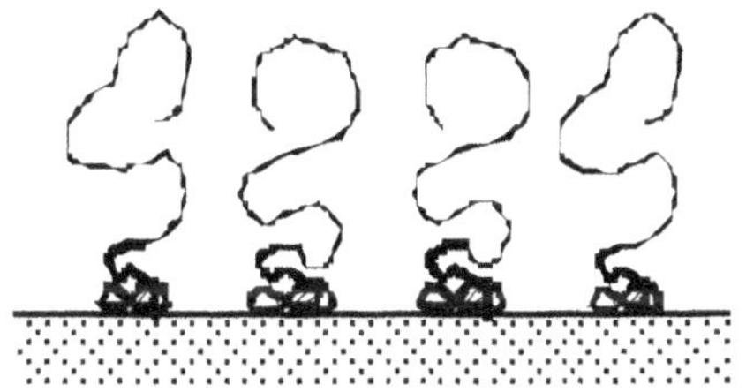

Fig. 2 Scheme of copolymer brushes adsorbed from D_2O solution. poly(isobutyl vinyl ether) blocks are collapsed near the surface and 2-methyl-2-oxazoline blocks are stretched out toward the solution

extended into the solution phase. The adsorbed chains adopted brush structures for a IBVE content above 20%, as shown by the high values of the amounts adsorbed (up to 8 mg m^{-2}) and the thicknesses corresponding to extended POXZ layers (Fig. 2). The grafting densities follow qualitatively the theoretical model of Marques and Joanny [7] for the adsorption of copolymers in selective solvents in the presence of micellar aggregates.

Adsorption in a selective solvent of the PIBVE block

Cyclohexane is a selective solvent of PIBVE blocks. The amounts adsorbed and the thicknesses obtained for the different copolymers are reported in Table 2. Both Γ and h are larger for copolymers than for PIBVE. These results cannot be accounted for by the direct adsorption of the

Table 2 Polymer adsorption in C_6D_{12}

Isobutyl vinyl ether (%)	M_w (g mol^{-1})	Γ (mg m^{-2})	*h* (Å)
30	1,690	3.6	120
37	1,830	3.8	78
46	3,130	2.9	70
50	5,000	3.0	110
100	2,600	1.2	46

Fig. 3 Scheme of copolymer micelles adsorbed from C_6D_{12} solution. The micellar core is made of 2-methyl-2-oxazoline

copolymers via their large POXZ block on the surface. The direct adsorption of micelles on the silica surface, by their PIBVE blocks, or the formation of bilayers could explain the observed behavior (Fig. 3).

Conclusion

Different interfacial architectures of amphiphilic diblock copolymers have been obtained by varying the selectivity of the solvent versus the copolymer composition. In a nonselective solvent, homopolymer-type adsorbed layers were obtained. In water, the insoluble block collapsed on the surface and brush structures were obtained. Their grafting densities increased with the IBVE molar ratio (up to 50%). The dense layers obtained for adsorption from cyclohexane were attributed to direct micellar adsorption or to the formation of bilayers.

References

1. Bijsterbosch HD, Cohen Stuart MA, Fleer GJ (1998) Macromolecules 31:9281
2. Gutberlet T, Wurlitzer A, Dietrich U, Politsch E, Cevc G, Steitz R, Lösche M (2000) Physica B 283:37
3. Volet G, Cheradame H (1999) Polym Int 48:307
4. Cosgrove T, Heath TG, Ryan K, Crowley TL (1987) Macromolecules 20:2879
5. Auvray L, Auroy P (1991) Neutron, X-ray and light scattering. Elsevier, Amsterdam
6. Auroy P, Auvray L, Léger L (1991) Macromolecules 24:2523
7. Marques C, Joanny JF, Leibler L (1988) Macromolecules 21:1051

Progr Colloid Polym Sci (2004) 126: 93–96
DOI 10.1007/b93984

R. Tsekov
E. Evstatieva

A fractal classification of the drainage dynamics in thin liquid films

R. Tsekov
Department of Physical Chemistry,
University of Sofia, 1126 Sofia, Bulgaria

E. Evstatieva
Department of Water Technology,
ITC, Karlsruhe Research Center,
76021 Karlsruhe, Germany

Present address: R. Tsekov (✉)
Institute of Physical Chemistry,
Physical Chemistry of Condensed Matter,
University of Karlsruhe,
76128 Karlsruhe, Germany
e-mail: eroumen@email.com

Abstract It is demonstrated that the dynamic structure is very important for the rate of drainage of a thin liquid film and it can be effectively taken into account by a dynamic fractal dimension. It is shown that the latter is a powerful tool for description of the film drainage and classifies all the known results from the literature. The general expression obtained for the thinning rate is a heuristic one and predicts a variety of drainage models, which are even difficult to simulate in practice. It is a typical example of a scaling law, which explains the origin of the complicated dependence of the thinning rate on the film radius.

Keywords Thin liquid films · Drainage · Fractal dynamics

Introduction

The drainage in thin liquid films is an important process in foams, emulsions and suspensions. Scheludko [1] has adapted to free films the Reynolds formula for the velocity of approach of two parallel rigid discs separated by a liquid slit:

$$V = \frac{2h^3(p_\sigma - \Pi)}{3\mu R^2}, \qquad (1)$$

where h is the film thickness, R is the film radius, μ is the liquid viscosity, p_σ is the capillary pressure in the meniscus and Π is the disjoining pressure. This equation is valid for films with tangentially immobile and plane-parallel surfaces. Deviations from Eq. (1) due to tangential surface mobility are reported. Since the films are usually stabilized by surfactants, the film drainage causes an interfacial gradient of the surfactant adsorption. Thus, a gradient of the surface tension appears and the corresponding Marangoni effect slows down the interfacial flow. In this case, the theory [2] predicts a thinning rate according to the Reynolds formula from Eq. (1) but with a reduced viscosity,

$$\eta = \mu\frac{1 + aE/D\mu}{4 + aE/D\mu},$$

where a, E and D are the adsorption length, the Gibbs elasticity and the diffusion coefficient of the surfactant, respectively. The thickness nonuniformity is another important reason for deviations from Eq. (1). Some experiments [3] have detected rates of thinning depending much less on the film radius compared to the Reynolds formula. Such behavior cannot be explained by the surface mobility, the effect of which is not expected to alter the functional dependence of the thinning rate on the film radius. Equation (1) requires strictly symmetrical drainage between two parallel flat surfaces, which is supplied in the Reynolds case by the rigidity of the solid interfaces. In contrast, the shape of free films is determined via the balance of the viscous, capillary and surface forces. Experimental investigations [4] have shown that large films are nonhomogeneous in thickness. Owing to the film geometry, the liquid in the film center flows more slowly compared to that in the film periphery and in this way a characteristic shape is formed, which is known under the name of dimple. The thinning rate of dimpled films is larger than the prediction of the Reynolds law [5].

Hydrodynamics of the liquid film thinning

In the frames of the lubrication approximation, the dynamics of the film drainage is described by the following linearized equation [2]:

$$12\eta R^2 \partial_t H = h^3 \Delta P, \tag{2}$$

where H is the local film thickness, h is the average film thickness, P is the driving pressure and Δ is a two-dimensional Laplace operator. A relationship between the driving pressure and the local film thickness is further supplied via the normal force balance on the film surfaces:

$$P = p - \sigma \Delta H / 2R^2. \tag{3}$$

Here p is the mean driving pressure averaged along the film area and σ is the surface tension. Note that p is equal to the difference between the capillary pressure in the meniscus, p_σ, and the disjoining pressure, Π. Since $h < < R$, the films are relatively flat; hence, in Eq. (3) the local capillary pressure is proportional to the film curvature and the disjoining pressure contributes by its mean value only. In general, the evolution of the film thickness is governed by Eqs. (2) and (3) and a set of an initial and four boundary conditions. The latter depend, however, substantially on the film formation. For this reason, there are plenty of different and even contradictive observations of the thinning rate. The scope of the present paper is to classify the laws of thinning by introducing a fractal scale of the dynamics of the film drainage.

Let us begin by the derivation of the Reynolds formula. Supposing the film surfaces are parallel, the local thickness, H, in the film is everywhere equal to the mean thickness, h. In this case, Eq. (2) acquires the form

$$-12\eta R^2 V = h^3 \Delta P, \tag{4}$$

where $V = -\partial_t h$ is the mean thinning rate. Integrating Eq. (4) twice under the constraint of zero driving pressure at the firm rim, one obtains $P = (3\eta R^2 V/h^3)(1-x^2)$, where $x = r/R$ is the dimensionless distance from the film center. Combining this result with Eq. (3) and integrating once again leads to

$$\partial_x H = \left(pR^2/\sigma\right)x - \left(3\eta R^4 V/2\sigma h^3\right)\left(2 - x^2\right)x. \tag{5}$$

Hence, it follows immediately from the boundary condition $(\partial_x H)_1 = 0$ that the rate of drainage obeys the Reynolds formula. Integrating Eq. (5) by remembering that h is the mean film thickness, one obtains the film profile:

$$H = h + \left(pR^2/12\sigma\right)\left(2 - 6x^2 + 3x^4\right). \tag{6}$$

As seen, Eq. (6) describes a dimple with a thicker part in the film center and a thinner ring at the film border. Of course, the film is no longer planar and to obtain the range of validity of the Reynolds formula one should require at least a positive thickness in the film thinnest part, i.e. the barrier rim. Using Eq. (6) one can transform the inequality $H(1) \geq 0$ to

$$R \leq \sqrt{12\sigma h/(p_\sigma - \Pi)}. \tag{7}$$

Therefore, the Reynolds formula is valid only for small films. This is the conclusion of many experiments [3], which clearly show that the Reynolds formula is applicable to sufficiently small films, which are relatively homogeneous in thickness. Note that all the films approach their equilibrium state with the Reynolds rate of thinning, because the inequality in Eq. (7) is always fulfilled in the limit $p_\sigma \rightarrow \Pi$. This is not surprising since the equilibrium films are flat.

The problem now is to describe the drainage of films with radii out of the range of validity of the inequality in Eq. (7). Experimental results [3] indicate an increase of the amplitude of thickness nonhomogeneity with increasing film size. This corresponds to increasing divergence from the idealized models of a planar film profile. Introducing a dimensionless variable, τ, via the relation

$$d\tau = \left(\sigma h^3/24\eta R^4\right)dt \tag{8}$$

and substituting Eq. (3) in Eq. (2), one obtains the following differential equation:

$$\Delta\Delta H = -\partial_\tau H. \tag{9}$$

Applying the Laplace transform, this equation changes to

$$\Delta\Delta \tilde{H} = -s\tilde{H} + h_0, \tag{10}$$

where $\tilde{H}$ is the Laplace image of the local film thickness and the initial profile of the film is set to be flat at thickness h_0. If the film is axisymmetric, the solution of Eq. (10) is the sum of the Bessel functions of zero order:

$$\tilde{H} = h_0/s + AJ_0\left(\sqrt[4]{-s}x\right) + BI_0\left(\sqrt[4]{-s}x\right) + CK_0\left(\sqrt[4]{-s}x\right) + DY_0\left(\sqrt[4]{-s}x\right). \tag{11}$$

The functions K_0 and Y_0 are not proper solutions since they posses peculiarity at $x = 0$. For this reason, the corresponding coefficients have to be set equal to zero, $C = D = 0$. The other two coefficients can be determined by the boundary condition $\left(\partial_x \tilde{H}\right)_1 = 0$ of the minimum at the barrier rim and the definition of the mean thickness to obtain

$$A = \frac{\left(s\tilde{h} - h_0\right)\sqrt[4]{-s}}{4sJ_1\left(\sqrt[4]{-s}\right)}, \quad B = \frac{\left(s\tilde{h} - h_0\right)\sqrt[4]{-s}}{4sI_1\left(\sqrt[4]{-s}\right)}.$$

Hence, the Laplace image of the film profile acquires the form

$$\tilde{H} = \frac{h_0}{s} + \left(s\tilde{h} - h_0\right)\frac{\sqrt[4]{-s}}{4s}\left[\frac{J_0(\sqrt[4]{-s}x)}{J_1(\sqrt[4]{-s})} + \frac{I_0(\sqrt[4]{-s}x)}{I_1(\sqrt[4]{-s})}\right], \qquad (12)$$

where the only unknown function is the Laplace image, $\tilde{h}$, of the mean thickness. It can be determined from the equilibrium pressure balance at the barrier rim:

$$\sigma\left(\Delta\tilde{H}\right)_1/2R^2 = p_\sigma/s. \qquad (13)$$

For the sake of simplicity, sufficiently thick films are considered here, where the disjoining pressure is negligible compared to the capillary pressure in the meniscus. Substituting Eq. (12) in Eq. (13), one obtains the relation

$$s\tilde{h} - h_0 = \frac{8p_\sigma R^2}{\sigma\sqrt{-s}\sqrt[4]{-s}}\frac{J_1(\sqrt[4]{-s})I_1(\sqrt[4]{-s})}{J_1(\sqrt[4]{-s})I_0(\sqrt[4]{-s}) - J_0(\sqrt[4]{-s})I_1(\sqrt[4]{-s})}. \qquad (14)$$

Thus, the Laplace image of the film profile acquires the form

$$\tilde{H} = \frac{h_0}{s} + \frac{2p_\sigma R^2}{\sigma s\sqrt{-s}}\frac{I_1(\sqrt[4]{-s})J_0(\sqrt[4]{-s}x) + J_1(\sqrt[4]{-s})I_0(\sqrt[4]{-s}x)}{J_1(\sqrt[4]{-s})I_0(\sqrt[4]{-s}) - J_0(\sqrt[4]{-s})I_1(\sqrt[4]{-s})}. \qquad (15)$$

Unfortunately, it is impossible to invert analytically the Laplace image from Eq. (15) to obtain the local film thickness. Since the main interest here is the evolution of the mean film thickness, the latter can be expressed from Eq. (14) in the form

$$h - h_0 = L^{-1}\left[\frac{8p_\sigma R^2}{\sigma s\sqrt{-s}\sqrt[4]{-s}}\frac{J_1(\sqrt[4]{-s})I_1(\sqrt[4]{-s})}{J_1(\sqrt[4]{-s})I_0(\sqrt[4]{-s}) - J_0(\sqrt[4]{-s})I_1(\sqrt[4]{-s})}\right]. \qquad (16)$$

where the operator L^{-1} indicates the inverse Laplace transformation. A favorite circumstance is that for microscopic films τ is a very small number. Hence, the Laplace images with large s are mostly important. Expanding the Laplace image in Eq. (16) for large s allows explicit inversion of the Laplace transformation to obtain

$$h_0 - h = 8\left(R^2 p_\sigma/\sqrt{2}\sigma\right)L^{-1}[s^{-7/4}] = 32R^2 p_\sigma \tau^{3/4}/3\sqrt{3}\sigma. \qquad (17)$$

This equation can be solved with respect to τ:

$$\tau = 9\left(\sigma/32R^2 p_\sigma\right)^{4/3}(h_0 - h)^{4/3}. \qquad (18)$$

A detailed analysis of Eq. (18) shows that it implies the contribution of the initial film profile on the rate of drainage. This effect is not desired since many experiments show that there are stages in the drainage when the film has already forgotten the initial condition. Fortunately, the dependence of τ on the film thickness in Eq. (18) is on the power 4/3, which is close to 1. Hence, the following physical approximation of Eq. (18) is plausible,

$$\tau = 12\left(\sigma/32R^2 p_\sigma\right)^{4/3}\left(h_0^{4/3} - h^{4/3}\right). \qquad (19)$$

which eliminates the disadvantage already mentioned. Substituting Eq. (19) in Eq. (8), one obtains the velocity of drainage of the films:

$$V = \frac{1}{3\eta}\sqrt[3]{\frac{h^8 p_\sigma^4}{2\sigma R^4}}. \qquad (20)$$

As seen, the velocity of drainage of large films differs substantially from the Reynolds law. It is inversely proportional to the 4/3 power of the film radius and this weaker dependence on R is in accordance with the experimental observations. Note that for film radii out of the range of validity of the inequality in Eq. (7) the thinning rate, V, from Eq. (20) is always larger than the Reynolds estimation.

Fractal dynamics of the liquid film drainage

The rate of drainage in Eq. (20) is valid for axisymmetric films only. Sometimes, this model does not correspond well to the real film shape, especially in large films formed by quick initial withdrawal of liquid. More complex phenomena [6] take place in such films and the radial symmetry of the thinning process is violated. The asymmetric film drainage is quicker than the axisymmetric one owing to the peristaltic pumping effect. Owing to the stochastic nature of the film shape in this case, the description of the film drainage is more complicated and requires a statistical approach. A scaling analysis of Eqs. (2) and (3) leads to the following equations

$$12\eta R^2 V = h^3 q^2 \delta P, \qquad (21)$$

$$\delta P = \left(\sigma h/2R^2\right)q^2, \qquad (22)$$

where q is the wave number of the thickness perturbations in the film and δP is the corresponding changes of the driving pressure. Substituting Eq. (22) in Eq. (21), one obtains an expression for the characteristic wave number:

$$q = (R/h)\sqrt[4]{24\eta V/\sigma}. \qquad (23)$$

Since the thickness perturbations are generated by the drainage, q increases with increasing of the thinning rate. Equation (23) also demonstrates that the larger the film is the more thickness perturbations there are. To find out the drainage rate of the films one can employ a scaling law for the driving pressure distribution. Introducing a fractal dimension, α, one can relate the pressure changes to the corresponding wave numbers via the expression

$$\delta P/p = (4/q)^{\alpha}, \quad (24)$$

where the factor 4 corresponds to the average distance between the zeros of the Bessel functions. Equation (24) is a typical example of a power law and implies a geometric correlation between the pressure perturbations. Substituting Eq. (24) in Eq. (22) and introducing Eq. (23) in the result yields the velocity of drainage of the films:

$$V = \frac{32\sigma h^4}{3\eta R^4}\left(\frac{R^2 p}{16\sigma h}\right)^{4/(2+\alpha)}. \quad (25)$$

The law from Eq. (25) discriminates between different types of drainage by the corresponding dynamic fractal dimension. For instance, if the film possesses two solid interfaces the latter induce strong correlation in both directions. In this case $\alpha = 2$ and Eq. (25) reduces to

$$V = \frac{2h^3(p_\sigma - \Pi)}{3\eta R^2}. \quad (26)$$

Equation (26) is in fact the Reynolds law from Eq. (1). If the drainage is strictly axisymmetric but not confined between solid interfaces, the radial direction is free for relaxation and a dimple occurs. In this case $\alpha = 1$ and Eq. (25) reduces to

$$V = \frac{1}{3\eta}\sqrt[3]{\frac{h^8(p_\sigma - \Pi)^4}{2\sigma R^4}}. \quad (27)$$

Equation (27) is a generalization of Eq. (20) including the effect of the disjoining pressure. Finally, one can imagine a completely random film broken down to many uncorrelated subdomains. In this case $\alpha = 0$ and Eq. (25) acquires the form

$$V = \frac{h^2(p_\sigma - \Pi)^2}{24\eta\sigma}. \quad (28)$$

Note that the rate of drainage of such stochastically corrugated films does not depend on the film radius since any spatial correlation of the dynamics is missing. In a previous paper [5], it was demonstrated that some experiments are well described by the following rate of drainage:

$$V = \frac{1}{6\eta}\sqrt[5]{\frac{h^{12}(p_\sigma - \Pi)^8}{4\sigma^3 R^4}}. \quad (29)$$

One can easily check that Eq. (29) corresponds to Eq. (25) for $\alpha = 1/2$. Hence, the conclusion is that the films described are neither planar nor strictly axisymmetric. The typical shapes of such films possess some kind of regularity [3] but it is far from a strong radial symmetry. For this reason the corresponding fractal dimension is a number between those of axisymmetric and random films.

Conclusion

The present analysis demonstrates that the morphology of the film profiles is very important for the rate of drainage. It can be effectively taken into account by the dynamic fractal dimension, which affects substantially the thinning rate as described in Eq. (25). As shown, the latter is a powerful law for description of the film drainage and reproduces all the known results from the literature. Equation (25) is a heuristic one and predicts a variety of drainage behaviors, which are even difficult to simulate in practice. It is a typical example of a scaling law, which explains the origin of the complicated dependence of the thinning rate on the film radius. Additional phenomena, such as interfacial friction [2] and marginal regeneration [7], can also be included by proper generalization of the effective viscosity, which, in this case, will become a thickness-dependent quantity.

Acknowledgement The authors are grateful to the Alexander von Humboldt Foundation for fellowships.

References

1. Scheludko A (1967) Adv Colloid Interface Sci 1:391
2. Ivanov IB (ed) (1988) Thin liquid films. Dekker, New York
3. Radoev B, Scheludko A, Manev E (1983) J Colloid Interface Sci 95:254
4. Manev E, Tsekov R, Radoev B (1997) J Dispersion Sci Technol 18:769
5. Tsekov R (1998) Colloids Surf A 141:161
6. Joye J, Hirasaki G, Miller C (1996) J Colloid Interface Sci 177:542
7. Nierstrasz VA, Frens G (2001) J Colloid Interface Sci 234:162

Progr Colloid Polym Sci (2004) 126: 97–101
DOI 10.1007/b94022

C. Regnaut
S. Amokrane

Hard-sphere mixture in the colloidal range: a critical analysis of approximate virial pressure binodals

C. Regnaut (✉) · S. Amokrane
Groupe de Physique des milieux denses, Faculté des Sciences et Technologie, Université Paris XII, 61 avenue du Général de Gaulle, 94010 Créteil Cedex, France
e-mail: regnaut@univ-paris12.fr
Tel.: +33-1-45171565
Fax: +33-1-45171579

Abstract Highly asymmetric binary mixtures of hard spheres are known to exhibit a metastable fluid–fluid phase separation. The route using the virial pressure is the simplest way to compute the binodal, provided that accurate contact values of the pair distribution functions are known. To this end, various empirical modifications of the Boublik–Mansoori–Carnahan–Starling–Leland equation of state have been proposed that take into account some self-consistency rules or constraints, especially in the colloidal range. The present comparative study shows that binodals can be obtained from a variety of equations of state but emphasises the difficulty to correlate the occurrence of the binodal with specific constraints on the pair distribution functions at contact.

Keywords Hard-sphere mixture · Equation of state · Phase transition · Colloids

Introduction

A phase instability in asymmetric mixtures of hard particles was suspected since the early work on the depletion effect [1] in colloid polymer mixtures. This field, however, saw renewed interest only when Biben and Hansen [2] suggested, at the beginning of the last decade, a spinodal instability in highly asymmetric binary mixtures of hard spheres (HS). The prediction made from integral equation theories was later confirmed by the simulation study of the full phase diagram of Dijkstra et al. [3]. For the big spheres, the salient features are a wide two-phase region of coexisting fluid and solid and a two-fluid-phase region (F–F) (concentrated and dilute big spheres) in metastable equilibrium. While these features of the phase diagram now seem well established, one important question from a practical point of view is the possibility of predicting the F–F transition from an analytical equation of state (EOS). This question underlies many recent studies of HS mixtures, since the Boublik–Mansoori–Carnahan–Starling–Leland (BMCSL) EOS [4] – the standard reference for many HS-like systems – does not predict any phase transition. The BMCSL EOS is also inadequate for studying structural quantities of HS-like colloids. Indeed the contact values of the pair distribution functions (PDFs) $g_{ij}(r=d_{ij})$ (for convenience these will be denoted as g_{ij}) predicted from this EOS become poor as the diameter ratio $R=d_2(\text{solute})/d_1$ (solvent) increases. The failure is particularly severe in the so-called colloidal limit in which the concentration, x_2, of the large particles vanishes and R goes to infinity. To overcome this drawback, virial expansions can be considered [5], but remain quantitatively uncertain. The free-volume approach of Lekkerkerker and coworkers [6, 7] could constitute a simple alternative. The incorrect trend with the size ratio [8, 9] predicted with the original expression of the free volume can be corrected but it remains sensitive to the precise model used for this quantity [9]. Another method is the calculation of the EOS directly from the virial expression of the pressure and the determination of ad hoc corrections to the BMCSL g_{ij}. This is the method initiated by Henderson and coworkers [10]. In this approach, the modified expressions must obey some exact constraints in the one-component and colloidal limits, and their value are controlled by comparison with available simulation data for the g_{ij} values in the colloidal range. The work of Henderson and coworkers focuses on contact properties at infinite dilution of the solute: the wall-particle limit (or contact theorem) connecting $g_{11}$and g_{12} and the asymptotic behaviour $\ln g_{22} \approx R$. The relevance of these two criteria has been discussed in our recent work [11], by studying various combinations of standard expressions of g_{ij} and the ones of Henderson and coworkers. It was concluded that although the contact theorem seems a sensitive criterion for improving ad hoc expressions, a precise parameterisation of g_{22} remains a challenge. However, as discussed by Santos et al. [12], seemingly better approximations of the g_{ij} values do not always lead to a more accurate EOS. Therefore, comparison of different extensions of the BMCSL ansatz is not obvious. These studies indeed differ on various points: the constraints and self-consistent relations retained, the initial quantity that is considered (compressibility factor or g_{ij} values), the symmetrical or asymmetrical treatment of g_{ij}, the universality of the proposal for g_{ij} values (i.e. extension to n-component mixtures and to other dimensions). In this paper, we extend our previous analysis [11], and present a comparative discussion of these various approaches and their ability in predicting both a F–F binodal and contact values in agreement with recent simulation data.

Recent models of g_{ij}, the Gibbs free energy and binodals

For HS, the virial pressure, P, is entirely expressible in terms of the contact values, g_{ij}, the (here additive) hard-core diameters, $d_{ij} = \frac{(d_i+d_j)}{2}$, the total number density, $\rho = N/V$, and the concentrations, $x_i = \frac{N_i}{N}$, with N the total number of particles and V the volume of the mixture. The compressibility factor is given by

$$Z = \frac{\beta P}{\rho} = 1 + \frac{2\pi\rho}{3}\sum_{i,j} x_i x_j d_{ij}^3 g_{ij}, \tag{1}$$

with $\beta = 1/k_B T$. The Gibbs free energy, G, can be determined by integrating the pressure as

$$\beta\frac{G}{N} = \int_0^{\eta(P)} \frac{Z(\eta') - 1}{\eta'} d\eta' + Z[\eta(P)] + \beta\frac{F_{id}}{N}, \tag{2}$$

where $\eta = \frac{\pi}{6}\rho\sum x_i d_i^3$ is the total packing fraction and F_{id} the ideal gas contribution. At equilibrium between two phases I and II, the pressure is constant and the chemical potential of each component is the same in the two phases. Consequently, the isobaric $G(P, x_2)$ versus x_2 curve (here we consider a binary mixture) has a common tangent at concentration x^{I}_2 and x^{II}_2, which we obtained graphically. The binodal is the locus of the points (x^{I}_2, x^{II}_2) as P is varied.

Most recent theories intending to improve the BMCSL g_{ij} values to be used in Eqs. (1) and (2) verify some obvious constraints: one-component limit of the three PDFs, $d_{11} = d_{12} = d_{22}$ or $x_1 = 0$ or $x_2 = 0 \Rightarrow g_{11} = g_{12} = g_{22} = g_{\text{pure}}$, and point-solvent limit of the three PDFs, $d_1 = 0 \Rightarrow g_{12} = g_{11} = (1-\eta_2)^{-1}$ and $g_{22} = g_{\text{pure}}$. Only some of them verify either only one or both constraints ofthe colloidal limit: the contact theorem, $g^{\infty}{}_{12} = 1 + 4\eta_1 g^{\text{pure}}(\eta_1)$ and the behaviour for very big spheres at infinite dilution, $\ln(g_{22}) \approx R$.

One may separate these recent approaches in two classes: "symmetric" ones, which as BMCSL are formally defined for all size ratios (i.e. $R < 1$ and $R > 1$) and consider an arbitrary number of components, and "asymmetric" with $R > 1$ and binary mixtures only.

In the first class, it is useful to introduce the reduced average curvature [12] defined by

$$z_{ij}^{-1} = \frac{d_{ij}}{d_i d_j}\frac{\sum_i x_i d_i^3}{\sum_i x_i d_i^2}. \tag{3}$$

The BMCSL relations expressed in terms of $x = (1-\eta)^{-1}$, η and z_{ij} read

$$g_{ij}^{\text{BMCSL}} = x + \frac{3}{2}x^2\eta z_{ij} + \frac{1}{2}x^3\eta^2 z_{ij}^2. \tag{4}$$

Improvements of the BMCSL relations according to the D-dimensional ansatz of Santos et al. (SYL) and fulfilment of condition 3 for $D = 3$ lead to [13]

$$g_{ij}^{\text{SYL}} = g_{ij}^{\text{BMCSL}} + \frac{1}{2}x^3\eta^2 z_{ij}(z_{ij} - 1). \tag{5}$$

The Rosenfeld (R) scaling form [14] is interesting to consider since it takes into account criterion 4 but not criterion 3:

$$g_{ij}^{\text{R}} = x e^{\frac{3}{2}x\eta z}. \tag{6}$$

Also, the very recent approximation of Viduna and Smith (VS) [15] almost verifies criterion 3 but not criterion 4:

$$g_{ij}^{\text{VS}} = x + f_1(\eta) z_{ij} + f_2(\eta)\left(\frac{2}{3} + \frac{\xi_1\xi_3}{3\xi_2^2}\right) z_{ij}^2, \tag{7}$$

where the functions f_n and ω_n defined in Ref. [15] are invariant in the permutation of i and j.

Recent asymmetric expressions are

- The Barrio and Solana (BS) [16] modified BMCSL expressions which improve g_{22} and verify criterion 3 but not criterion 4:

$$g_{11}^{\text{BS}} = g_{11}^{\text{BMCSL}} + \theta x_2 d_1;\; g_{12}^{BS} = g_{21}^{BS} = g_{12}^{\text{BMCSL}} + \theta d_1^2 d_{12}^{-1};\; g_{22}^{\text{BS}} = g_{22}^{\text{BMCSL}} + \theta x_1 d_2. \tag{8}$$

The Henderson–Chan (HC) modified expressions [10] as used in Ref. [11] (g_{22} with an extra term in x^3) verify both criteria 3 and 4, where $\tilde{\eta} = \eta_1 + \frac{\eta_2}{R}$ and the damping factor $a = \left(\frac{\tilde{\eta}}{\eta}\right)^n$ is adjusted via n [10, 11]:

$$
\begin{aligned}
g_{11}^{\mathrm{HC}} &= g_{11}^{\mathrm{BMCSL}};\; g_{12}^{\mathrm{HC}} = g_{21}^{\mathrm{HC}} = g_{12}^{\mathrm{BMCSL}} \\
&\quad + \frac{R^2-1}{(R+1)^2}\tilde{\eta}^2 x^3 - \frac{R^3-1}{(R+1)^3}\tilde{\eta}^3 x^3, \\
g_{22}^{\mathrm{HC}} &= g_{22}^{\mathrm{BMCSL}} + a\left[\frac{3}{2}x^2(\eta - R\tilde{\eta}) + \frac{x^3}{2}(\eta^2 - R^2\tilde{\eta}^2)\right] \\
&\quad + \exp\left[-\frac{3}{2}ax^2(\eta - R\tilde{\eta})\right] - 1,
\end{aligned}
\tag{9}
$$

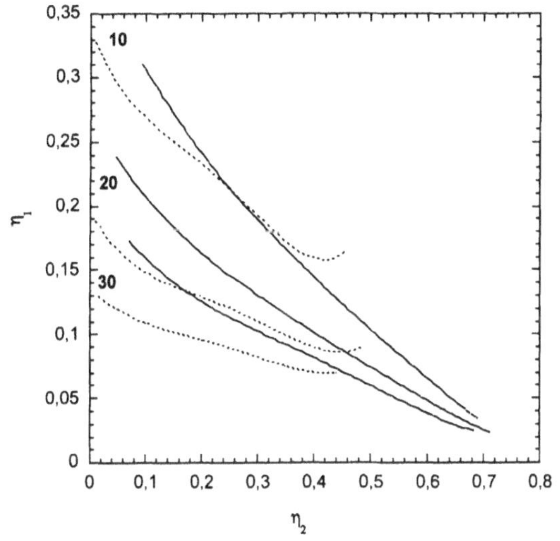

Fig. 1 Theoretical binodals obtained with the Rosenfeld scaling form [14] (*continuous line*) versus the simulation of the effective fluid [3] (*dotted line*). From *top* to *bottom*: size ratio $R = 10$, 20, 30

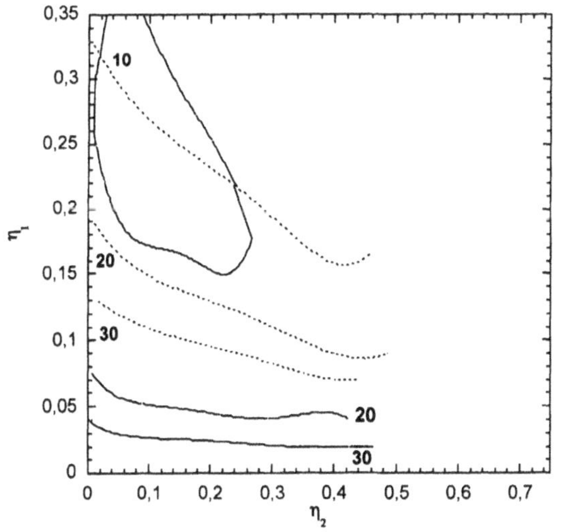

Fig. 2 Same as Fig. 1 with the model of Santos et al. [13]

Binodals and g_{22} contact values

A careful exploration of the Gibbs free energy in the P–x_2 plane and the search for the occurrence of a phase transition was performed for the models presented. The quantitative agreement may be judged from Figs. 1, 2, 3 and 4 for the binodals (the comparison is made with the simulation of Dijkstra et al. in the effective one-component model for $R = 10$, 20, 30), and in Table 1 for the contact values of the large spheres where comparison is made with Monte Carlo simulations at $R = 10$ [17] and also at $R = 20$ [18]. This last value provides a much more

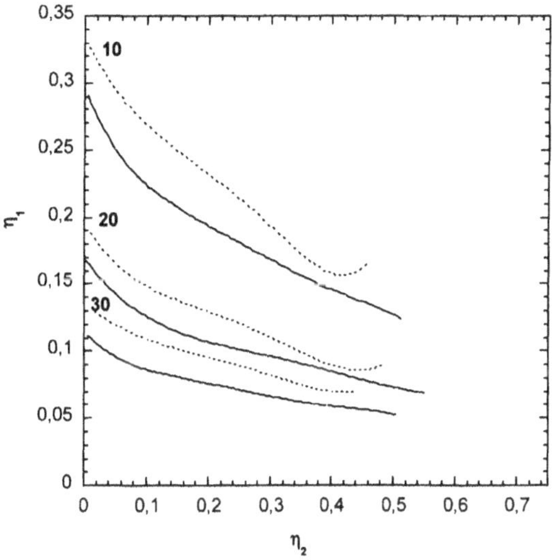

Fig. 3 Same as Fig. 1 with the model of Barrio and Solana [16]

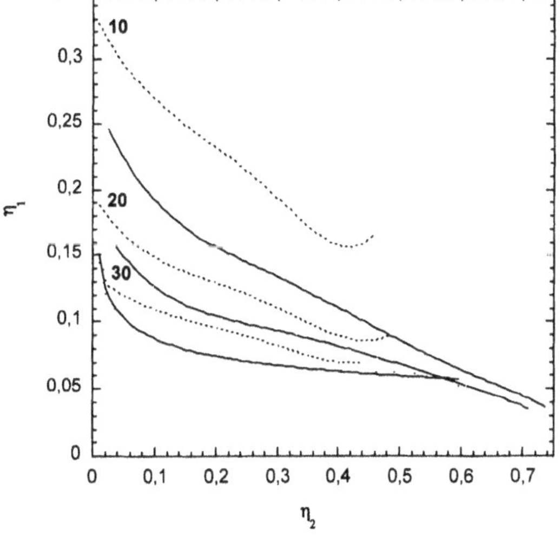

Fig. 4 Same as Fig. 1 with the Henderson–Chan modified expression [10] with $n = 0.5$ [11]

Table 1 Contact values of the pair distribution functions of the large spheres, g_{22}, at selected packing fractions η_1 (small spheres), η_2 (large spheres) and diameter ratio R. Equation-of-state model of Boublik–Mansoori–Carnahan–Starling–Leland (*BMCSL*) [4], Santos et al. (*SYL*) [13], Barrio and Solana (*BS*) [16], Rosenfeld (*R*) [14], Viduna and Smith (*VS*) [15] and Henderson–Chan (*HC*) [10] and Monte Carlo (*MC*) simulations [17, 18]

R	η_1	η_2	BMCSL	SYL	R	BS	VS	HC (n=5)	HC (n=8)	MC
10	0.026	0.524	11.4	11.1	30.3	11.9	14.3	11.0	11.1	16.0
	0.058	0.292	6.02	6.55	11.5	6.53	7.86	6.04	5.87	8.86
	0.049	0.5	14.9	15.9	59.7	15.9	20.2	14.9	14.4	19.9
	0.1	0.5	34.1	38.3	693	38.1	48.9	88.9	41.7	52.8
20	0.088	0.1	8.69	11.9	38.2	10.9	12.8	17.3	12.4	23.5
	0.123	0.1	14.6	19.6	180	19.3	22.2	85.4	46.9	65.0

stringent test of the exponential dependence with R than those made in other studies.

For all the previous symmetrical approaches, some improvement over the BMCSL approach in the colloidal range is indeed observed since they eventually predict binodals. However, the marked quantitative discrepancies between these models are difficult to correlate with the constraints of the colloidal limit and the contact values of the PDF. Indeed the Rosenfeld scaling form does not at all satisfy the contact theorem, but it actually predicts correct trends in the binodals with R (Fig. 1). It satisfies the exponential behaviour of g_{22}, but gives exceedingly large values not only for g_{22} (Table 1) but also for $g_{11}$and g_{12} at high packing fractions η_1 or η_2. The SYL model obeys exactly the contact theorem with quite acceptable g_{11} and g_{12} but is increasingly poor at large R for g_{22} and for the binodals (Fig. 2). More puzzling is the recent VS model [15]: the contact theorem is almost verified and all g_{ij} values look very accurate for R=10, but the binodal is very poor when compared with simulation since the critical point occurs [15] outside the domain of packing fractions ($\eta^{max}{}_2 \approx 0.50$) corresponding to a liquid.

When compared with the symmetrical models, the asymmetrical BS (Fig. 3) and HC (Fig. 4) ones show somewhat better qualitative agreement with the binodals from simulation. Both models verify the contact theorem exactly, and their predictions for g_{11} and g_{12} are in close agreement with available simulation data. More puzzling is the fact that the BS model, which does not reproduce g_{22} at large R (Table 1), is as good as the HC model as R increases (R=30). This remark is, however, not new, since our previous work [11] has shown that combining the HC $g_{12}$with the BMCSL g_{11} and g_{22} – a minimal modification of the BMCSL ansatz – is sufficient to obtain a binodal. This is not true with the combination g_{11} and g_{12} from BMCSL and g_{22} from HC, which does not predict a binodal.

Conclusion

Correlating some constraints and consistent rules for a specific g_{ij} with the existence of a binodal proves to be difficult. Indeed, its occurrence depends on the (non-trivial) behaviour of the second partial derivative at constant pressure of the Gibbs free energy with respect to the concentration. The condition that P is constant links the values of the three g_{ij} values via the relation $\eta_1 = \eta_1$ (P, η_2). This implies that any ad hoc change in one of the g_{ij} affects the value of the remaining ones, i.e. a self-consistent adjustment of the other g_{ij} would be necessary. This may explain why the crude expressions of Rosenfeld which treat the three g_{ij} on an equal footing do produce a binodal, although their value is rather inaccurate. The possibility of predicting both a correct binodal and accurate contact values of the PDFs from an analytical EOS remains therefore an unsolved problem.

References

1. (a) Asakura S, Oosawa F (1954) J Chem Phys 22:1255; (b) Vrij A (1976) Pure Appl Chem 48:471
2. (a) Biben T, Hansen JP (1991) Phys Rev Lett 66:2215; (b) Biben T, Hansen JP (1991) J Phys Condens Matter 3:65
3. Dijkstra M, van Roij R, Evans R (1999) Phys Rev E 59:5744
4. (a) Boublik T (1970) J Chem Phys 53:71; (b) Mansoori GA, Carnahan NF, Starling KE, Leland TW Jr (1971) J Chem Phys 54:1523
5. Coussaert T, Baus M (1998) J Chem Phys 109:6012
6. Lekkerkerker HNW, Stroobants A (1993) Physica A 195:387
7. Lekkerkerker HNW, Oversteegen SM (2002) J Phys Condens Matter 14:9317
8. Rosenfeld Y (1995) J Phys Chem 99:2857
9. Amokrane S, Regnaut C (1996) Phys Rev E 53:1990

10. (a) Henderson D, Chan KY (1998) J Chem Phys 108:9946; (b) Henderson D, Boda D, Chan KY, Wasan D (1998) Mol Phys 95:131, (c) Cao D, Chan KY, Henderson D, Wang W (2000) Mol Phys 98:619
11. (a) Regnaut C, Dyan A, Amokrane S (2001) Mol Phys 99:2055; (b) Erratum (2002) Mol Phys 100:2907
12. Santos A, Yuste SB, Lopez de Haro M (1999) Mol Phys 96:1
13. Santos A, Yuste SB, Lopez de Haro M (2002) J Chem Phys 117:5785
14. Rosenfeld Y (1988) J Chem Phys 89:4272
15. (a) Viduna D, Smith WE (2002) Mol Phys 100:2815; (b) Viduna D, Smith WE (2002) Mol Phys 100:2903
16. Barrio C, Solana JR (2000) J Chem Phys 113:10180
17. Lue L, Woodcock LV (1999) Mol Phys 96:1435
18. Malherbe J G, Amokrane S (2001) Mol Phys 99:355

Progr Colloid Polym Sci (2004) 126: 102–107
DOI 10.1007/b94455

P. Germain
M. Bouaskarne
J-G. Malherbe
C. Regnaut
S. Amokrane

Validity of the hard-sphere mixture model for sterically stabilised colloids: a critical viewpoint

P. Germain (✉) · M. Bouaskarne
J-G. Malherbe · C. Regnaut · S. Amokrane
Groupe de Physique des milieux denses,
Faculté des Sciences et Technologie
Université, Paris XII,
61 avenue du Général de Gaulle,
94010 Créteil Cedex, France
e-mail: germain@univ-paris12.fr
Tel.: +33-1-45171443
Fax: +33-1-45171579

Abstract The validity of the hard-sphere mixture model for sterically stabilised colloids is discussed. The inadequacy of the hard-sphere mixture model in the context of pure colloid–solvent systems is analysed both at the experimental and at the theoretical level. A simple model is presented to discuss the respective weights of the direct solute–solute interaction and the indirect ones mediated by the solvent. The range of the size ratio in which the indirect one is relevant is examined. In this regime, the adequacy of the hard-sphere mixture as a reference system for perturbation treatments is investigated.

Keywords Steric stabilisation · Colloid interactions · Potential of mean force · Phase transition

Introduction

The model of a binary mixture of hard sphere with a large difference in size has been extensively used in recent years for the analysis of the structure and the thermodynamics of sterically stabilised colloids. It indeed captures the well-known depletion effect [1], which has been regarded up to now as the dominant mechanism for explaining the phase behaviour of these systems. In the packing fraction representation, the phase behaviour predicted from this model depends on a unique parameter, the big to small sphere size ratio, R. Its most salient feature is the existence, for sufficiently high R, of a metastable fluid–fluid (F–F) transition, which is always pre-empted by the fluid–solid one (see Ref. [2] and references therein). As a consequence, this model predicts that a sizeable volume fraction of solutes could be dispersed in the fluid phase only in a gaseous "solvent".

This scenario may, however, be questioned in the light of several known experimental facts:

1. Fluid dispersions that are rich in solute particles actually exist in liquid solvents. For example, octadecyl chain coated silica particles can be dispersed in benzene up to a volume fraction $\eta = 0.28$ [3]. In microemulsions, volume fractions of the dispersed phase that are even higher are frequently considered [4].
2. The phase boundaries often depend on the temperature. Examples are colloidal silica particles [3] or reverse micelles [4]. The latter show, in addition, a lower consolute point in the temperature-volume fraction plane.
3. The phase behaviour can be different in different solvents [5] or with solutes of different surface composition [6], even when the size asymmetry remains the same.

These points being by definition foreign to the hard-sphere model, one needs to improve the modelling to restore the specificity of real systems. Of course, these known facts have already been analysed in the original papers [3, 4, 5, 6] from various models and theoretical methods. In recent papers, however, we showed in the framework of the effective fluid approach (using as input the effective interaction obtained from the integral equations of statistical mechanics [7]), that a simple understanding of points 1–3 may be gained by taking into account two essential physical effects. On the one hand, the attractive forces between the components of the mixture, especially the solvent–solvent and solute–solvent ones, strongly modify the effective interaction between the solutes with respect to the depletion potential [8]. These modifications lead, of course, to great changes in the phase behaviour [8]. On the other hand, the direct interaction between the solute particles must be considered at the same scale as the short-range structure in the indirect interaction (that mediated by the solvent). When both show a comparable spatial variation, the image of a nearly hard-sphere solute may become insufficient [9]. Thus, two different situations should be considered. The first one is that of pseudo-binary mixtures – such as colloids with added polymers or mixtures of differently sized colloids. In this case, the smaller species is itself a supramolecular object and is much bigger than the particles of the suspending medium. The depletion potential is due to the smaller objects and not to the solvent itself, whose role is assumed to be negligible. It is precisely in this case that the asymmetric binary mixture of hard spheres, with $R{\sim}10$, has been often invoked in the literature [10]. The second situation is that of colloidal particles suspended in a molecular solvent – say a reverse micelle or colloidal silica in an organic solvent – for which the size ratio ranges from $R{\sim}20$ to $R{\sim}100$. In order to estimate the relevance of the hard-sphere model for such a system, it is therefore important to understand the regime of size asymmetry in which the short-range structure is relevant, and conversely the situations in which it is offset by the direct interaction between the solutes. A clarification of this point is one of the objectives of this work.

Ultimately, we consider the regime for which the indirect interaction is relevant. We then investigate the question whether the hard-sphere mixture may be used at least as an appropriate reference to study more realistic models (i.e. with attractive forces) in the framework of perturbation treatments. This motivation is indeed frequently invoked to justify the persistent interest in this model for solvent–solute systems (see Ref. [11] and references therein). It is shown that the very narrow range of the effective interaction between solutes makes even this issue problematic [12].

This paper is organised as follows. A simple model of sterically stabilised colloidal particle is used to examine the respective weights of the direct solute–solute interaction and of the indirect one. The possible insufficiency of a perturbation treatment using the hard-sphere mixture as a reference system is then illustrated The paper ends with a brief conclusion.

Indirect potential versus direct potential

The designation "sterically stabilised colloid" refers to a wide variety of experimental systems [13]. A quantitative account of the direct interaction between the solute particles in a specific suspension and, in particular, that of the complex mechanism leading to steric stabilisation would require a detailed model of the material forming the particles, of its surface layer and of the suspending medium [13]. A detailed model of the latter, together with its interaction with the solute surface, is also necessary in order to compute the indirect interaction. An illustration of some aspects of this difficult task is given in Ref. [9] (water/AOT reverse micelles in decane). However, the purpose of the present short paper is neither to give a thorough analysis of all possible contributions to the total effective interaction, nor to model a specific system. What we are seeking is a simple way to estimate even in a very gross way the importance of the indirect interaction with respect to the direct one. We thus start with a very simple "generic" model, which incorporates in a minimal way the phenomenology of steric stabilisation, namely a short-range repulsion that compensates near contact the van der Waals attraction. To this end, we consider a spherical colloid viewed as a supramolecular object formed of Lennard-Jones-like particles, of total radius $R = R' + D$ (in units of the solvent radius) where R' is the core radius and D a length that controls the relative magnitudes of the repulsive and attractive terms. The total attractive and repulsive contributions to the direct potential are then computed by integrating the $-c_6/r^6$ and c_{12}/r^{12} interactions between the microscopic components over the volume of the two spheres of radius R' and R, respectively. The $-c_6/r^6$ term gives the well known Hamaker result [12] and explicit formulas for the repulsive c_{12}/r^{12} term are given in Ref. [14]. This leads to a direct potential:

$$\Phi^{\mathrm{dir}} = \Phi^{\mathrm{att}} + \Phi^{\mathrm{rep}}, \tag{1}$$

with

$$\Phi^{\mathrm{att}}(r) = -c\frac{\pi^2}{6}\left[\frac{2R'^2}{r^2 - (2R'^2)} + \frac{2R'^2}{r^2} + \ln\left(\frac{r^2 - (2R'^2)}{r^2}\right)\right] \tag{2}$$

and

$$\Phi^{\mathrm{rep}} = c\left(\frac{16}{4725}\pi^2 R^6\right) \times \frac{\sum_{n=0}^{8} P_{16-2n}(R) r^{2n}}{\left(\left[(r+R)^2 - R^2\right]\left[(r-R)^2 - R^2\right]\right)^7}, \tag{3}$$

where $P_n(R)$ is a polynomial of degree n [13]. In this expression, $c = c_6 = c_{12}$, which amounts to assuming that the diameter of the solvent is equal to that of the Lennard-Jones objects forming the colloid (otherwise one should define a different size ratio R)

It is stressed that taking different radii R and R' for the repulsive and attractive terms is just an ad hoc way to incorporate the idea that in sterically stabilised particles the repulsion should dominate near contact. Indeed, by varying the parameter $D << R$ one may mimic different physical situations in which the respective magnitudes of the repulsion are determined by different geometrical parameters. D may roughly be correlated with the length of the surface layer chains and their degree of overlap. Typical examples are silica particles covered with a dense layer of octadecyl chains or water-in-oil particles covered with less densely packed AOT surfactant chains.

As emphasised in the Introduction, the question of the relevance of the indirect part of the potential arises more particularly for very large R (solute–solvent system): in this case the short-range effects related to the discrete nature of the solvent are expected to be offset by the direct interaction between the big objects. To illustrate this point, we considered the case $R = 50$ and $c/k_BT = 1$ – one has for example $c/k_BT = 0.8$ for colloidal suspensions based on hydrocarbons [13]. The potential of mean force, Φ^{ind}, was computed from the Ornstein–Zernike equations at infinite dilution with the hypernetted chain (HNC) closure [7]. We did not include solvent–solvent and solvent–solute attractive forces, Φ^{ind} being thus the pure depletion potential. In what follows, the magnitude of the attractive well at contact is therefore overestimated with respect to real physical systems [8]. We mention here that Φ^{ind} computed with the HNC compares well with simulations for $R = 10$. No simulation data exist for higher values. However, the potential computed from the formula of Götzelman et al. [14] (Eq. 4) is close to the HNC one; therefore the HNC should give a reasonable estimate. Our results are shown for a solvent density $\rho_1^* = 0.4$ in Figs. 1 and 2 corresponding to two different definitions of the effective hard-sphere solute radius, R^{eff}, used to compute the depletion potential: in Fig. 1, we simply took $R^{eff} = R$, whereas in Fig. 2, R was determined so as to have $\Phi^{dir}(R^{eff})/k_BT = 3/2$ (still with $R^{eff} = 50$). In the latter case one has roughly $R \sim 49.5$ (the solvent particles do not penetrate the surface layer). In each case the potentials are computed for $D = 1$ (Figs. 1a, 2a) and $D = 0.5$ (Figs. 1b, 2b). We see that the effect of the depletion potential is never significant beyond a narrow shell close to contact, even in the case illustrated in Fig. 2 which is most favourable to the depletion mechanism. In Fig. 1, the depletion potential is almost completely offset by the direct one. In Figs. 1b and 2b, the direct potential has a deeper minimum than in Figs. 1a and 2a. This would mimic a situation with less efficient steric stabilisation. As a result, the depletion potential is more quickly offset when the solute particles are away from contact.

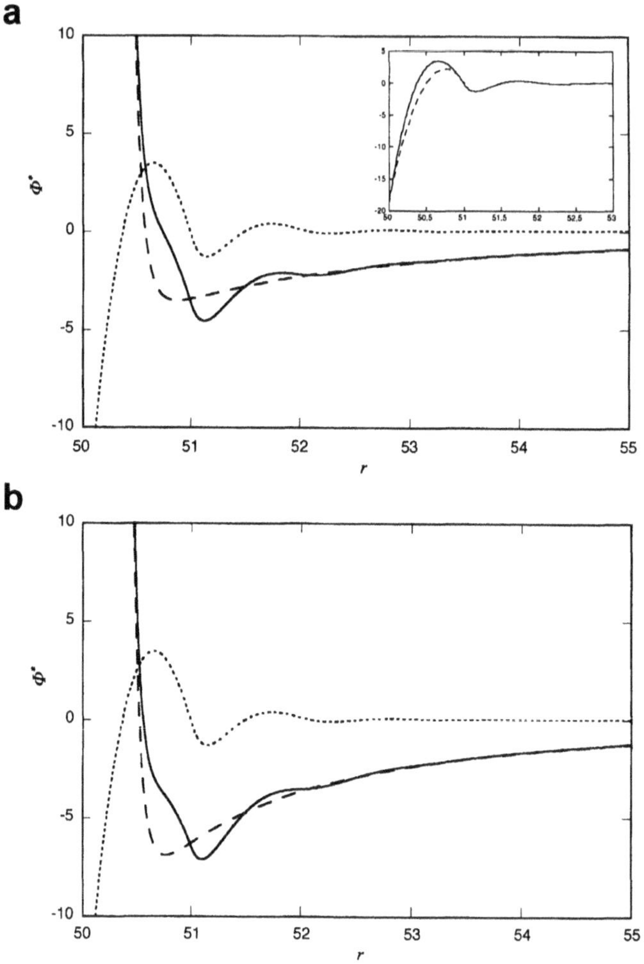

Fig. 1 Solute–solute interaction potential, Φ^*, in units of k_BT for $R = R^{eff} = 50$, and $\rho_1 = 0.4$ (in units of σ_1^3 with σ_1 the solvent diameter); r is the distance between the centres of the solutes (in units of σ_1): **a** $D = 1$, **b** $D = 0.5$. Hypernetted chain (*HNC*) depletion potential (*dotted line*); direct potential (*dashed line*); total potential (*solid line*). *Inset* in *a*: depletion potentials computed from the HNC closure (*solid line*) and from Eq. (4) (*dashed line*)

These results suggest that, in most solute–solvent systems with $R = 50$, the effect of the indirect potential should be rather marginal with respect to the direct one. Indeed, the situation described in Fig. 2 is the one most favourable to the depletion effect. First, we neglected the attractive forces for computing the indirect potential, which amounts to overestimating the depletion well. Second, the solute radius used to compute the indirect potential is larger than that involved in the direct one. Even in this case, the depletion well occurs in so tight a shell that its influence on the phase diagram should not be significant – except near close-packing. We also note that the usual statement according to which the solute–solvent systems with very large size asymmetry ($R > 100$) should be regarded as hard-sphere-like one-component

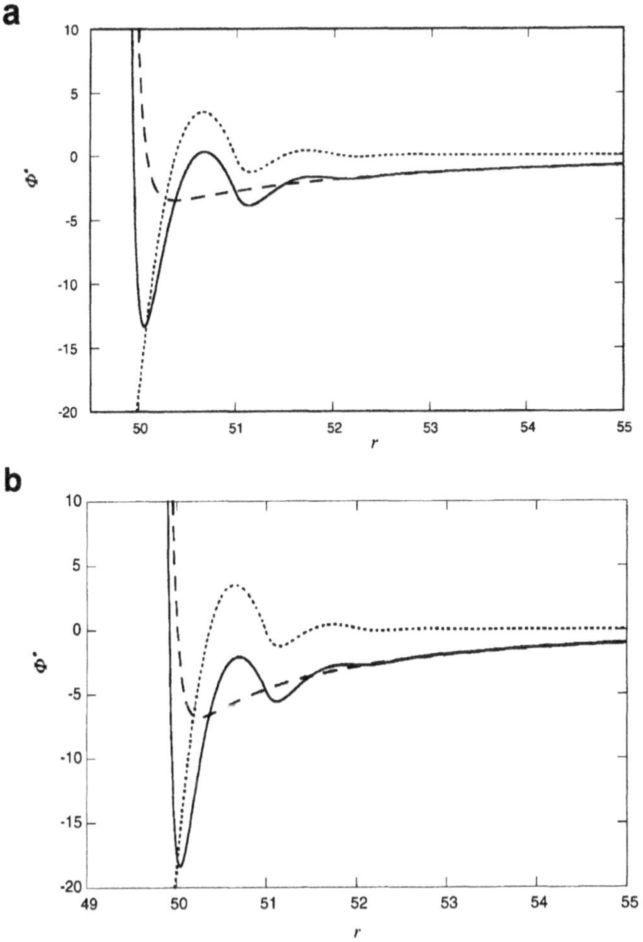

Fig. 2 Φ^* in units of k_BT for $\rho_1 = 0.4$ (in units of σ_1^3 with σ_1 the solvent diameter), $R^{\text{eff}} = 50$, and $\Phi^{\text{dir}}(R^{\text{eff}})/k_BT = 3/2$; r is the distance between the centres of the solutes (in units of σ_1): **a** $D = 1$, **b** $D = 0.5$. HNC depletion potential (*dotted line*); direct potential (*dashed line*); total potential (*solid line*)

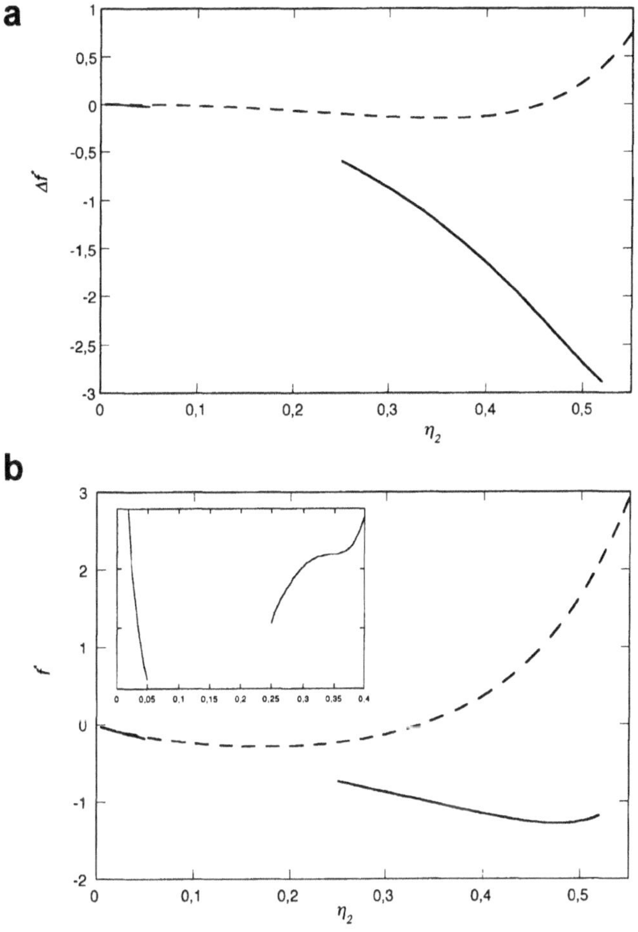

Fig. 3 a Comparison for $\rho_1^* = 0.12$ (in units of σ_1^3 with σ_1 the solvent diameter) between the reference HNC (*RHNC*) (*solid line*) and the perturbation (*dashed line*) reduced excess free energy, $\Delta f^* = \frac{\sigma_2{}^3 \Delta F/k_BT}{V}$, with σ_2 the solvent diameter and V the total volume. η_2 is the solute packing fraction. The missing part of the *solid curve* corresponds to the domain where the RHNC has no solution. **b** Reduced free energy, $f^* = \frac{\sigma_2{}^3 F/k_BT}{V}$. RHNC (*solid line*), perturbation (*dashed line*). The *inset* shows the reduced RHNC free energy after subsraction of the linear component

systems may be questioned in the light of the present results. They indeed show that the direct potential may have a sizeable attractive tail over a range of distances larger than what is usually implicitly assumed.

On the other hand, mixtures of two colloids with different sizes [10] belong to another category. For these systems, the variation of the direct potential of the bigger species has a much smaller range than that of the depletion well. This direct interaction could therefore be safely modelled as a hard-sphere one.

Finally, the inset shows for $R = 50$ the HNC depletion potential and that computed from the analytical expression of Götzelman et al. [14]:

$$\Phi^{\text{GED}}(r) = -\frac{R^{\text{eff}}+1}{2}\left[3x^2\eta_1^* + \left(9x + 12x^2\right)\eta_1^{*2} + \left(36x + 30x^2\right)\eta_1^{*3}\right], \tag{4}$$

where $x = r - R^{\text{eff}} - 1$ and $\eta_1^* = \frac{\pi}{6}\rho_1^*\sigma^3$ is the solvent packing fraction of the reservoir (see inset in Fig. 1a). The good agreement between the two potentials, especially at contact, is particularly interesting. Indeed, as Eqs. (1), (2), (3) and (4) are fully analytical, they provide a convenient way to get a quick estimation of the regime in which a given suspension is.

Validity of the perturbation theory with a hard-sphere mixture reference system

In this section, we investigate the case of solute–solvent mixtures, $R \sim 20$ (e.g. microemulsions). According to the previous discussion, the indirect solute–solute interaction should be relevant. The fine structure of the direct potential should also be significant in this regime [9]. However, to simplify the discussion, we consider here only the treatment of the solvent–solvent and solvent–solute attractive potentials, u_{11} and u_{12}, by perturbation, the direct potential being a hard-sphere one. Even in this case different combinations of u_{11} and u_{12} can give a large variety of physical situations. To determine whether the hard-sphere mixture system may be considered as an appropriate reference in a perturbation scheme, we took $R = 10$ and the following potentials: a weak solvent–solvent Lennard-Jones potential ($u_{11}(r) = 4\varepsilon_{11}(\frac{1}{r^{12}} - \frac{1}{r^6})$, with $\varepsilon_{11} = 0.1k_BT$, and r in units of the solvent diameter) and a stronger solvent–solute Yukawa one ($u_{12}(r) = \varepsilon_{12}^* \frac{\sigma_{12}}{r} \exp[\lambda_{12}(r - \sigma_{12})]$, with $\varepsilon_{12}^* = 2.36k_BT$, $\lambda_{12} = 2.5$ and $\sigma_{12} = (1 + R)/2$). This case is convenient because it provides an example of a F–F transition for which the perturbation theory with a hard-sphere mixture reference system can be tested numerically (see later). This transition was found in the effective one-component representation, using again the HNC closure to compute the potential of mean force Φ^{tot} (the free energy was computed with the very accurate optimised reference HNC, RHNC, theory [15]). We obtained a critical solvent density $\rho_{1c}^* \sim 0.1$ in the reservoir, whereas the depletion potential Φ^{dep} gives with the same method $\rho_{1c}^* \sim 0.51$. This strong reduction of ρ_{1c}^* is due to the very different structure of Φ^{tot} and Φ^{dep}, as Φ^{tot} has an attractive tail with much longer range than Φ^{dep}.

The perturbation free energy was computed in the following way [12]. First, we computed the reference free energy, F^{dep}, corresponding to the depletion potential with the RHNC theory [15]. Second, we computed the excess free energy, ΔF^{pert}, due to the attractive forces by applying the perturbation treatment to $\Delta\Phi = \Phi^{tot} - \Phi^{dep}$:

$$\frac{\Delta F^{pert}}{V} = \frac{\rho_2}{2} \int_V g^{dep}(r) \Delta\Phi(r) dr, \qquad (5)$$

where ρ_2 is the solute density, V the total volume and $g^{dep}(r)$ the pair distribution function computed for the depletion potential Φ^{dep}. The perturbation free energy is then $F^{pert} = F^{dep} + \Delta F^{pert}$ (the justification for this calculation is given in the second reference of Ref. [12]). It appears from Eq. (5) that the validity of the perturbation theory with a hard-sphere mixture reference system is closely related to the solute correlation function: if the latter is not too strongly modified by the perturbation potential, the corresponding excess free energy is expected to be satisfactorily estimated by Eq. (5), while one may expect more severe problems in the opposite case. Unfortunately, it is the second situation that occurs more frequently for systems with very short range interactions, such as those considered here [12] (see also Ref. [16] for a discussion of the van der Waals approach to very short ranged interactions). This leads to a poor estimation of the excess free energy term as observed in Fig. 3a, where we compare the RHNC and the perturbation excess free energies just above the F–F transition line ($\rho_1^* = 0.12$). The consequences of this large discrepancy between F^{pert} and F^{RHNC} for the phase transition is shown in Fig. 3b: for the same density, the perturbation fails to predict the F–F transition. The failure is even complete for the potentials considered here since the F–F transition is not found in the perturbation treatment up to $\rho_1^* = 0.5$. The perturbation treatment is not possible for higher values of ρ_1^* because of the problems of numerical convergence of the RHNC equations of the reference system close to its critical density. These numerical problems also prevented us from testing the perturbation theory for other F–F transitions predicted by the RHNC for other choices of u_{11} and u_{12} as these occur at much higher density (see third reference of Ref. [8]). The accuracy of the perturbation excess free energy, ΔF, could nevertheless be tested at low bulk density (second reference of Ref. [12]). A similar pathology is found – which worsens when ρ_1^* increases. This confirms that the hard-sphere mixture is not, for many solvent sterically stabilised colloid systems, a good reference system. On the contrary, it might still remain useful in the case of pseudo binary mixtures for which the direct interaction is of very small range with respect to the size of the objects. This point, however, should be examined further (see Refs. [12, 16] for a partial discussion).

References

1. (a) Asakura S, Oosawa F (1954) J Chem Phys 22:1255; (b) Vrij A (1976) Pure Appl Chem 48:471
2. Dijkstra M, van Roij R, Evans R (1999) Phys Rev E 59:5744
3. Duits MHG, May RP, Vrij A, de Kruif CG (1991) Langmuir 7:62
4. Chen SH, Rouch J, Sciortino F, Tartaglia P (1994) J Phys Condens Matter 6:10855
5. Robertus C, Joosten JGH, Levine YK (1990) J Chem Phys 93:7293
6. (a) Lemaire B, Bothorel P, Roux D (1983) J Phys Chem 87:1023; (b) Brunetti S, Roux D, Bellocq AM, Fourche G, Bothorel P (1983) J Phys Chem 87:1028.
7. Hansen JP, McDonald IR (1986) Theory of simple liquids. Academic, New York
8. (a) Clément-Cottuz J, Amokrane S, Regnaut C (2000) Phys Rev E 61:1692; (b) Amokrane S, Malherbe JG (2001) J Phys:Condens Matter 13:7199; (c) Malherbe JG, Amokrane S, Regnaut C (2002) Phys Rev E 66 (in press)
9. (a) Amokrane S, Bouaskarne M (2000) J Chem Phys 112:11107; (b) Bouaskarne M Amokrane S, Regnaut C (2001) J Chem Phys 114:2442

10. Imhof A, Dhont JKG (1995) Phys Rev Let 75:1662
11. (a) Matyushov D, Ladanyi B (1997), J Chem Phys 107:5815; (b) Regnaut C, Dyan A, Amokrane S (2001) Molec Phys 99:2055; (c) Viduna D, Smith WR (2002) Mol Phys 100:2815
12. (a) Germain P, Amokrane S (2002) Phys Rev E 65:31109; (b) Germain P, Amokrane S Phys Rev E (to be published)
13. Hunter R (1995) Foundations of colloid science. Oxford Science, Oxford
14. Henderson D, Duh D-M, Chu X, Wasan D (1997) J Colloid Interface Sci 185:265
14. Götzelman B, Evans R, Dietrich S (1998) Phys Rev E 57:6785.
15. (a) F. Lado (1973) Phys Rev A 8:2548; (b) Lado F (1982) Phys Lett 89A:196; (c) Lado F, Foiles SM, Ashcroft NW (1983) Phys Rev A 4:2374
16. Louis AA (2001) Philos Trans R Soc Lond 359:939

Progr Colloid Polym Sci (2004) 126: 108–113
DOI 10.1007/b94018

Jan Sefcik
Marcel Verduyn
Giuseppe Storti
Massimo Morbidelli

Charging and colloidal stability of electrostatically stabilized latex particles

J. Sefcik · M. Verduyn · G. Storti
M. Morbidelli (✉)
Institut für Chemie- und Bioingenieurwissenschaften, Swiss Federal Institute of Technology Zurich, ETH Hönggerberg/HCI, 8093 Zurich, Switzerland
e-mail: morbidelli@chem.ethz.ch
Tel.: +41-1-6323034
Fax: +41-1-6321082

Abstract In this work we study how the aggregation kinetics of primary particles in polymeric latexes stabilized by an anionic surfactant is related to surface charge and solution ionic strength. Polystyrene latex particles stabilized with sodium dodecyl sulfate have been investigated under conditions of slow irreversible aggregation, with stability ratios between 10^7 and 10^9. Using standard models for colloidal stability we analyze relations between solution salt concentration, surface charge and aggregation kinetics of primary latex particles. Apparent surface charge densities estimated from aggregation kinetics are in qualitative disagreement with those previously determined from electrophoretic mobility data. Possible strategies to arrive at qualitatively correct and predictive models for the colloidal stability of these systems are discussed.

Keywords Colloids · Stability · Derjaguin–Landau–Verwey–Overbeek · Charge · Surfactant

Introduction

Colloidal particles move around owing to Brownian motion in macroscopically static liquids, and attractive interactions due to dispersion forces drive their aggregation when they come in contact. On the other hand, charging of the particle surface results in formation of diffuse electrical double layers, which overlap as two particles approach each other, thus leading to repulsive forces. If one considers an idealized system where monodisperse spherical solid particles at concentration N aggregate when coming in contact by Brownian motion in a fluid with viscosity μ at temperature T, then the rate of doublet formation due to aggregation of primary particles was derived in the classical work of Smoluchowski as $\beta_S N^2$, with the ideal Smoluchowski rate constant $\beta_S = 8kT/3\mu$ [1]. When interparticle forces and hydrodynamic interactions are taken into account, the rate of doublet formation can still be expressed as βN^2, but now with a general aggregation rate constant β. It is customary in the colloidal literature to relate β to β_S (obtained in the absence of interparticle interactions) by a multiplicative factor called the stability ratio, $W = \beta_S/\beta$ [2], given by

$$W = 2a \int_{2a}^{\infty} \frac{\exp[U(r)]}{G(r)r^2} \mathrm{d}r. \quad (1)$$

Here a is the particle radius, U is the dimensionless interparticle interaction energy (expressed in the units of kT) and G is the hydrodynamic interaction function [3], accounting for additional hydrodynamic friction of particles moving towards each other at distances comparable to their size.

In many electrostatically stabilized colloidal suspensions W has been found experimentally to depend linearly on the solution salt concentration, c, when plotted in log–log coordinates, i.e. $\log W = k_1 \log c + k_2$, in a given concentration range below the critical coagulation concentration. This was first observed for iodide sols [4] and has since been documented for various polymer latex systems [5,6]. In order to explain this behavior, one has torelate interparticle forces to the solution salt concentration through an appropriate theory. Reering and Overbeek invoked the Derjaguin–Landau–Verwey–Overbeek (DLVO) theory to explain their observations. The DLVO theory of colloidal stability assumes that the attractive dispersion force and the repulsive electrostatic force between two colloidal particles are additive and that no other significant interaction forces are present [7]. Accordingly, the total energy of interaction between two primary particles, U, is expressed as the sum of the attractive dispersion contribution, U_{dis}, and the repulsive electrical double-layer contribution, U_{edl}. The attractive energy of interaction U_{dis} can be explicitly calculated assuming that intermolecular dispersion forces are pairwise additive and integrating over the interior volume of the two interacting particles. The corresponding expression for two equal size spherical particles is available [8].

As already mentioned, the electrostatic repulsive interaction between two primary particles is caused by the overlap of their diffuse double layers. In order to calculate the energy of repulsion for the pair of particles, we need the diffuse-layer electrostatic potential, ψ_{d}. When using the Stern model of the inner part of the electrical double layer, it was found that the double-layer interaction energy is bracketed by two limiting regimes, where either ψ_{d} or the surface charge density, $-\sigma_{\mathrm{d}}$, remains constant as two aggregating particles are approaching each other at given conditions. Situations intermediate between these two limiting cases are described by the charge regulation regime, which takes into account the shift of ionic equilibria during the particle approach [9]. Here we thus consider the two limiting cases – constant potential (CP) and constant charge (CC) – in order to bracket the true surface charging behavior, whatever it might be.

In the literature there are a number of expressions available for the energy of repulsion of a pair of colloidal particles due to overlap of their electrical double layers (for overviews see Refs. [8, 10]). Their ranges of applicability depend on the dimensionless thickness of the electrical double layer, κa, where κ is the inverse of the Debye length, and the dimensionless diffuse-layer potential, $e\psi_{\mathrm{d}}/kT$, of the particles. A linear theory suitable for small potentials ($|e\psi_{\mathrm{d}}/kT| < 1$) is based on the Debye–Hückel approximation (linearization of the Poisson–Boltzmann equation) to calculate the exact repulsive interaction energy for two flat plates. Then one can use the Derjaguin approximation to calculate the corresponding interaction between spheres at separations small compared to their radius ($r/a = h << 1$) by expanding the potential as a series in powers of $1/\kappa a$ (so that it is required that κa is large [10]). For the case of CP this approach was used by Hogg et al. [11]. However, as potentials become larger, the linear theory becomes less accurate compared to the full solution of the nonlinear Poisson–Boltzmann equation. Ohshima and Kondo developed [12] a more accurate approximation of the nonlinear energy of interaction for the case of equal size spheres (again, it is required that $h << 1$ and $\kappa a >> 1$), where in the limit of $|\psi_{\mathrm{d}}| << kT/e$ the linear formula of Hogg et al. is recovered. For the case of CC we use an analytical formula recently introduced by Ohshima [13] based on approximates for the energy of interaction between two flat plates accurate at short distances ($\kappa h < 2$). The energy of interaction between two equal spheres is then computed by numerical integration of the Derjaguin formula.

ψ_{d} is related to σ_{d} through the Gouy–Chapman equation, which for 1–1 electrolytes is as follows:

$$\sigma_{\mathrm{d}} = -\frac{\varepsilon kT\kappa}{e}\left[2\sinh\left(\frac{e\psi_{\mathrm{d}}}{2kT}\right) + \frac{4}{\kappa a}\tanh\left(\frac{e\psi_{\mathrm{d}}}{4kT}\right)\right]. \quad (2)$$

This equation gives the surface charge density within 5% of an accurate numerical solution of the Poisson–Boltzmann equation for any value of ψ_{d} and $\kappa a > 0.5$ [10].

Although the DLVO theory has been able to qualitatively and sometimes quantitatively explain various aspects of colloidal stability [14], it has often been noted that surface charging and its effect on colloidal stability could not be consistently described while trying to reconcile results from various experimental characterizations [6]. In this work we investigate electrostatically stabilized polystyrene latexes using standard DLVO models for colloidal stability in order to analyze relations between solution salt concentration, surface charge and aggregation kinetics of primary latex particles.

Experimental

Polystyrene latexes with nearly monodisperse particle size distribution were prepared by a standard emulsion polymerization procedure as described previously [15] using sodium dodecyl sulfate (SDS) as the emulsifier and sodium persulfate as the initiator. The mean hydrodynamic radius, a, of the synthesized particles was determined by dynamic light scattering using a Zetasizer 5000 (Malvern Instruments) and the solid weight fraction was determined by gravimetry using an HG53 halogen moisture analyzer (Mettler Toledo). The particle number concentration, N^0_1, was then calculated from the measured mean radius (a) and the solid weight fraction using a polymer density of 1,050 kg/m^3. The bare surface charge corresponding to the surface concentration of dissociable groups, assuming that no counterions are bound, was calculated as $\sigma_0 = -F\Gamma$, where F is the Faraday constant and Γ is the surface coverage of the emulsifier (surfactant) determined previously for this system [16].

The experimental procedure for measuring aggregation kinetics was as follows. First, 10 ml of a concentrated NaCl solution was added to 50 ml of the original latex dispersion so that the resulting mixture (after a short time of stirring to achieve aqueous phase homogeneity) reaches the desired NaCl concentration. The aggregation kinetics was followed by monitoring the time evolution of the mean hydrodynamic radius of the aggregating suspension, R_{DLS}. Particle concentrations in the dispersions were too high for on-line measurements of the scattered light intensity. Therefore, samples were withdrawn from the aggregating system and diluted with doubly distilled water to stop aggregation. Since aggregation of primary particles is irreversible, dilution does not affect any aggregates present. There was no observable change in aggregate size on the time scale of our experiments. Dynamic light scattering was performed on these diluted samples with the standard red light laser wavelength $\lambda = 632.8$ nm at the scattering angle $\theta = 90°$, corresponding to the scattering vector amplitude $q = (4\pi n/\lambda)\sin(\theta/2)$, where n is the refractive index of the dispersant (water).

Results and discussion

We list the relevant characteristics of the two latexes used in the experiments in Table 1, while Table 2 reports the characteristics of each experimental run. These include the aqueous phase ionic strength, I, the dimensionless Debye length, κa, the aggregation rate constant, β, and the stability ratio, W. The values of β were obtained as follows. In each experiment, as a function of time, we calculated the scaled mean hydrodynamic radius

$$f(R_{\mathrm{DLS}}) = \frac{R_{\mathrm{DLS}}(t)/R_{DLS}(0) - 1}{1 - BR_{\mathrm{DLS}}(t)/R_{\mathrm{DLS}}(0)}, \tag{3}$$

where $R_{\mathrm{DLS}}(0) = a$ is the measured (mean) hydrodynamic radius of primary particles and B is the ratio of singlet and doublet hydrodynamic radii taken as 1/1.38 [17]. In the early stage of aggregation when only primary particles and doublets are present, $f(R_{\mathrm{DLS}})$ changes linearly with time and the corresponding slope is equal to $2S_2(q)\beta N^0{}_1$, where $S_2(q) = [1 + \sin(2aq)/(2aq)]/2$ is the structure factor of the doublet of solid primary particles, q is the scattering vector amplitude and $N^0{}_1$ is the initial concentration of primary particles. Since $S_2(q)$ and $N^0{}_1$ are known (Tables 1, 2), from the measurement of this slope we obtain an estimate of β.

For each experimental system we then solved the equation

$$\frac{4kT}{3\mu\beta a} = \int_{2a}^{\infty} \frac{\exp\left[U_{\mathrm{vdW}}(r) + U_{\mathrm{edl}}(r)\right]}{G(r)r^2}\,\mathrm{d}r, \tag{4}$$

for the unknown diffuse-layer potential in the electrostatic interaction energy expression, U_{edl}, using expressions for the CP [12] and the CC [13] limit, respectively. For polystyrene we used the value of the Hamaker constant $A = 6.5 \times 10^{-21}$ J in the expression of the dispersion attractive interaction [8]. Finally, we calculated the corresponding diffuse-layer charge from the Gouy–Chapman equation (Eq. 2). The results obtained are summarized in Table 2. It can be seen that in the case of CC the calculated diffuse-layer potentials and the corresponding charges are 10–15% lower in absolute value than those calculated for the CP case, while following the same qualitative trends.

Let us now see how we can predict the value of the stability ratio using the DLVO theory. For this we need to add to Eqs. (1) and (2) one more piece of information in order to compute ψ_{d}, σ_{d} and W (or β). For this we compare the following three approaches.

1. Surface potential independent of ionic strength. We assume that the diffuse-layer potential of isolated particles is not dependent on the ionic strength. In this approach we cannot a priori estimate ψ_{d} and therefore we set it equal to the average of the experimental data in Table 2, i.e. ψ_{CP} for latexes B and F is −44.5 and −37.2 mV, respectively. The corresponding stability ratio can be calculated from Eq. (1), using appropriate expressions for the electrostatic repulsion, for ionic strengths up to the critical coagulation concentration.
2. Surface charge independent of ionic strength. We assume that the surface charge of isolated particles is not dependent upon ionic strength, and using the data in Table 2 we set it equal to the average of experimental values, i.e. σ_{CP} for latexes B and F is −52.8

Table 1 Properties of the latexes used in the aggregation experiments: particle radius, a, number concentration, $N^0{}_1$, surfactant coverage, Γ, bare surface charge, σ_0, doublet structure factor, $S_2(q)$

Latex	a (nm)	$N^0{}_1$ (10^{14}/ml)	Γ (μmol/m^2)	σ_0 (mC/m^2)	$S_2(q)$
B	37.5	5.67	1.95	−188	0.851
F	57	1.51	0.64	−62	0.699

Table 2 Stability ratio, diffuse-layer potential and particle surface charge, calculated from aggregation kinetics (constant potential, CP, and constant charge, CC, limits), and from the counterion binding model

Latex	I (mM)	κa	Logβ	log$W$	ψ_{CP} (mV)	ψ_{CC} (mV)	σ_{CP} (mC/m^2)	σ_{CC} (mC/m^2)	σ_{mod} (mC/m^2)
B1	183	52.9	−25.7	8.8	−45.0	−40.4	−50.3	−44.2	−36.4
B2	194	54.5	−25.4	8.5	−44.8	−40.4	−51.5	−45.4	−36.0
B3	217	57.6	−24.9	8.0	−44.5	−40.4	−54.0	−48.0	−35.3
B4	240	60.6	−24.1	7.2	−43.7	−39.9	−55.5	−49.7	−34.5
F1	147	72.1	−24.8	7.9	−37.5	−33.3	−36.1	−31.5	−20.0
F2	158	74.7	−24.3	7.4	−37.1	−33.1	−36.9	−32.4	−19.7
F3	170	77.5	−24.0	7.1	−37.1	−33.2	−38.3	−33.7	−19.3

and $-37.1\ mC/m^2$, respectively. Since the bare charges, σ_0, calculated from the surfactant coverage Γ assuming no counterions bound are -188 and $-62\ mC/m^2$ (Table 1), this implies a degree of counterion binding of $\alpha = 0.72$ and 0.40 for latexes B and F, respectively. The corresponding diffuse-layer potential can be calculated from the Gouy–Chapman equation (Eq. 2), and the stability ratio from Eq. (1), using appropriate expressions for the electrostatic repulsion, for ionic strengths up to the critical coagulation concentration.

3. Surface charging model. In this approach we add a piece of new physical information, which relates to the fact that the surface charge can be computed on the basis of the bare surface charge and the counterion binding equilibria. We consider in the following the model developed earlier [16] to interpret data from electrophoretic mobility measurements for the same latexes as used in this study. The surface charge in this model, σ_{mod}, is expressed in terms of the bare charge $\sigma_0 = -F\Gamma$ (where F is the Faraday constant) and the degree of counterion binding, α:

$$\sigma_{mod} = -F\Gamma(1 - \alpha), \quad (5)$$

$$\alpha = \frac{K_{Na} I \exp\left(-\frac{e\psi_d}{kT}\right)}{1 + K_{Na} I \exp\left(-\frac{e\psi_d}{kT}\right)}. \quad (6)$$

Here K_{Na} is the counterion binding constant and I is the bulk solution ionic strength, which in our case is equal to the bulk sodium concentration. For SDS adsorbed on polystyrene latex particles we found previously that the best agreement with electrophoretic mobility measurements was obtained using for the surfactant coverage an adsorption isotherm determined through independent experiments, while fitting the counterion binding constant $K_{Na} = 6\ dm^3/mol$, where the shear plane position was related to the Debye length, $1/\kappa$, as $\Delta_{shear} = 0.2/\kappa$. Again, the corresponding diffuse-layer potential can be calculated from the Gouy–Chapman equation (Eq. 2) and the stability ratio from Eq. (1), for ionic strengths up to the critical coagulation concentration.

The predicted stability ratio for latex B using approaches 1–3 is plotted in Fig. 1. Since in approaches 1 and 2 we were fitting the experimental data, the calculated lines are near the experimental data points, but the corresponding slopes are not in agreement. On the other hand, in approaches 3 we used an independently determined surface charging model and the predicted stability ratios are several orders of magnitude lower than those measured. Discrepancies between double-layer potentials and the corresponding charges determined from electrokinetic and aggregation measurements are commonly observed [8] and could in some cases be justified by considering that the position of the shear plane is shifted outward from the outer Helmholtz plane [6]. However, in this case the shear plane position was already carefully considered when the surface charging model used in approach 3 was developed [16].

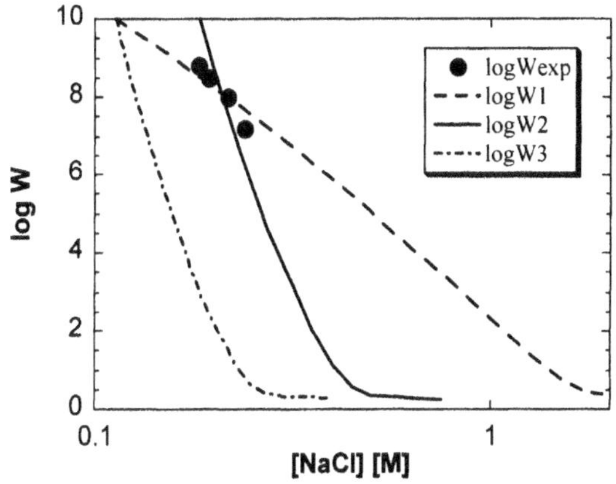

Fig. 1 Comparison of the experimental stability ratio values for latex B as a function of ionic strength with various models: surface potential independent of ionic strength (*W1*), surface charge independent of ionic strength (*W2*), counterion binding model (*W3*)

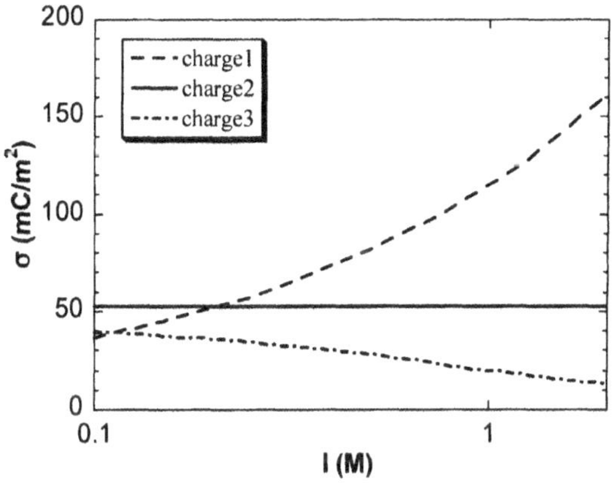

Fig. 2 Surface charge of latex B as a function of ionic strength according to various charging models: surface potential independent of ionic strength (*charge 1*), surface charge independent of ionic strength (*charge 2*), counterion binding model (*charge 3*)

In order to evaluate and contrast these three approaches we plotted the calculated diffuse-layer charges for latex B as a function of the solution ionic strength for each approach (Fig. 2). Qualitatively different behavior of surface charge versus ionic strength can be seen for the three approaches. For the case of the surface potential fixed at a certain value independent of the ionic strength (approach 1) we get the surface charge increasing with ionic strength. The standard electrical double-layer theory can lead to the surface charge either decreasing or increasing with the ionic strength, depending on whether the bulk electrolyte is a binding or an inert ion. When the binding ion concentration is constant and the ionic strength increases owing to an inert ion (as is the case for carboxylate latexes at a constant pH [6]), the increased shielding causes a decrease in the absolute value of the diffuse-layer potential. Therefore the degree of counterion binding decreases and thus the surface charge increases with increasing ionic strength. On the other hand, if the binding ion is the same as in the bulk electrolyte (as is the case for the carboxylate latex with a strong acid added as the only electrolyte), the degree of counterion binding increases with increasing ionic strength and thus the surface charge decreases.

In our case, sulfate groups at the surface can, in principle, bind both sodium and proton counterions, but at relatively high ionic strengths owing to NaCl at neutral pH, the bulk concentration of sodium is about 6 orders of magnitude larger than that of protons and we do not expect a significant degree of proton binding compared to that of sodium cations in our system. Therefore the degree of (sodium) counterion binding increases and the surface charge decreases with increasing ionic strength. Since this implies that the surface potential must also decrease with increasing ionic strength, assuming the surface potential to be fixed over a wider range of ionic strength would be unphysical. On the other hand, assuming the surface charge to be fixed independent of the ionic strength (approach 2) is also not physically sound, since it would imply that the counterion binding constant decreases with increasing ionic strength. For the last approach, which indeed appears to be more realistic, we can see from Fig. 2 that the measured stability ratio of the charge-stabilized colloidal particles considered here is about 6 orders of magnitude larger that that predicted by the DLVO theory using the surface charging model determined independently from electrophoretic mobility measurements. In principle one could adjust the Hamaker constant in order to obtain a better fit to the experimental stability ratio values; however, the required Hamaker constant would have to be significantly lower than the value we used here, $A=6.5\times10^{-21}$ J, which is already at the low end of Hamaker constant values reported in the literature for polystyrene particles [6]. This indicates that the apparent degree of counterion binding necessary to explain aggregation kinetic data using the DLVO theory is much lower than that needed to explain electrophoretic mobility measurements. Although not shown here, the same conclusion is reached by analyzing the data for latex F.

Thus, we can conclude that none of the treatments of surface charging leads to even qualitative consistency between surface charging deduced from electrical double-layer theory and electrophoretic mobility measurements, and aggregating kinetics of primary latex particles stabilized by sulfate surfactant described by the classical DLVO theory.

Conclusions

Aggregation rate constants of primary particles were measured for various combinations of particle size, surfactant coverage and electrolyte (NaCl) concentration which lead to relatively large Fuchs stability ratios of the order of 10^7—10^9. From such values we calculated the corresponding diffuse-layer potential using an expression of the particle interaction energy which accounts for dispersion attraction and diffuse double-layer repulsion in a DLVO framework. The diffuse-layer potentials and the corresponding surface charge densities were calculated for both CP and CC limiting regimes, using appropriate solutions of the nonlinear Poisson–Boltzmann equation. The surface charge densities calculated this way are in qualitative disagreement with those previously determined from electrophoretic mobility data. In particular, it is found that, at least in the frame of this work, the surface charge estimated from the aggregation measurements is significantly larger than that computed from electrophoretic mobility data. These data can be reconciled following several possible strategies. One would be to account for additional repulsive interactions, such as those due to ionic dehydration of binding counterions [18] or to dipolar interactions between bound surfactant–counterion pairs [19], which augment the DLVO interaction energy; however, appropriate experiments would have to be designed in order to quantify such interactions. On a more fundamental level, the basic assumption in the DLVO theory about the independence and additivity of dispersion and electrostatic interactions could be revisited [20], considering the large surface charge and ionic strength values considered in this work.

Acknowledgement We gratefully acknowledge the financial support by the Swiss National Science Foundation (grant NF 2000-061883).

References

1. Smoluchowski M (1917) Z Phys Chem 92:129
2. Fuchs N (1934) Z Phys 89:736
3. Honig EP, Roeberse GJ, Wiersema PH (1971) J Colloid Interface Sci 36:97
4. Reerink H, Overbeek JTG (1954) Discuss Faraday Soc:74
5. Maroto JA, de las Nieves FJ (1998) Colloid Polym Sci 276:453
6. Behrens SH, Christl DI, Emmerzael R, Schurtenberger P, Borkovec M (2000) Langmuir 16:2566
7. Vervey EJW, Overbeek JTG (1948) Theory of the stability of lyophobic colloids. Elsevier, New York
8. Elimelech M, Gregory J, Jia X, Williams RA (1995) Particle deposition and aggregation. Butterworth-Heinemann, Woburn, MA
9. Behrens SH, Borkovec M (1999) J Phys Chem B 103:2918
10. Russel WB, Saville DA, Schowalter WR (1989) Colloidal dispersions. Cambridge University Press, Cambridge
11. Hogg R, Healy TW, Fürstenau DW (1966) Trans Faraday Soc 62:1638
12. Ohshima H, Kondo J (1988) J Colloid Interface Sci 122:591
13. Ohshima H (2000) J Colloid Interface Sci 225:204
14. Hunter RJ (2001) Foundations of colloid science, 2nd edn. Oxford University Press, Oxford
15. Verduyn M (1999) Aggregation of polymer particles in emulsion. PhD thesis. ETH, Zurich
16. Sefcik J, Verduyn M, Storti G, Morbidelli M (2003) Langmuir 19:4778
17. Holthoff H, Egelhaaf SU, Borkovec M, Schurtenberger P, Sticher H (1996) Langmuir 12:5541
18. Israelachvili JN (1992) Intermolecular and surface forces, 2nd edn. Academic, London
19. Manciu M, Ruckenstein E (2001) Langmuir 17:7061
20. Bostrom M, Williams DRM, Ninham BW (2001) Phys Rev Lett 87:168103

Progr Colloid Polym Sci (2004) 126: 114–116
DOI 10.1007/b93985

Lionel Foret
Alois Würger

Long-range interactions of surface charges on electrolyte films

L. Foret · A. Würger (✉)
CPMOH, Université Bordeaux 1,
Unité Mixte de Recherche CNRS 5798
351cours de la Liberation,
33405 Talence, France
e-mail: a.wuerger@cpmoh.u-bordeaux1.fr

Abstract A surface charge at an electrolyte–insulator interface and its counterion cloud form an electric dipole, thus giving rise to a long-range interaction that has been identified as the usual dipolar interaction. In this paper, we show that the picture of interacting dipoles ceases to be valid for a thin film with a thickness smaller than the Debye screening length. Indeed, in this case, both parallel and antiparallel dipoles repel each other, with a force much stronger than expected from a dipolar interaction. This originates from the delocalization of the screening clouds in very thin films.

Complex problems from soft matter and biophysics have stimulated interest in membrane model systems, such as black soap films or lamellar lyotropic phases with added proteins or colloidal particles. They give rise to fundamental problems like crystallization protein [1] protein–membrane interaction [2], and protein [3]. The properties of charged molecules and electrolyte solution are, to a large extent, controlled by screened electrostatic interactions that depend strongly on the interfacial geometry.

A charged particle adsorbed on an electrolyte–insulator interface forms, with its screening cloud, an electric dipole (Fig. 1a). In his pioneer experimental work on 2D colloidal crystallization at a water–air interface, Pieranski [4] observed a long-range repulsive force which he interpreted in terms of a dipolar energy. From a theoretical point of view, it has been confirmed that the potential energy of screened surface charges at the interface of a semiinfinite electrolyte is given by a modified dipolar interaction [5]. Similar behavior has been found theoretically for a classical Coulomb plasma near a wall [6]. Very recently, this model has been refined, by taking into account the finite size of the colloidal particle and, in particular, the dipole moment in the insulating half-space arising from the particle segment below the interface [7, 8].

In this paper, we study the screened electrostatic interaction of charges at the interfaces confining an electrolyte layer. Generalizing Hurd's work on semiinfinite electrolytes to films of finite thickness, we calculate the interaction potential of two charges on the same interface, and on opposite interfaces. In particular, we discuss the validity of the picture of interacting dipoles that was developed for adsorbed colloidal particles.

Our model consists of an electrolyte confined by two planar interfaces, which contains some point charges, Q, that induce a nonuniform distribution of the positive and negative salt and counterions, $n_\pm(\mathbf{r})$. We start from the free-energy functional of these bulk ion densities, $n_\pm$:

$$F = k_B T \sum_\pm \int d^3 r n_\pm(\mathbf{r}) \left[\ln \frac{n_\pm(\mathbf{r})}{n_s} - 1 \right] + \frac{e}{2} \int d^3 n_t(\mathbf{r}) \phi(\mathbf{r}), \quad (1)$$

the first term being the ideal gas entropy and the second one the electrostatic energy. Here, $n_t(\mathbf{r}) = n_+ - n_- + (Q/e) \sum_\alpha \delta(\mathbf{r} - \mathbf{r}_\alpha)$ denotes the total charge number density and ϕ the total electrostatic potential.

In a classical way, charge density and potential are related by the Poisson–Boltzmann equation. In the present work, we use its linearized version, i.e., we resort to the Debye–Hückel approximation (DHA) [9]. The

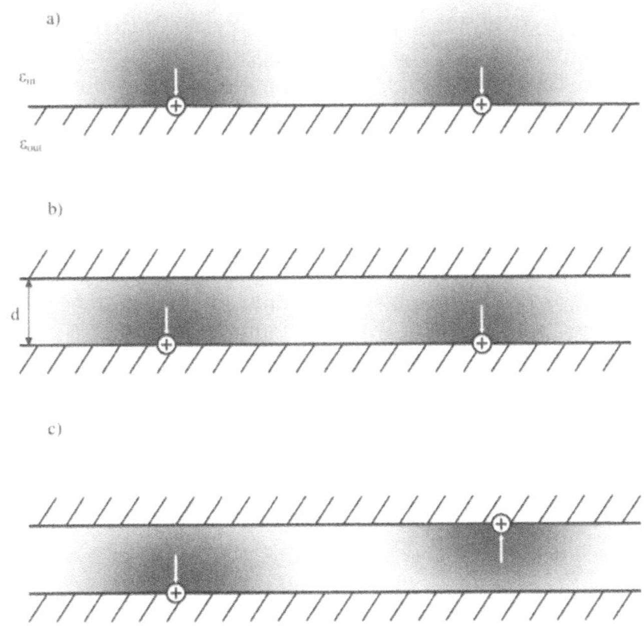

Fig. 1a, b Schematic view of two surface charges and their screening clouds. The resulting dipole moments are indicated as *white arrows*. **a** Semiinfinite electrolyte, **b** and **c** confined electrolyte

potential and the charge density can, thus, be written as a superposition of the potentials and screening clouds created by the discrete charges. Inserting the equilibrium solution $n_\pm(\mathbf{r})$ obtained from the DHA into Eq. (1), the free energy for fixed surface charge positions, $\{\mathbf{r}_\alpha\}$, becomes [10]

$$F(\{r_\alpha\}) = \frac{Q}{2}\sum_{\alpha,\beta} \varphi\beta(r_\alpha), \tag{2}$$

where $\varphi_\beta(\mathbf{r})$ is the potential created by the point charge at $\mathbf{r}_\beta$ and its counterion cloud. Thus, in the DHA, $V(\mathbf{r}_\alpha, \mathbf{r}_\beta) = Q\varphi_\beta(\mathbf{r}_\alpha)$ is the interaction energy of two surface charges at $\mathbf{r}_\alpha$ and $\mathbf{r}_\beta$. The DHA remains valid while the potential energy is smaller than the thermal energy $Q\varphi << k_B T$, i.e., while the Bjerrum length, $l_B = Qe/4\pi\varepsilon k_B T$, is smaller than the screening length and the separation of the lateral charges [9]. In the physical systems of interest, since l_B is very small ($\simeq 7$ Å) the preceding restrictions are satisfied; the DHA turns out to be of particular relevance in the long-range interaction studies.

In the remainder of this paper we consider one charge at the origin, with the screened potential $\varphi(\rho, z)$, and a second charge at lateral distance ρ, either on the same interface, $z=0$, or on the opposite one, $z=d$. The interaction energy is given by $V = Q\varphi(\rho,0)$, or $V_{\uparrow\downarrow} = Q\varphi(\rho,d)$ respectively. The potential $\varphi(\rho,z)$ is a solution of the linearized Poisson–Boltzmann equation with usual electrostatic boundary conditions, and depends strongly on the dielectric constant inside, ε_{in}, and outside, ε_{out}, the film; in most cases, that of the aqueous core, $\varepsilon_{\text{in}} \simeq 80\varepsilon_0$, is much larger than $\varepsilon_{\text{out}} \simeq \varepsilon_0$. We give the solution in terms of its 2D in-plane Fourier transform,

$$\varphi(\rho, z) = Q\int \frac{\mathrm{d}q}{2\pi} q \frac{A_+ \mathrm{e}^{-p(z-d)} + A_- \mathrm{e}^{-p(d-z)}}{A_+^2 \mathrm{e}^{pd} - A_-^2 \mathrm{e}^{-pd}} J_0(q\rho) \tag{3}$$

with the zero-order Bessel function, $J_0(q\rho)$, and the shorthand notation

$$A_\pm = \varepsilon_{\text{out}} q \pm \varepsilon_{\text{in}} p \quad \text{and} \quad p = \sqrt{q^2 + \lambda_D^{-2}}.$$

The Debye screening length, λ_D, gives the size of the counterion distribution and depends on the electrolyte concentration, n_s, as $\lambda_D = (2n_s e^2/k_B T \varepsilon_{\text{in}})^{-1/2}$. The Fourier integral (Eq. 3) can be performed in some limiting cases only. We are interested in lateral distances much larger than the Debye length, $\rho >> \lambda_D$. In this limit, we treat successively the cases of large and small thickness, as compared to the Debye length. ($d >> \lambda_D$ and $d << \lambda_D$).

The limit of thick films ($d >> \lambda_D$) is equivalent to a semiinfinite electrolyte; since the counterion distribution is hardly affected by the second interface, the corrections due to finite d are negligible. This situation applies to cold plasmas or colloids adsorbed on a wall [4, 5, 6]. These works show that the interaction decays algebraically with the third power of the inverse distance [5],

$$V = \frac{\varepsilon_{\text{out}}}{2\pi\varepsilon_{\text{in}}^2} \frac{Q^2 \lambda_D^2}{\rho^3}, \quad \text{if} \quad \rho \gg \lambda_D \tag{4}$$

which has been taken as a signature of a dipole–dipole interaction. Indeed, this picture is confirmed when noting that a discrete charge and its screening cloud give rise to a dipole moment $Q\lambda_D$ (Fig. 1a). We will show, however, that this picture ceases to be valid as the thickness of the aqueous core decreases.

For thin films ($d << \lambda_D$), the screening cloud of a given charge is squeezed by the opposite interface. We evaluate the screened interaction, $Q\varphi(\rho,z)$, with a second charge that may be located on the same interface ($z=0$), or on the opposite one ($z=d$). Noting the inequalities $d << \lambda_D << \rho$, we find the integral (Eq. 3) is dominated by q values smaller than $1/\rho$. We thus have $q\lambda_D << 1$ and $qd << 1$, and obtain

$$\varphi(\rho, 0) = \varphi(\rho, d) = Q\lambda_{D^2} \int \frac{\mathrm{d}q}{2\pi} \frac{qJ_0(q\rho)}{\varepsilon_{\text{in}} d + 2\varepsilon_{\text{out}} q \lambda_D^2}. \tag{5}$$

This integral can be evaluated analytically [11],

$$\varphi(\rho, 0) = \varphi(\rho, d) = \frac{Q}{4\pi\varepsilon_{\text{out}}\rho}\left[1 - \frac{\pi}{2}\Xi\left[H_0(\Xi) - y_0(\Xi)\right]\right]. \tag{6}$$

Hence the interaction potential is given by the bare interaction of two surface charges at lateral distance ρ, multiplied by a screening factor that involves the zero-order Struve function, H_0, and the zero-order Bessel function of the second kind, Y_0, with the dimensionless parameter

$$\Xi = \frac{\varepsilon_{in}}{2\varepsilon_{out}} \frac{\rho d}{\lambda_D{}^2}. \tag{7}$$

Before evaluating Eq. (6) further, we discuss general aspects related to the picture of interacting dipoles. As shown in Fig. 1, each adsorbed charge and its screening cloud give rise to a finite dipole moment of magnitude $Q\lambda_D$ and for $d >> \lambda_D$ and $Qd/2$ for $d << \lambda_D$. This picture is confirmed in the former case, since Eq. (6) is the interaction of two parallel dipoles $Q\lambda_D$ at a distance ρ.

For thin films, $d << \lambda_D$, we consider the two situations shown in Fig. 1. From the dipole picture, one expects a positive interaction energy for two charges on the same interface ($z=0$), and a negative energy for charges on opposite interfaces ($z=d$). The actual potential (Eq. 6), however, is the same in both cases. We may conclude that the dipole picture developed in Refs. [4, 5, 6] for $d=\infty$ fails for the interaction of screened charges on a thin film.

We are going to study the limits $\Xi >> 1$ and $\Xi << 1$, where Eq. (6) reduces to simple power laws. Since the film thickness, d, can hardly decrease below a few nanometers, and since $\varepsilon_{out} << \varepsilon_{in}$ for most systems, Ξ is, in general, larger than unity. Thus, for small thickness (around a few nanometers) and large distances, we have $\Xi >> 1$ and the interaction energies $V = Q\varphi(\rho,0)$ and $V_{\uparrow\downarrow} = Q\varphi(\rho,d)$ can be approximated by the first term of the expansion of Eq. (6) in powers of $1/\Xi$:

$$V = V_{\uparrow\downarrow} = \frac{\varepsilon_{out}}{\pi\varepsilon_{in}^2} \frac{Q^2\lambda_D{}^4}{d^2\rho^3} \quad \text{if} \quad \Xi >> 1 \tag{8}$$

Despite the power law $V \sim \rho^{-3}$, this is clearly not a dipolar interaction. Note that two screened charges, on the same or on opposite interfaces, are subject to a repulsive force in a lateral direction. (Antiparallel dipoles as in Fig. 1c would attract each other.) Moreover, Eq. (6) is significantly larger, by a factor of $\lambda D^4/d^4 >> 1$, than the interaction energy of two dipoles of magnitude $Qd/2$.

In physical terms, this stronger interaction arises from the spreading of the screening cloud in the confined geometry. In thin films, both the electrostatic potential, φ, and the net charge density, $n = n_+ - n_-$, decay algebraically at large distances, i.e., the counterions are no longer localized in the vicinity of the surface charge. As a consequence, even at large distances, the potential energy of two screened charges is not given by the dipolar interaction. The screening of the point charges, Q, being incomplete, their repulsive interaction exceeds the dipole energy. As is obvious from the factor $1/d^2$ in Eq. (7), this effect increases with decreasing d.

For very thin films, one reaches $\Xi << 1$. Although this limit is difficult to obtain in real systems, we note its particularly simple form of the interaction energy. For small Ξ, the second term, in braces, in Eq. (6) is negligible, resulting in

$$V = V_{\uparrow\downarrow} = \frac{Q^2}{4\pi\varepsilon_{out}\rho} \quad \text{if} \quad \Xi << 1. \tag{9}$$

Thus the interaction is the bare electrostatic potential energy of two surface charges, Q, as if the film together with the counterions had disappeared. Indeed, the counterions are smeared out well beyond the distance ρ, and hardly contribute to the electrostatic energy. From this picture it seems natural that the energy (Eq. 6) is the same for charges on the same side and on opposite sides of the film.

In summary, we have calculated the long-range interaction ($\rho >> \lambda_D$) between charges adsorbed on the interfaces of an electrolyte confined between two planar walls. The effect of confinement is found to be important when the film thickness is smaller than the typical size of the counterion distribution ($\lambda_D \geq d$). In this case, the interaction is significantly stronger than the dipolar energy. This observation can be traced back to the fact that screening of surface charges is weakened in thin films.

References

1. Petkova V et al (2002) Biophys J 82:541
2. Lhert F et al (2002) Langmuir 18:512
3. May S et al (2000) Biophys J 79:1747
4. Pieranski P (1980) Phys Rev Lett 45:569
5. Hurd AJ (1985) J Phys A 18:L1055
6. Jancovici B (1982) J Stat Phys 29:263
7. Aveyard R et al (2002) Phys Rev Lett 88:246102
8. Aveyard R (2000) Langmuir 16:1969
9. Safran SA (1994) Statistical thermodynamics of surfaces, interfaces, and membranes. Addison-Wesley, Reading, MA
10. Foret L, Kühn R, Würger A (2002) Phys Rev Lett 89:156102 (2002)
11. Gradshteyn IS, Ryshik IM (1965) Table of integrals, series, and products. Academic, New York

Progr Colloid Polym Sci (2004) 126: 117–120
DOI 10.1007/b94005

R. Tsekov
E. Evstatieva
K.W. Stöckelhuber
P.G. Smirniotis

Stability of TiO_2 suspensions in reactors for degradation of toxic pollutants

R. Tsekov
Department of Physical Chemistry,
University of Sofia, 1126 Sofia, Bulgaria

E. Evstatieva
Department of Water Technology,
ITC, Karlsruhe Research Center, 76021
Karlsruhe, Germany

K.W. Stöckelhuber
Max Planck Group for Colloids and
Surfaces, TU Bergakademie,
09599 Freiberg, Germany

P.G. Smirniotis
Department of Chemical Engineering,
University of Cincinnati, Cincinnati,
OH 45221, USA

Present address: R. Tsekov (✉)
Institute of Physical Chemistry,
Physical Chemistry of Condensed Matter,
University of Karlsruhe, 76128 Karlsruhe,
Germany
e-mail: eroumen@email.com

Abstract The stability of TiO_2 suspensions in photocatalytic reactors for degradation of pollutants is described as a result of competition between the electrostatic repulsion and the attraction due to the van der Waals and hydrophobic interactions. A simple expression for the hydrophobic force is derived, which depends substantially on the pollutant concentration and the adsorption length. The stability analysis performed yields a power dependence of the critical pollutant concentration on the electrolyte concentration, above which the suspensions are unstable. The scaling parameters of this power law depend significantly on the adsorption length and they are tabulated in the paper.

Keywords Suspensions · Stability · Hydrophobic Force · Coagulation

Introduction

Photocatalysts based on TiO_2 are often employed for purification and advanced treatment of water [1] since they are appropriate for decomposition of organic molecules to water, carbon dioxide and mineral ions [2]. The aim of our project is to set up an effective photocatalytic reactor for degradation of toxic pollutants operating on the basis of the band-gap model [3] by the use of TiO_2 photocatalysts. Since the catalytic medium consists of TiO_2 particles suspended in water, it is important to analyze the suspension stability in order to prevent coagulation of TiO_2 particles and segregation from the medium.

The problem of binding of two TiO_2 particles is equivalent to the problem of rupture of the liquid film dividing the particles. When the film ruptures, the particles affix to each other and form a larger particle. If the process advances, the gravitational force prevails over the Brownian motion and the suspension separates into two phases. According to colloidal science, the film stability is a result of competition of interfacial forces mediated by hydrodynamic resistance. The classical Derjaguin–Landau–Verwey–Overbeek theory considers only the van der Waals and electrostatic interactions. It has been recognized further that there are additional macroscopic forces in films due to hydration, hydrophobic and other specific interactions [4]. The knowledge of all these components is important to answer the questions how do TiO_2 particles interact and what is the role of other substances in the coagulation process.

Theory

Following the well-known Derjaguin approach, the potential energy of interaction of two identical spherical particles with radius R can be related via the following expression

$$U = \pi R \int_{r}^{\infty} W \mathrm{d}h \tag{1}$$

to the free energy excess, W, per unit area of a flat film with thickness h as compared to the corresponding bulk liquid. As mentioned in the Introduction, there are several distinguished components of W. In the case of a TiO_2 suspension in water $W = W_{VW} + W_{EL} + W_{HP}$, where W_{VW}, W_{EL} and W_{HP} are the van der Waals, the electrostatic and the hydrophobic components, respectively. The van der Waals interaction is universal, while the electrostatic component appears because the TiO_2/water surface is charged. The hydrophobic interaction is induced by the organic pollutants, which adsorb on the TiO_2/water surface and, thus, they change its specific properties.

The simplest and relatively exact expression for the van der Waals energy excess is

$$W_{VW} = -A/12\pi h^2, \tag{2}$$

where $A = 2.6 \times 10^{-19}$ J is the Hamaker constant for the $TiO_2/H_2O/TiO_2$ system [5]. Since the suspended TiO_2 particles are chemically identical, the films between them are symmetric and the resulting van der Waals force is attraction. Hence, the van der Waals interaction is the principal destabilizing force in the catalytic suspension of TiO_2. In the frames of validity of Eq. (2) the attraction energy goes to infinity when the distance between the particle surfaces approaches zero. Thus, the van der Waals force is also important for the practically irreversible attachment of the particles on each other. Owing to the van der Waals interaction the activation energy of detachment tends to infinity and the process is very slow.

The classical expression for electrostatic energy at constant surface potential, ϕ, on the TiO_2/water interface is given by the following expression [6]

$$W_{EL} = 2\varepsilon_0\varepsilon[(4kT/ze)\tanh(ze\phi/4kT)]^2 \exp(-h/d)/d, \tag{3}$$

where $d = \sqrt{\varepsilon_0\varepsilon kT/c}/ze$ is the Debye length, $\varepsilon_0\varepsilon$ is the dielectric permittivity of the suspension, e is the electron charge and c is the concentration of z charged electrolyte ions. There are two different kinds of electrolyte effects on the electrostatic interaction. An increase of the concentration of ions not adsorbing specifically on the film interfaces decreases the Debye length and thus suppresses the electrostatic repulsion. Ions, adsorbing specifically on the film interfaces, change the surface potential and depending on their sign can lead to an increase or a decrease of W_{EL}. A favorite circumstance is that ϕ on the TiO_2/water interface varies in the range 100–150 mV at different electrolyte concentrations [7]. These values are larger that the thermal potential $4kT/ze$ and for this reason the term $\tanh(ze\phi/4kT)$ in Eq. (3) is always close to 1. Hence, Eq. (3) can be simplified to

$$W_{EL} = 32\varepsilon_0\varepsilon(kT/ze)^2 \exp(-h/d)/d, \tag{4}$$

where W_{EL} becomes independent of the surface potential. The electrostatic force is obviously repulsive since the two TiO_2 particles are equally charged; hence, it is the basic stabilizing effect of the suspension. Since the electrostatic repulsion decreases by increasing the electrolyte concentration, the latter is very important for the coagulation process.

The hydrophobic force is the most complicated one. It is induced by the adsorption of species on the TiO_2 particle surface. A general measure of the free-energy excess W_{HP} is the difference between the adsorption parts of the surface tension in the film and on a single interface [8]. If we restrict the consideration to the case of relatively low saturated interfacial layers, the free-energy excess can be calculated from the difference of adsorptions via the following relation

$$W_{HP} = 2kT(\Gamma - \Gamma_h), \tag{5}$$

where Γ_h is the adsorption of pollutants on the TiO_2/water interfaces of a symmetric film with thickness h and $\Gamma = \Gamma_{h\to\infty}$. The factor 2 in Eq. (5) takes into account the effect of the two particles.

The further elaboration of Eq. (5) requires an expression for the thickness dependence of the adsorption. Thus, one can employ the Maxwell relation $2(\partial_h\Gamma_h)_\mu = (\partial_\mu\Pi)_h = C - C_h$ [8], where $\Pi = p - p_h$ is the disjoining pressure, μ is the chemical potential, C_h is the pollutant concentration in the film with thickness h and $C = C_{h\to\infty}$ is the average concentration of pollutants in the reactor. In the case of low concentrations the adsorption can be related to the pollutant concentration via Henry's law, $\Gamma_h = aC_h$, where a is the equilibrium constant of adsorption or the adsorption length. Substituting Henry's law in the Maxwell relation, one obtains a differential equation for the h-dependence of the adsorption:

$$2\partial_h\Gamma_h = C - \Gamma_h/a. \tag{6}$$

The solution of Eq. (6) under the condition of stick adhesion $\Gamma_0 = 2\Gamma$ is $\Gamma_h = aC[1 + \exp(-h/2a)]$. Now, introducing the adsorption in Eq. (5), the free-energy excess acquires the form

$$W_{HP} = -2kTaC \exp(-h/2a). \tag{7}$$

As is to be seen, W_{HP} is negative and the attractive hydrophobic force is another destabilizing effect in the suspensions. Indeed, coagulation of suspensions induced by addition of polymer molecules has been observed experimentally [9].

The total excess of free energy is equal to the sum $W = W_{VW} + W_{EL} + W_{HP}$. Introducing here the expressions for the free-energy components from Eqs. (2), (4) and (7), respectively, and substituting the result in Eq. (1), the latter after performing the integration changes to

$$U = \pi R\left[-A/12\pi r + 32\varepsilon_0\varepsilon(kT/ze)^2\exp(-r/d) - 4kTa^2C\exp(-r/2a)\right]. \tag{8}$$

Obviously, some parameters in Eq. (8) can be considered as general for our reactors. Here we can mention the particle radius $R = 0.1$ μm, the Hamaker constant $A = 2.6\times10^{-19}$ J, the dielectric permittivity $\varepsilon = 80$ and the temperature $T = 298$ K. Since usually CO_3^{2-}, SO_4^{2-} and HPO_3^{2-} ions are generated at the degradation of toxic pollutants, one can assume that $z = 2$. Hence, only three variable parameters remains in Eq. (8): the electrolyte concentration, c, included in d, the pollutant concentration, C, and the adsorption length, a. The latter reflects the specific properties of the pollutants undergoing degradation.

Application

In general, the potential energy from Eq. (8) exhibits a maximum and a minimum. The parameters of the energy profile depend substantially on the concentrations of pollutants and electrolyte in the reactor. The potential-energy profiles at various c and C are presented in Figs. 1 and 2, where the realistic value of $a = 10$ nm is employed [10]. As seen in Fig. 1, the increase of the electrolyte concentration decreases the energy barrier owing to the screening of the electrostatic repulsion. At c of the order of 0.6 mM the barrier approaches zero and fast coagulation takes place. Hence, if the reactor operates at $C = 1$ mM it is necessary to guarantee special conditions that the electrolyte concentration does not exceed 0.5 mM. In Fig. 2 the case of constant electrolyte concentration is presented. It shows that the increase of the pollutant concentration leads to destabilization of the suspension as a result of an increase of the hydrophobic attraction. At C above 2 mM fast coagulation occurs and this is the maximal allowed working concentration of pollutants when $c = 0.2$ mM.

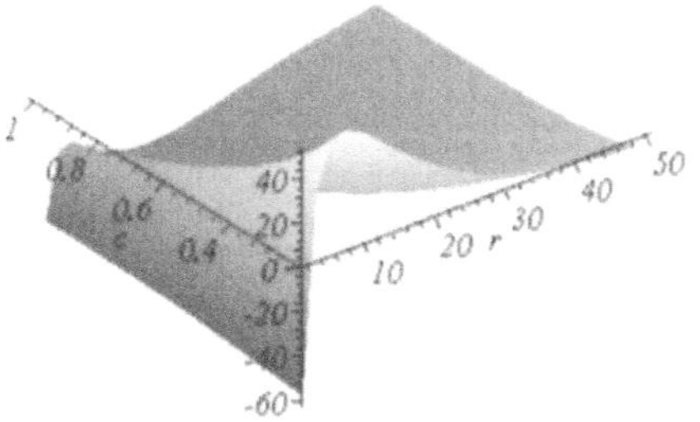

Fig. 1 Potential-energy curves U (kT) versus r (nm) at $a = 10$ nm, $C = 1$ mM and different electrolyte concentrations c (mM)

In general, there is a relation between the concentrations of pollutants and electrolytes indicating the region of the suspension stability. According to the theory [6], the fast coagulation starts when the maximum of the potential energy approaches zero. Mathematically this is equivalent to the following system of two equations:

$$U = 0, \quad \partial_r U = 0. \tag{9}$$

Substituting here the potential energy from Eq. (8), one obtains a system of two transcendental equations with three variables. Using numerical calculations it is possible to exclude the critical distance from Eq. (9). Thus, one obtains the dependence of the critical concentration of pollutants from the electrolyte concentration and the adsorption length, which is presented in Fig. 3. The surface represents the maximal value of the pollutant concentration in order to prevent coagulation of the medium. Obviously, the critical pollutant concentration decreases on increasing the electrolyte concentration. An approximate analysis of these numerical calculations shows that a good interpolation of the critical dependence of C versus c is given by the following power law

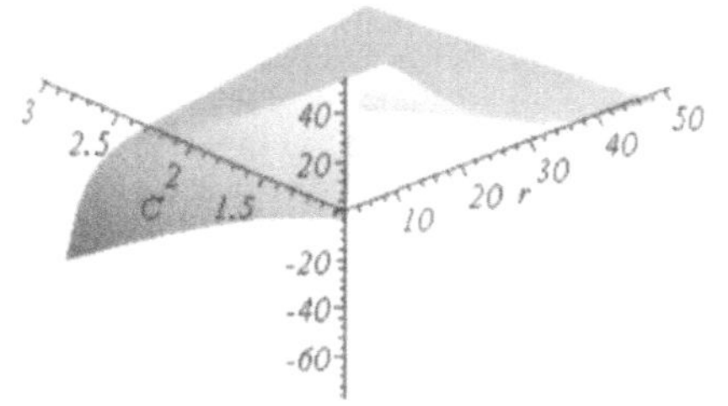

Fig. 2 U (kT) versus r (nm) at $a = 10$ nm, $c = 0.2$ mM and different pollutant concentrations C (mM)

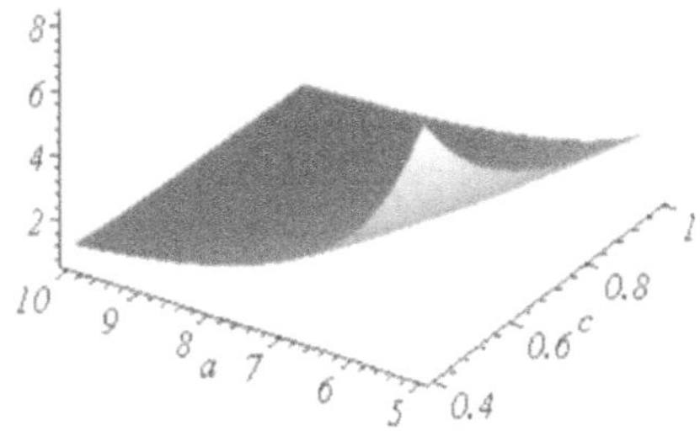

Fig. 3 The dependence of the critical pollutant concentration C (mM) on the electrolyte concentration c (mM) and the adsorption length a (nm)

Table 1 Values of the specific parameters in Eq. (10) at different adsorption lengths

a (nm)	5	10	20
n	1.23	0.67	0.41
B (mM)	1.59	0.79	0.31

$$C/B = (B/c)^n, \qquad (10)$$

where n and B are constants depending on a. The values of n and B for different realistic values [10] of a are summarized in Table 1. As seen, both parameters are inversely proportional to a in the frames of the considered region. Thus, it is easy to predict the value of the critical concentration of pollutants when the specific value of a and the electrolyte concentration are known.

Conclusions

The balance between the attraction, due to van der Waals and hydrophobic interactions, and the electrostatic repulsion determines the stability of photocatalytic suspensions of TiO_2 particles in water. Since the van der Waals force is universal, only the other two forces are accessible for manipulation in order to achieve stable suspensions. A reasonable parameter to control the electrostatic force is the electrolyte concentration, while the surface electrostatic potential on the TiO_2/water interface is very high. The increase of the electrolyte concentration leads to screening of the electrostatic repulsion and thus induces fast coagulation. In the present paper a simple expression for the hydrophobic force was obtained, where the main parameters are the concentration of the pollutants in the bulk and the adsorption length. Thus, the stability analysis reveals a dependence of the critical concentration of pollutants on the electrolyte concentrations at a given adsorption length. Using the power law derived one is able to determine the desirable maximal pollutant concentration at which the photocatalytic reactor still operates. An important point here is that in some reactors the concentrations could change in time and, for this reason, at any step the stability of the reactor should be proved. If the reactors operate in a turbulent flow the temperature in the present consideration has to be effectively increased by inclusion of the dispersion of the hydrodynamic velocity fluctuations.

Acknowledgements R.T. and P.G.S. wish to thank NATO for a grant (SfP-974209). E.E. is thankful to the Alexander von Humboldt Foundation for a Stability Pact fellowship. K.W.S. is grateful to DFG SFB 285 for financial support.

References

1. Linsebigler A, Lu G, Yates J (1995) Chem Rev 95:735
2. Rajeshwar K (1996) Chem Ind 12:454
3. Legrini O, Oliveros E, Braun A (1993) Chem Rev 93:671
4. van Oss C (1994) Interfacial forces in aqueous media. Dekker, New York
5. Horn RG, Clarke DR, Clarkson MT (1988) J Mater Res 3:413
6. Verwey E, Overbeek J (1948) Theory of the stability of lyophobic colloids. Elsevier, Amsterdam
7. Healy TW, White LR (1978) Adv Colloid Interface Sci 9:303
8. Tsekov R, Schulze HJ (1997) Langmuir 13:5674
9. Fröberg J, Rojas O, Claesson P (1999) Int J Miner Process 56:1
10. Hato M (1996) J Phys Chem 100:18530

Progr Colloid Polym Sci (2004) 126: 121–123
DOI 10.1007/b94003

J.-B. Salmon
A. Colin
S. Manneville
D. Roux

Dynamical rheology of a lyotropic lamellar phase near an out-of-equilibrium transition

J.-B. Salmon (✉) · A. Colin
S. Manneville · D. Roux
CRPP, avenue Albert Schweitzer,
33600 Pessac, France
e-mail: jean-baptiste.salmon@espci.fr
Tel.: +33-5-56845619
Fax: +33-5-56845600

Abstract This work deals with sustained oscillations of a complex fluid near a shear-induced transition. Rheological measurements of the onion texture near a layering transition are performed in a Couette cell. This complex fluid exhibits different regimes of sustained oscillations of the shear rate on a large time scale (500 s) at controlled stress. Transient dynamics strongly suggest the presence of a doubling period transition to chaos and the presence of a Hopf bifurcation. Using dynamical system theory, we analyse the different asymptotic temporal series of the shear rate and show that the dynamics do not strictly correspond to classical three-dimensional chaos. This may due to spatial organization of the fluid in the gap of the Couette cell.

Keywords Complex fluid · Lamellar phase · Rheology · Chaos

Introduction

Complex fluids (foams, liquid crystals, pastes, etc.) are known to exhibit different unusual behaviour when submitted to a shear flow [1, 2]. Such behaviour is due to a coupling between the structure and the flow. Recently, it has been observed that several complex fluids display oscillations of the shear rate in rheological experiments [3, 4, 5]. The measured periods of these oscillations are very large compared to the mechanical times of the experiments (rotation of the Couette cell, for instance). These oscillations do not arise from a simple elastic instability, but from the complex relation between the structure and the flow [4, 5]. Such dynamics may be described using dynamical system theory and may display chaotic behaviour [3, 5]. Theoretically, it has been shown that some models of soft glassy materials under shear present chaotic dynamics [6]. In the first part of this work, we show rheological measurements performed on the onion texture of a lyotropic lamellar phase near a shear-induced transition. In the second part, we present transient and asymptotic dynamics which suggest the presence of low-dimensional chaotic states. We also try to evidence the presence of a Hopf bifurcation in the recorded dynamics.

Rheology of the sodium dodecyl sulphate/octanol/brine lamellar phase

Experimental

The complex fluid we studied is a lyotropic lamellar phase composed of sodium dodecyl sulphate (SDS) (6.5 wt%), octanol (7.8 wt%) and brine (20 g l^{-1} NaCl). At equilibrium, this phase is made of surfactant bilayers stacked regularly and separated by the aqueous solvent [7]. The distance between the lamellae, i.e. the smectic period, was about 15 nm. In order to measure the viscosity, the lamellar phase was confined in Mooney–Couette cells (gaps 1 or 0.5 mm and outer radius 25 mm). A stress-controlled rheometer recorded the shear rate as a function of time. The Couette cell was transparent and a laser (He–Ne,15 mW) gave the diffraction pattern of the fluid undergoing shear, allowing us to determine the microscopic structure of the lamellar phase.

The layering transition

It has been shown in previous work that shear controls the texture of the lamellar phase. At a typical shear rate of 1 s^{-1}, the texture changes radically: the membranes are wrapped in multilamellar vesicles of the order of microns (called onions), with a close compact organization. The corresponding diffraction pattern is a homogeneous ring, indicating that onions are disordered. At a typical shear rate of 15 s^{-1}, six fuzzy peaks appear on the ring. This indicates an organization of two-dimensional layers of onions with hexagonal order sliding on each other. This transition is called the layering transition [8].

Dynamics of the shear rate

Flow curves and dynamical regimes

For temperatures above 27 °C, it is difficult to define a stationary viscosity in the vicinity of the layering transition. A typical flow curve obtained in a Couette flow is shown in Fig. 1(see Ref. [5] for details).

Our major result is the existence of different dynamical behaviours. For instance, three different dynamics of the shear rate corresponding to branches BC and BE are shown in Fig. 2. Such dynamics have a mean period of the order of 500 s and amplitude that slightly evolves under applied stress. Note that the mean period does not depend on σ. With the help of the recorded diffration patterns, we observed that the oscillations of the shear rate are correlated to a structural change in the fluid: the shear rate oscillates between the disordered state (high viscosity) and the layered state (low viscosity) [4, 5].

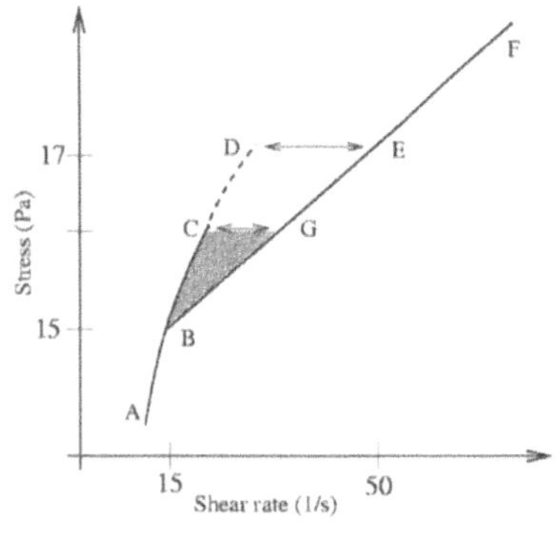

Fig. 1 Typical flow curve in the vicinity of the layering transition. Branches *AB* and *EF* are stationary states. The other branches correspond to different dynamical regimes; the *dotted line* corresponds to metastable states. Branch *AD* corresponds to disordered onions, *FB* to the layered state

Analysis using dynamical system theory

The dynamics shown in Fig. 2a displays strong similarities with a chaotic state because it is an aperiodic state; therefore a transition to chaos should be present, showing, in particular, a Hopf bifurcation leading to a periodic state [9]. At the precision $\delta\sigma$ of our setup, we never observed a periodic state. This strongly suggests that the Hopf bifurcation is subcritical (see Fig. 3). At critical stress σ_1, a periodic state appears with a finite amplitude as in our experiments and coexists with the stationary state. Between N and P a transition to chaos appears on the periodic state. At stress σ_2, the stationary state is no longer stable and a chaotic state is observed. If the applied stress is decreased from this aperiodic state, hysteretic behaviour should be present. Such behaviour has not been observed; this means that $||\sigma_2-\sigma_1|| << \delta\sigma$.

Stochastic resonance

Figure 2c shows typical dynamics observed approaching the aperiodic state displayed in Fig. 2b. It corresponds to a noisy state with small oscillations. In the vicinity of the out-of-equilibrium transition (the layering transition),

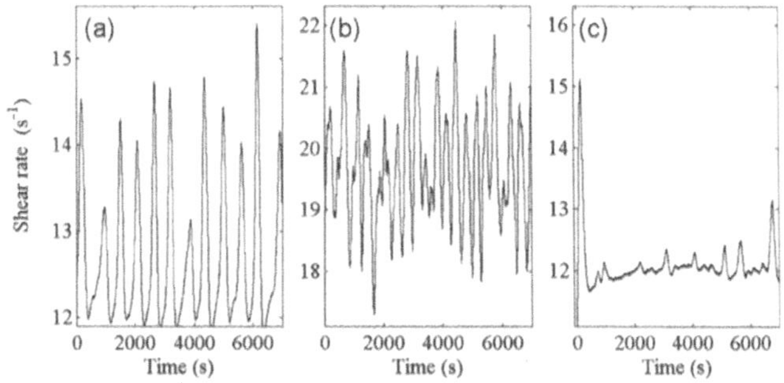

Fig. 2 Different dynamical states in the vicinity of the layering transition: **a** aperiodic state at point *C*, **b** complex dynamics in branch *BE* and **c** stochastic resonance in branch *BC*

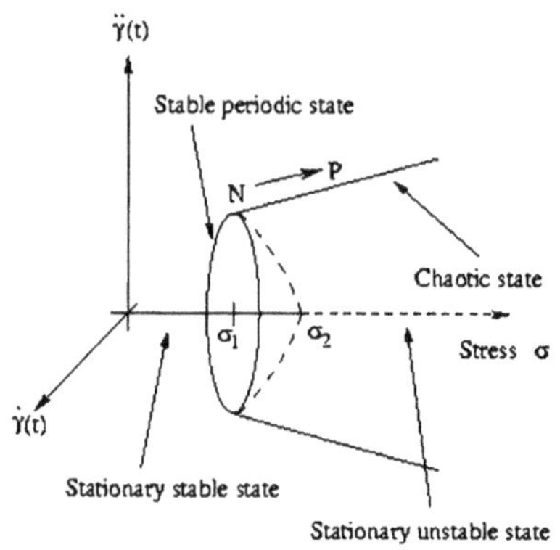

Fig. 3 Subcritical Hopf bifurcation; between *N* and *P* transition to chaos occurs

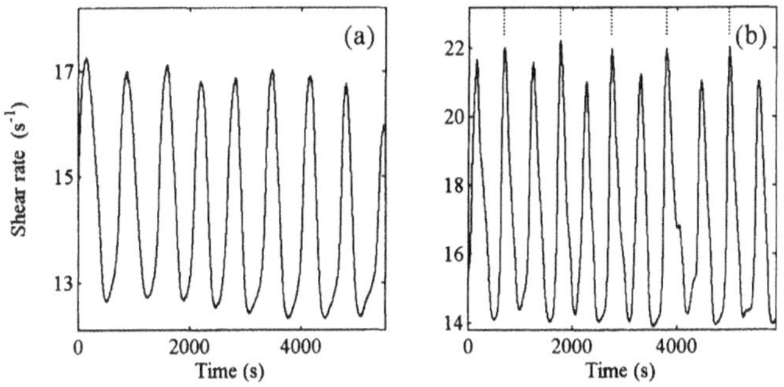

Fig. 4 Different transient dynamics observed in the vicinity of the layering transition: **a** periodic state of the shear rate ($T \approx 500$ s); **b** periodic state with period $2T$, *dotted lines* indicate this period

noise appears and reveals the presence of a periodic state. Such a phenomenon is called stochastic resonance and is characteristic of the presence of a Hopf bifurcation [10].

Period doubling: a possible transition to chaos?

The transient states displayed in Fig. 4 were recorded in the vicinity of the layering transition near point B on the branch BG. Figure 4a corresponds to a transient periodic state with a period of 500 s and Fig. 4b corresponds to a transient periodic state of period $2T$. Such observations strongly suggest the presence of a Hopf bifurcation and a transition to chaos via a period doubling scenario [9].

Conclusion

The rheology of the lamellar phase under study presents a lot of dynamical regimes. Using basic tools of dynamical systems theory, we have evidenced the presence of a subcritical Hopf bifurcation leading to an aperiodic state. However, a more quantitative study has revealed that the aperiodic state shown in Fig. 2b does not strictly correspond to simple three-dimensional chaos [5]. This is probably due to a coupling with spatial effects. In the present system, rheological chaos has not been fully demonstrated despite strong indications of a Hopf bifurcation and a transition to chaos via a period doubling transition.

References

1. Cates ME, Evans MR (eds) (2000) Soft and fragile matter: non equilibrium dynamics metastability and flow. Institute of Physics, Bristol
2. McLeish T (ed) (1997) Theoretical challenges in the dynamics of complex fluids. Kluwer, Dordrecht
3. Bandyopadhyay R, Basappa G, Sood AK (2000) Phys Rev Lett 84:2022
4. Wunenburger A-S, Colin A, Leng J, Arnéodo A, Roux D (2001) Phys Rev Lett 86:1374
5. Salmon J-B, Colin A, Roux D (2002) Phys Rev E 66:031505
6. Cates ME, Head DA, Ajdari A (2002) Phys Rev E 66:025202
7. Herve P, Roux D, Bellocq A-M, Nallet F, Gulik-Krzywicki T (1993) J Phys II 3:1255
8. Diat O, Roux D, Nallet F (1995) Phys Rev E 51:3296
9. Guckenheimer J, Holmes P (1984) Nonlinear oscillations, dynamical systems, and bifurcations of vector fields. Springer, Berlin Heidelberg New York
10. Gang H, Ditzinger T, Ning CZ, Haken H (1993) Phys Rev Lett 71:807

Progr Colloid Polym Sci (2004) 126: 124–129
DOI 10.1007/b94023

J. Oberdisse
F. Boué

Rheology–structure relationship of a model nanocomposite material

J. Oberdisse (✉) · F. Boué
Laboratoire Léon Brillouin,
CEA/CNRS, CEA Saclay,
91191 Gif sur Yvette, France
e-mail: oberdisse@cea.fr

Abstract We report the structure and rheological properties of solvent-cast silica-filled latex films, completing the data presented in previous work. The structure of the filled latex films has been measured by small-angle neutron scattering. The scattered intensity varies considerably with the physicochemical parameters, indicating important structural modifications. Our analysis allows us to quantify the degree of aggregation of the silica in the matrix, which is studied as a function of pH, silica volume fraction and bead type. It is found that the pH has a dominant effect on the structure of the samples, unlike the silica volume fraction. Rheological properties of the samples are tested by means of uniaxial strain experiments and expressed in terms of a strain-dependent reinforcement factor. We show that the degree of aggregation of the filler particles is correlated with the low deformation response of the films, and that the rheological properties of the resulting material can thereby be tuned over a large range.

Keywords Colloid · Latex · Small-angle neutron scattering · Composite

Introduction

Elastomers filled with small and hard particles are of great importance for the rubber industry, where carbon black and silica are commonly used fillers. These improve the mechanical properties of polymeric material like the elastic modulus or resistance to abrasion [1]. In this article, we present some elements of a study of the relationship between the filler structure and the rheological properties of a model nanocomposite material. The data presented here are intended to complete the picture of the structure [2] and rheological properties [3] of our model system [4, 5]. It consists of (hard) nanosilica beads embedded in a soft polymeric matrix, which is itself formed from nanolatex particles. Our system is solvent-cast, i.e. it is controlled by physicochemical manipulations in solution. The sensitivity to pH (and also electrolyte concentration) is due to the silanol and carboxyl groups on the surface of the beads. The repulsion between particles in solution can therefore be controlled through the pH or the electrolyte concentration. The advantages of our system are that the silica–latex surface interactions are always the same, independent of the final structure of the filler in the matrix, and that the constituents, latex and silica particles, can be studied individually. The latex forms a continuous and homogeneous matrix, and we treat the silica–latex films essentially as "frozen colloids" of silica in a matrix, and discuss their structure accordingly. Note that the nonergodic nature of these colloids does not influence the structure measured in a (static) small-angle neutron scattering (SANS) experiment, which averages over macroscopic parts of the samples.

Experimental

The starting components were aqueous colloidal suspensions of silica from Akzo Nobel (two sizes: Bindzil 30/220, hereafter called

B30, and Bindzil 40/130, hereafter called B40) and nanolatex polymer beads (Rhodia). It is a core–shell latex of poly(methyl methacrylate) (PMMA) and poly(butylacrylate) (PBuA), with a hydrophilic shell containing methacrylic acid. Colloidal stock solutions of silica and nanolatex were brought to desired concentration and pH, mixed, and degassed under a primary vacuum in order to avoid bubble formation. Nanocomposites were formed by evaporating the solvent at 65 °C. SANS experiments were performed at LLB on beamline PACE and at ILL on beamline D11, using standard configurations [2]. Data treatment was done with a self-written program following standard procedures, with H_2O as a calibration standard. Samples for stress–strain isotherms were brought to constant thickness using sandpaper. They were stretched up to rupture in a controlled constant-rate deformation. The force, $F(\lambda)$, where $\lambda = L/L_0$ denotes the elongation with respect to the initial length, L_0, was measured with an HBM Q11 force transducer, and converted to the (real) stress inside the material, σ. The film was assumed to deform homogeneously, and to be incompressible.

Results

Characterization of colloidal systems

The intensity scattered by colloidal solutions of B30 and B40 at 1% in D_2O (containing about 2% of H_2O, the initial solvent) is shown in Fig. 1. This is to illustrate that these constituents are well-defined colloidal particles, with comparable radii of gyration. A more detailed analysis of the form factors [2] gives the following parameters (R_0, σ) of a log–normal size distribution: $R_0 = 76.9$ Å, $\sigma = 0.186$ (B30); $R_0 = 92.6$ Å, $\sigma = 0.279$ (B40); $R_0 = 138.9$ Å, $\sigma = 0.243$ (nanolatex). In Fig. 1, the intensity of a 10% solution of B40 is also plotted. Owing to the electrostatic repulsion between silica beads in aqueous solution, the beads are expected not to be aggregated. Indeed, the spectrum displays a peak with a nice structure around 1.8×10^{-2} Å^{-1}, which indicates good dispersion. One can calculate the average aggregation number from the peak position, q_0, of the structure factor by means of a simple cubic-structure model:

$$N_{agg} = \Phi_{silica}(2\pi/q_0)^3/V_{si}. \tag{1}$$

In Eq. (1), Φ_{silica} denotes the volume fraction of silica, and V_{si} is the volume of an average silica bead, determined previously (for B40 $V_{si} = 4.72 \times 10^6$ Å^3). We find $N_{agg} \approx 1$.

Structure of latex–silica films

We now turn to the structure of nanocomposite films as seen by SANS. The intensity scattered by latex–silica films made with a nanolatex and the bigger silica beads (B40) is shown in Fig. 2. The pH of the solution before solvent evaporation was set to 7.5. The silica volume fraction was set to 6, 9 and 15%. It can be seen from the almost identical shape of the curves that the structure does not change much as the silica volume fraction increases.

All samples show the same high-q behaviour, which is of course due to the identical silica beads present in the film. A small-q increase is also present in all spectra, which indicates that there is some aggregation on larger scales, apparently independent of volume fraction. Finally, all three intensities display a structure factor peak around 0.006–0.007 Å^{-1}. Assuming a simple cubic model (Eq. 1) for the structure, the average aggregation number, N_{agg}, can be deduced; it is given in Table 1. The increase in N_{agg} is relatively weak, in the sense that it is less strong than the increase in volume fraction. In other words, as the volume fraction increases, aggregates increase both in size and in number. This is different from the situation at pH 9 (with the same silica B40, discussed in Ref. [2]), where the aggregation number is

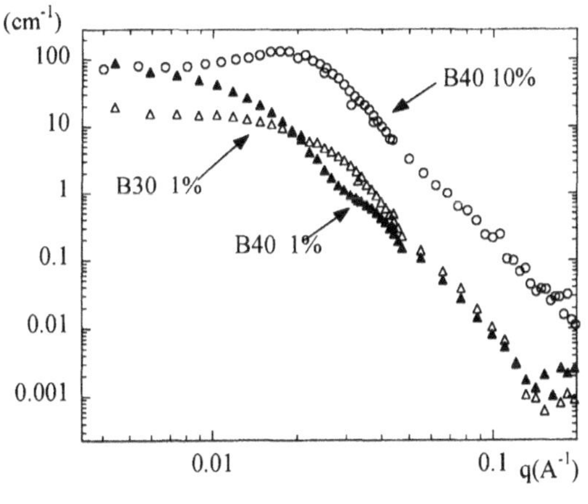

Fig. 1 Scattered intensity, I, as a function of wavevector, q, for colloidal solutions of silica (Bindzil 30/220, *B30*, and Bindzil 40/130, *B40*) in D_2O at 1% and for B40 at 10% (H_2O)

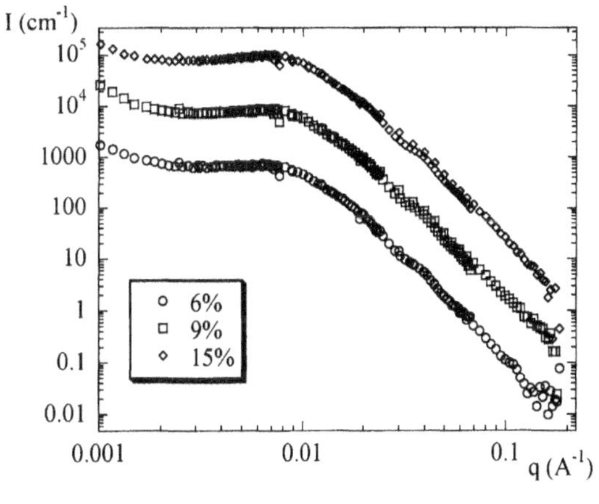

Fig. 2 I as a function of q for nanocomposites (pH 7.5, $\Phi_{silica} = 6$, 9, and 15%). For clarity, the intensity is multiplied by 10 (or 100) for the sample at 9% (or 15%)

Table 1 Average aggregation number of the system (Bindzil 40/130, pH 7.5, $\Phi_{silica} = 6, 9, 15\%$)

Φ_{silica} (%)	N_{agg}
6	≈12
9	≈17
15	≈22

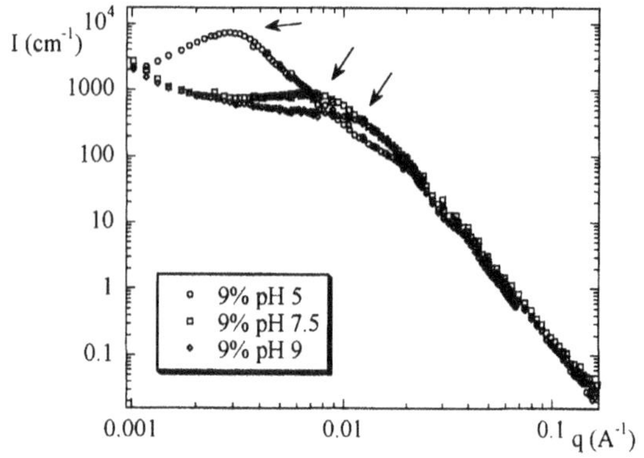

Fig. 3 *I* as a function of *q* for nanocomposites made from solutions at pH 5, 7.5 and 9 (B40, $\Phi_{silica} = 9\%$). The peak position is indicated by the *arrows*

essentially fixed (or even decreases with volume fraction). The effect of pH is illustrated in Fig. 3, where we compare spectra of B40–latex nanocomposites at identical volume fraction ($\Phi_{silica} = 9\%$) and pH 5, 7.5 and 9.

The large increase in the intensity for the low-pH samples can clearly be seen. The intensity at the maximum ($q \approx 0.008$ Å^{-1}) at intermediate pH, for example, is about 7 times smaller than the corresponding intensity at acidic pH. The structure factor peak is thus more and more pronounced as the pH decreases. From its position, the aggregation number is found to increase considerably, from about 1 (pH 9) to 16 (pH 7.5) and finally to about 200 (pH 5). Naively, one might be surprised that the intensity does not increase as much as the average aggregation number. This is due to the enhanced decrease with *q* of the form factor of bigger aggregates, as shown by numerical simulation [2].

As can be seen in these examples, the structure of silica–latex nanocomposites is essentially determined by the pH of the solution before evaporation. At a fixed silica volume fraction, $\Phi_{silica,}$ low pH leads to large aggregates. Φ_{silica} is found to have little influence on the aggregation number, or at least considerably less than the pH. To further illustrate the importance of the pH in a different system, we now turn to samples made with the smaller silica beads, B30. A series of samples at high silica volume fraction, $\Phi_{silica} = 15\%$, were prepared. The results of the SANS experiments are shown in Fig. 4, where we plot the scattered intensities from nanocomposite films made with B30, at different solution pH (only three spectra are shown for clarity). At low wavevectors,

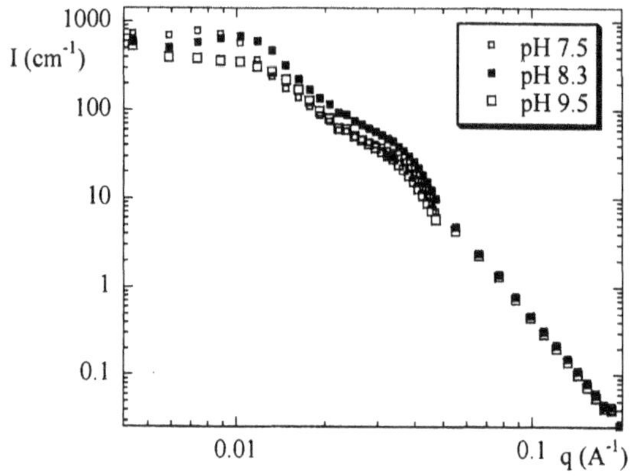

Fig. 4 *I* as a function of *q* for nanocomposites containing 15% of B30. The pH varies from 7.5 to 9.5 as indicated

Table 2 Average aggregation number (Bindzil 30/220, $\Phi_{silica} = 15\%$, pH 7.5–9.5)

pH	N_{agg}
7.5	42 ± 10
7.8	25 ± 5
8.3	12 ± 3
9	≈10
9.5	≈10

the intensity decreases as the pH increases. In parallel, the structure factor peak shifts to higher wavevectors. The average aggregation number deduced from the peak (or shoulder) position according to Eq. (1) is given in Table 2.

In previous work, we also measured spectra at a lower silica volume fraction of 5%. The pH-dependence of the aggregation number which we extracted from the scattering patterns is shown in Fig. 5. We have superimposed the data from Table 2 corresponding to the higher silica volume fraction of 15%. The data are found to fit nicely into the 5% set, with slightly higher aggregation numbers. In the inset, a zoom of the low aggregation number tail is shown.

Strain-dependent reinforcement factor

We have measured the stress–strain isotherms of pure nanolatex films [3], and the main results are summarized here. The rheological properties of these films are found to depend strongly on pH. In order to determine the reinforcement due to the silica structures only, we have introduced a different way of presenting the data. It consists in dividing the stress, $\sigma(\lambda)$, of the nanocomposite by the stress, $\sigma_{latex}(\lambda)$, of the nanolatex film made at the same pH. The resulting reinforcement factor is strain-dependent, and it tells us how much stronger the

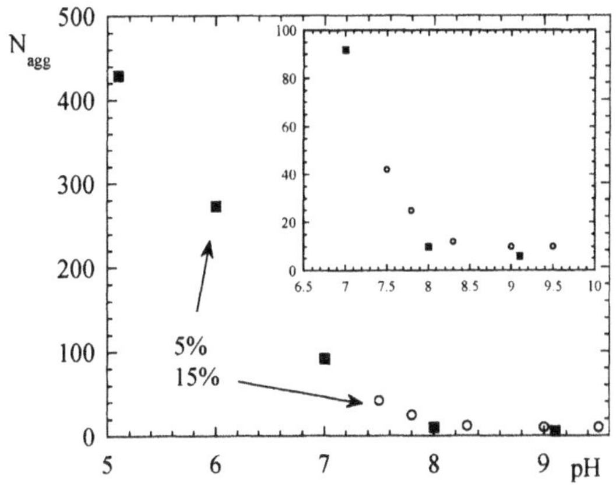

Fig. 5 Average aggregation number, N_{agg}, as a function of pH (B30, $\Phi_{silica} = 5$ and 15%). In the *inset* a zoom of the low-N_{agg} region is shown

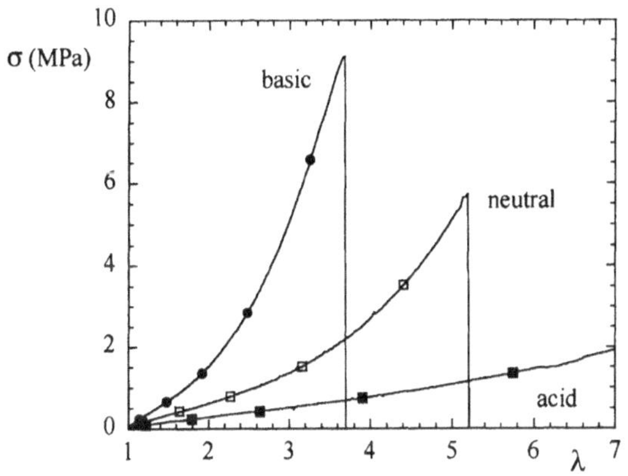

Fig. 6 Stress–strain isotherm of a pure nanolatex made from solutions at different pH

nanocomposite is with respect to the pure latex film. This is of special importance in our system, because the nanolatex itself has pH-dependent properties. An example of stress–strain isotherms of a nanolatex film at acidic, neutral and basic pH in solution before film formation is shown in Fig. 6.

The reinforcement factor, $\sigma(\lambda)/\sigma_{latex}(\lambda)$, of the system (B30, $\Phi_{silica} = 5\%$) is plotted in Fig. 7. The strong increase in reinforcement at low pH is clearly visible. At larger strain, the reinforcement decreases considerably, and seems to reach a plateau value at very large strain. This indicates that (1) some filler structure is probably destroyed during the deformation, and (2) that the changes in reinforcement due to different silica–polymer interactions at different pH are negligible with respect to the changes in reinforcement due to the structure.

Concerning point 1 of the preceding paragraph, we tried to parameterize the strong decrease in the reinforcement factor as a function of strain in terms of a power-law behaviour [6]. To this purpose, we subtracted

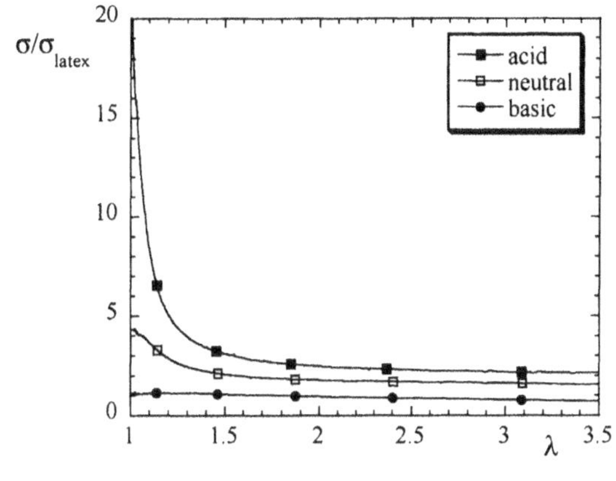

Fig. 7 Reinforcement factor, $\sigma(\lambda)/\sigma_{latex}(\lambda)$, of the system (B30, $\Phi_{silica} = 5\%$) for different pH values

the high-deformation plateau value in reduced stress $(\sigma/\sigma_{latex})_0$. It is then found that the stress is well described as a sum of two power laws in λ:

$$\sigma(\lambda)/\sigma_{latex}(\lambda) - (\sigma/\sigma_{latex})_0 = A\lambda^{\alpha} + B\lambda^{\beta}. \qquad (2)$$

Note that the procedure is empirical, and that the plateau value, different for each pH, is chosen in order to obtain a power-law behaviour. The results are that the quick decay corresponds to values of α between 9 and 11, the slow decay corresponds to values of β around 2.4, and that the ratio A/B, which characterizes the relative importance of the fast and the slow decrease in reinforcement, decreases from about 3.5 to 2 when increasing the pH from 4 to 7.

For the time being the parameterization (Eq. 2) of the data is purely descriptive. Its aim is to find general laws describing the evolution of reinforcement with strain. It is hoped that the strong decrease expressed by a high-power α (around 10!) can be correlated with the evolution of the structure under strain [6].

Another parameter of special interest is Young's modulus, E. The small deformation reinforcement factor can be estimated in a first approximation by the classical Einstein equation [7, 8]:

$$E/E_{latex} = 1 + 2.5\Phi_{silica} + \ldots \qquad (3)$$

The reinforcement factor of Young's modulus, E/E_{latex}, is plotted in Fig. 8 as a function of pH for the same samples as in Fig. 5 ($\Phi_{silica} = 5\%$). Note that the volume fraction is identical for all samples, and Eq. (3) would predict a constant reinforcement. The experimental result, however, shows a large increase in reinforcement as the pH decreases.

This discrepancy is of course due to the limitations of the Einstein equation, which works well for dilute systems of individual beads. As can be seen in Fig. 5, there is aggregation in this system, and the reinforcement factor, Fig. 8, seems to be well correlated with the

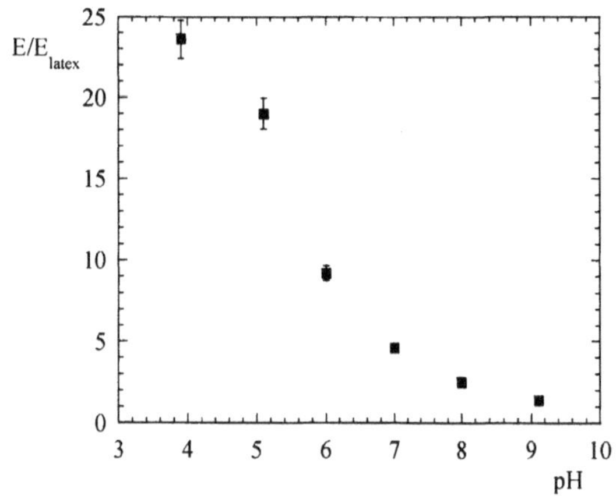

Fig. 8 Reinforcement factor, E/E_{latex}, of the system (B30, $\Phi_{\mathrm{silica}} = 5\%$) for different pH values

aggregation number. Note that several alternative equations for the reinforcement factor as a function of filler volume fraction exist in the literature [9]. A first possibility is to take the aggregate volume fraction, which includes contributions of bound rubber instead of the silica volume fraction [10]. Then a second-order term is sometimes added to Eq. (3), in order to account for pairwise interactions [11, 12]. Finally, a percolation term may be used to describe a divergence of the modulus with increasing filler volume fraction [9, 13]. Let us mention that other, less empirical, equations exist, which usually stem from some theoretical description of the filler network [14]. All these points are discussed in Ref. [3], where we analysed two series in volume fraction at fixed pH.

Discussion and conclusion

In the first part, we presented experimental results on the structure of model nanocomposite samples, as seen by SANS. From the intensity scattered by dilute solutions of silica beads, we know that the individual components come from a well-defined, nonaggregated colloidal system, and that their sizes are in a range interesting for reinforcement applications. The analysis of the structure of the nanocomposite films along the different lines in the phase diagram demonstrates the strong dependence of N_{agg} on the pH, and its rather weak dependence on Φ_{silica}. This can be seen nicely in Fig. 5, where results from different pH and different silica volume fractions are superimposed.

Another interesting point is the influence of the silica bead size. The smaller silica beads, B30, are found to aggregate more than the bigger, B40, beads. Initially, no such comparison was planned, but in some cases samples with almost identical concentrations were prepared. This is the case for the samples with $\Phi_{\mathrm{silica}} = 15\%$ at pH 7.5: the aggregation number is about twice as high in the case of B30 ($N_{\mathrm{agg}} = 42$) than in the case of B40 ($N_{\mathrm{agg}} = 22$). The same differences are observed at low pH [2]: at pH 5, the 5% sample of B30 has an average aggregation number of about 430, whereas we find only $N_{\mathrm{agg}} \approx 170$ for the closest B40 sample (6%). As a last example, also taken from Ref. [2], comparison of a sample series in silica volume fraction at pH 9 demonstrates that the average number of aggregation is very small for B40 (between 1 and 2, even slightly decreasing with Φ_{silica}), and about 10 for B30. Note that all these observations are mutually consistent, but we still lack a satisfying explanation for this behaviour.

The rheological properties of the nanocomposite films were tested by means of stress–strain isotherms. The stress of the pure latex films being also pH-dependent, we presented the data in terms of the reduced stress. It is found that highly structured fillers (big aggregates) give a strong reinforcement effect in the low-deformation range. Young's modulus of the nanocomposites, for example, has a pH dependence which is closely correlated to the aggregation number as measured by SANS. At higher elongation, the reinforcement factor of "acid" samples, initially very high, weakens. Because smaller reinforcement seems to be correlated with smaller aggregation numbers, we associate this decrease with the destruction of the filler structure during elongation, thereby resulting in the formation of smaller aggregates. In a purely empirical way, we have tentatively described the observed decrease in reinforcement by power laws. The sum of a quick decrease ($\propto \lambda^{10}$) and a more moderate one ($\propto \lambda^{2.4}$) gives a very satisfying though empirical description of the data.

Acknowledgements We are indebted to Bruno Demé (ILL) who helped us with the experiment on D11, and to Akzo Nobel and Rhodia for the silica and the latex beads, respectively.

References

1. Mark JE, Erman B, Eirich FR (eds) (1994) Science and technology of rubber. Academic, San Diego
2. Oberdisse J Demé B (2002) Macromolecules 35:4397
3. Oberdisse J (2002) Macromolecules 35:9441
4. Rharbi Y, Joanicot M, Vacher A, Cabane B, Boué F (1999) Europhys Lett 46:472
5. Rharbi Y, Boué F, Joanicot M, Cabane B (1996) Macromolecules 29:4346
6. Oberdisse J, Boué F (2002) Proceedings of Groupe Français de Rhéologie St Etienne
7. Einstein A (1906) Ann Phys 19:289
8. Smallwood HM (1944) J Appl Phys 15:758–766
9. Ahmed S, Jones FR (1990) J Mater Sci 25:4933–4942
10. (a) Medalia AI (1974) Rubber Chem Technol 47:411; (b) Polmanteer KE, Lenz CW (1975) Rubber Chem Technol 48:795; (c) Vidal A, Donnet JB (1987) Prog Colloid Polym Sci 75:201
11. Guth E, Gold O (1938) Phys Rev 53:322
12. Guth E (1944) J Appl Phys 16:20
13. Mooney M (1951) J Colloid Sci 6:162
14. Witten TA, Rubinstein M, Colby RH (1993) J Phys II 3:367

Progr Colloid Polym Sci (2004) 126: 130–133
DOI 10.1007/b94008

Markus Plaschke
Jörg Rothe
Melissa A. Denecke

Characterization of Gorleben groundwater colloids by scanning transmission X-ray microscopy

M. Plaschke (✉) · J. Rothe
M.A. Denecke
Forschungszentrum Karlsruhe,
Institut für Nukleare Entsorgung,
Postfach 3640, 76021 Karlsruhe, Germany
e-mail: plaschke@ine.fzk.de
Tel: +49-7247-824747,
Fax: +49-7247-823927

Abstract The role of groundwater colloids influencing mobilization or immobilization of actinides in natural aquatic systems is addressed in the context of safety aspects of nuclear waste disposal. Groundwater colloids from the Gorleben site (Lower Saxony, northwest Germany) are characterized in the presence of Eu(III) as a homologue for trivalent actinides. Scanning transmission X-ray microscopy (STXM) measurements are performed with samples in contact with aqueous solution to ensure that colloids and their aggregates are in the native hydrated state. At least five regions with different C K near-edge X-ray absorption fine structure (NEXAFS) spectral features can be detected in the STXM micrographs of the Eu(III)-saturated humic aggregates. The C K NEXAFS measurements indicate segregation of carbon species into regions with different contents of carboxylic, aromatic, phenolic or aliphatic chemical functional groups.

Keywords Humic acid · Groundwater colloids · Aggregation · Scanning transmission X-ray microscopy

Introduction

Aggregated colloids in Gorleben groundwater are investigated by synchrotron-based scanning transmission X-ray microscopy (STXM). Employing Fresnel zone plates as focussing optics, STXM enables soft X-ray spectromicroscopy measurements, simultaneously providing in situ morphological and chemical-state information on aqueous suspensions of organic matter within the "water window" (i.e., between the C K and O K absorption edges at 284 and 537 eV, respectively). STXM transmission images are recorded by scanning samples through a focussed X-ray beam, allowing visualization of structures with dimensions ranging from submicron to several microns. By varying the energy of the incident radiation, the absorption contrast can be adjusted to reveal the spatial distribution of a selected element, as well as its chemical state from the near-edge X-ray absorption fine structure (NEXAFS) [1]. Differentiation of inorganic from organic components and visualization of their interaction is possible through comparison of STXM images recorded at photon energies above and below the C K absorption edge [2]. Additionally, C K NEXAFS in the soft X-ray regime provides a tool for direct speciation of carbon-containing macromolecules from characteristic spectral transition resonances [3]. Humic substances have already been the subject of X-ray microscopy investigations e [4, 5]. In previous STXM studies the aggregation of Aldrich humic acid (HA) induced by Eu(III) was accompanied by the segregation into zones with different optical densities [6].

Experimental

In this study a groundwater sample from the Gorleben aquifer (Lower Saxony, northwest Germany), designated as Gohy-2227, was investigated. The groundwater samples were previously characterized in detail, where an original pH of 7.7, an ionic strength of 44.0 mmol/l and a dissolved organic carbon content of 81.9 mg C/l was found [7]. Sample solutions were prepared by adjusting the original groundwater samples to a pH of 5.6 and adding the desired amount of a Eu(III) stock solution (ICP-AES standard, Johnson Matthey, Karlsruhe, Germany, diluted to 10^{-3} mol/l; the pH of the stock solution was adjusted to 4) to a final Eu(III) concentration of 6.5×10^{-5} mol/l. Slightly acidic pH conditions are appropriate to ensure that Eu(III) is available for complexation.

STXM investigations were performed at the National Synchrotron Light Source (NSLS), beamline X1A outboard-STXM, Brookhaven, USA [8]. All the samples were investigated in the fully hydrated state. Thin films of colloid suspensions were enclosed between silicon nitride membranes as described elsewhere [9]. Diffraction lenses (Fresnel zone plates) focussed the undulator beam at the X1A outboard-STXM into a submicron spot of soft X-rays. The zone plate used in this study had a diameter of 160 μm and a width of 45 nm of the outermost ring segment, giving a spatial resolution (Rayleigh criterion) of 55 nm [10]. For acquisition of transmission images the sample was scanned in the x/y-plane perpendicular to the beam direction. The STXM endstation provides a flux of about 10^7 photons/s, with an energy bandwidth of about 0.1 eV at the C K edge [11]. The spherical grating monochromator was calibrated against the C K absorption threshold resonance of carbon dioxide at 292.76 eV [12].

In the spectromicroscopy mode, stacks of images from selected sample regions were recorded as a function of the incident photon energy, E. C K NEXAFS spectra were extracted through the analysis of the variation of the transmitted photon flux at regions of interest. The absorption was calculated as $\mu(E)\times d=\ln[I_0(E)/I(E)]$, where $\mu(E)$ and d are the linear absorption coefficient and the sample thickness, respectively. $I_0(E)$ was extracted from image regions free of particles; the transmitted intensity, $I(E)$, was derived from regions containing carbon-rich structures. For a review of the stack data analysis procedure, see Ref. [13]. The C K NEXAFS spectra were obtained from a stack of images recorded in 1-eV steps at 275–280 eV, 0.1-eV steps at 280–292 eV, 0.5-eV steps at 292–300 eV and 1-eV steps at 300–305 eV. In order to enable quantitative comparison of the absorption measurements, a linear preedge background (275–280 eV) was subtracted from the spectra, followed by normalization (μd = 1) at 302 eV.

Results and discussion

The sizes of the humic colloids in the original groundwater taken from the Gorleben aquifer are reported to be in the range of a few nanometers [7, 19]. For characterization with STXM larger aggregates are needed. It is known from atomic force microscopy (AFM) studies that aggregation of humic substances is induced by multivalent cations like Eu(III) [19]. Addition of Eu(III) cations to the Gorleben groundwater Gohy-2227 induces the formation of carbon-rich aggregates with zones of strongly varying optical density. The morphologies observed in the STXM images in Fig. 1 are dominated by diffuse, rather weakly absorbing, "cloudy" aggregates with diameters of several microns. Dense, particulate structures (designated in the following as "nodules" and "sticks") with lateral dimensions down to the resolution limit of about 100 nm are embedded into these cloudy structures (Fig. 1a, b). The cloudy structures appear prior to the much denser nodules and sticks upon increasing the incident X-ray energy, while recording spectral image stacks. Figure 1c shows a rather large aggregate, where zones of different optical densities are interweaved. A segregation process is evident, where dark, dense patches of organic material dispersed in lighter, less dense regions with about one third the optical density of the denser material are observed. Both structures in the aggregate in Fig. 1c exhibit no

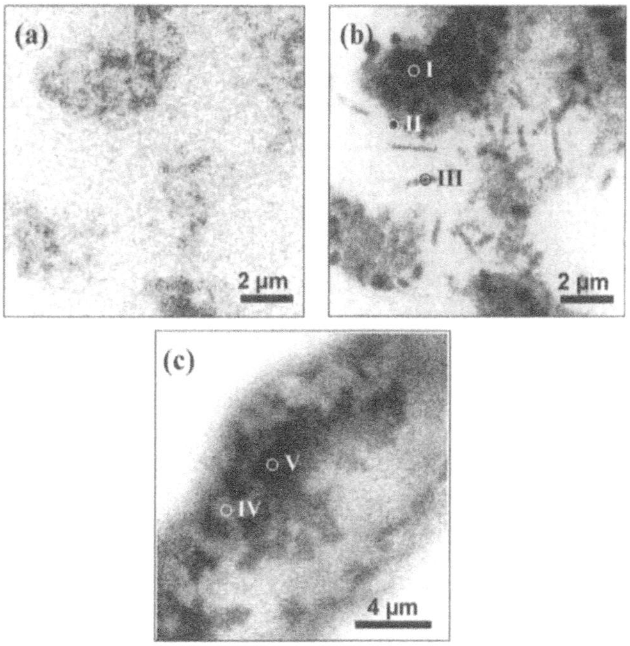

Fig. 1a, b Scanning transmission X-ray microscopy (*STXM*) images of Eu(III)-induced Gohy-2227 aggregates recorded at the energy of the aromatic (**a** 284.8 eV) and the carboxyl (**b** 288.2 eV) C K near-edge X-ray absorption fine structure (*NEXAFS*) resonances. **c** STXM image of a Eu(III)-induced Gohy-2227 aggregate with different morphology recorded at 284.8 eV. The *Roman numerals* mark different morphological structures chosen for the extraction of C K NEXAFS spectra in Fig. 2 (pH 5.6, Eu(III) 6.5×10^{-5} mol/l)

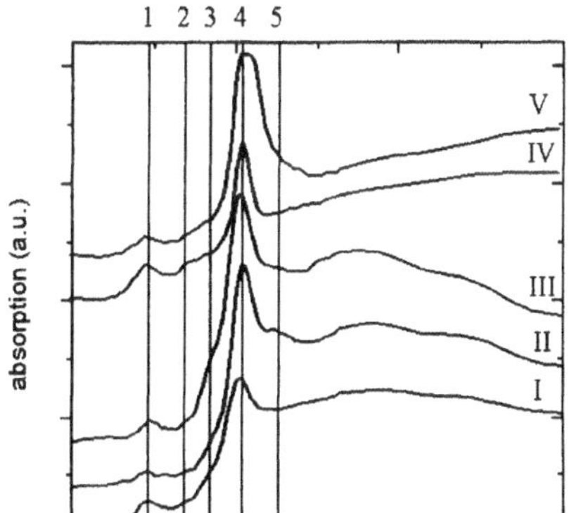

Fig. 2 C K NEXAFS spectra derived from the STXM stack images of Gohy-2227 aggregates extracted from regions marked in Fig. 1b and c with Roman numerals. The *numbers* refer to energy positions of transition bands assigned in Table 1 (pH 5.6, Eu(III) 6.5×10^{-5} mol/l)

Table 1 Mean energies and assignments of C $1s \rightarrow \pi^*$/Rydberg transitions of selected carbon groups as compiled in Ref. [3]. The *numbers* in *parentheses* refer to positions marked in Fig. 2

C–C aromatic (1)	C–OH phenol (2)	C–H aliphatic (3)	C=O (COOH) (4)	C–OH alcohol (5)
284.8 eV	287.1 eV	287.5 eV	288.5 eV	289.5 eV

particulate fine structure, although recorded with the same x/y step width of 100 nm as the image in Fig 1a and b. The normalized C K NEXAFS spectra extracted from different sample sites (designated by I–V) are compared in Fig. 2. All the morphologies in Fig. 1 exhibit distinct variations of the C K NEXAFS spectral signature. Peaks and shoulders at the rising edge and broad resonances at the crest of the C K NEXAFS spectra of these humic aggregates can be identified with electronic transitions to unoccupied molecular orbitals, i.e., C $1s \rightarrow \pi^*$/Rydberg states below and C $1s \rightarrow \sigma^*$ states above the ionization threshold. On the basis of spectra of well-characterized organic compounds reported in the literature [3, 14–16], the assignment of the spectral features to carbon functional groups is possible (Table 1).

The strong resonance, observed at 284.8 eV, indicates that the cloudy zones (I in Fig. 1b) are formed by humic macromolecules with a high content of aromatic carbon groups (energy position 1 in Fig. 2). The spectrum extracted from the image stack is similar to that observed for the dense fraction of the Aldrich HA aggregates reported in Ref. [6]. In contrast to this, the C K NEXAFS spectra indicate that the particulate structures (II and III in Fig. 1b) are dominated by carbon species rich in carboxyl functional groups (energy position 4 in Fig. 2), with only minor contributions from aromatic carbon. Additionally, these spectra exhibit a shoulder at about 289.5 eV typical for C–OH alcohol groups (energy position 5 in Fig. 2), followed by markedly pronounced features in the region of the σ^* resonances (at around 292 eV). The further shoulder at around 287.5 eV (energy position 3 in Fig. 2), pronounced in the particulate aggregates (II and III in Fig. 1b), is typically assigned to aliphatic carbon [3, 17]. Similarly, this energy band is enhanced in the dense zones of Eu(III)-induced Aldrich HA aggregates compared to uncomplexed HA [18]. Since aliphatic groups are noncomplexing, the pronouncedness of this feature may be indicative of a functional group involved in Eu(III)/HA complex formation (e.g., a shifted band of carboxylic or phenolic carbon). The light zones (IV) in Fig. 1c are characterized by more intense absorption in the spectral region assigned to aromatic and phenolic groups (energy positions 1 and 2 in Fig. 2), followed by a relatively sharp carboxyl resonance (energy position 4 in Fig. 2). In contrast, the darker zones (V) in Fig. 1c exhibit less absorption in the aromatic and phenolic region, but a remarkably intense and broad resonance in the region assigned to carboxylic groups (centered around 288.5 eV).

These results show that the organic matter in Gorleben groundwater is characterized as a heterogeneous dispersion of species with markedly distinct C K NEXAFS spectral features pointing to the chemical functional groups involved in the Eu(III)-induced aggregation. The observed segregation into zones with different optical densities and morphologies is complementary to recent AFM results, where different types of morphologies are found, for example, spherical particles and aggregates, fibrous structures and structures appearing to be hollow [19]. The segregation process as depicted in the STXM micrographs, combined with the C K NEXAFS-spectra, supports a new concept of humic substances on the basis of recent NMR results [20]. According to this concept, humic substances can be described as a complex mixture of plant components at various stages of humification. They may be associations of degradation products of, for example, polysaccharides, polypeptides or aromatic lignin fragments, varying in chemical composition and microstructure. A more detailed description of the chemical characteristics of these species and their cation complexation properties will be obtained by focussing on the comparison of the C K NEXAFS-spectra of the segregated regions with model compounds containing corresponding chemical functional groups.

Acknowledgements We are grateful for beam-time allotment by BNL/NSLS. All data taken using the X-1A STXM developed by the group of Janos Kirz and Chris Jacobsen at SUNY Stony Brook, with support from the Office of Biological and Environmental Research, U.S. DoE under contract DE-FG02-89ER60858, and the NSF under grant DBI-9605045.

References

1. Rothe J, Kneedler EM, Pecher K, Tonner BP, Nealson KH, Grundl T, Meyer-Ilse W, Warwick T (1999) J Synchrotron Rad 6:359
2. Rothe J, Denecke MA, Dardenne K (2000) J Colloid Interface Sci 231:91
3. Scheinost AC, Kretschmar R, Christl I, Jacobsen C (2001) Spec Publ R Soc Chem 273:39
4. Myneui SCB, Brown JT, Martinez GA, Meyer-Ilse W (1999) Science 12:1335
5. Schmidt C, Thieme J, Neuhäusler U, Schulte-Elbert U, Abbt-Braun G, Specht C, Jacobsen C (2000) Association of particles and structures in the presence of organic matter, Proceedings of the 6th international conference on X-ray microscopy, Berkeley. American Institute of Physics, Woodbury, NY, p 313
6. Plaschke M, Rothe J, Schäfer T, Denecke MA, Dardenne K, Pompe S, Heise K-H (2002) Colloids Surf A 197:245
7. Artinger R, Kienzler B, Schüßler W, Kim JI () In: Buckau G (ed) Effects of humic substances on the migration of radionuclides: complexation and transport of actinides. Wissenschaftliche Berichte, FZKA 6124. Forschungszentrum Karlsruhe, Karlsruhe, p 23
8. Jacobsen C, Williams S, Anderson E, Browne MT, Buckley CJ, Kern D, Kirz J, Rivers M, Zhang X (1991) Opt Commun 86:351
9. Neuhäusler U, Abend S, Jacobsen C, Lagaly G (1999) Colloid Polym Sci 277:719
10. Spector S, Jacobsen C, Tennant D (1998) In: Thieme J, Schmahl G, Umbach E, Rudolph D (eds) X-ray microscopy and spectromicroscopy. Springer, Berlin Heidelberg New York, pp
11. Winn B, Ade H, Buckley C, Howells M, Hulbert S, Jacobsen C, Kirz J, Mc Nulty I, Miao J, Oversluizen T, Pogorelsky I, Wirick S (1996) Rev Sci Instrum 67:1
12. Hitchcock AP, Mancini DC (1994) J Electron Spectrosc 67:1
13. Jacobsen C, Flynn G, Wirick S, Zimba C (2000) J Microsc 197:173
14. Ishii I, Hitchcock AP (1988) J Electron Spectrosc Relat Phenom 46:55
15. Francis JT, Hitchcock AP (1992) J Phys Chem 96:6598
16. Ade H, Winesett DA, Smith AP, Qu S, Ge S, Sokolov J, Rafailovich M (1999) Europhys Lett 45:526
17. Cody GD, Ade H, Wirick S, Mitchell GD, Davis A (1998) Org Geochem 28:441
18. Rothe J, Plaschke M, Denecke MA (2003) J Phys IV 104:421
19. Plaschke M, Römer J, Kim JI (2002) Environ Sci Technol 36:4483
20. Simpson AJ, Kingery WL, Hayes MHB, Spraul M, Humpher E, Dvortsak P, Kerssebaum R, Godejohann M, Hofmann M (2002) Naturwissenschaften 89:84

Progr Colloid Polym Sci (2004) 126: 134–138
DOI 10.1007/b94024

Rossitza G. Alargova
Kaoru Tsujii

Behavior of colloids in supercritical water: an attempt to study diffusion coefficients using dynamic light scattering

R.G. Alargova · K. Tsujii
The DEEPSTAR Group,
Japan Marine Science and Technology
Center, 2–15 Natsushima-cho,
Yokosuka 237-0061, Japan

Present address: R.G. Alargova
Dept. Chem. Eng., NCSU,
Raleigh, NC 27606, US
e-mail: rgalargo@ncsu.edu
Tel.: +1-919-513-4648
Fax: +1-919-515-3465

Present address: K. Tsujii
Nanotechnology Research Center, RIES,
Hokkaido University
N21, W10 Kita-ku
Sapporo 001-0021, Japan

Abstract This paper deals with studying the diffusion coefficients of polystyrene and colloidal gold particles dispersed in water above its critical point (374.0 °C and 22.064 MPa) using dynamic light scattering (DLS). The possible contribution from pure water to the total intensity of the scattered light and to the intensity autocorrelation function (ACF) measured by DLS has been examined at the conditions of particle size measurements. The results show that in the case of colloidal dispersions the ACF reflects mainly the particle diffusion even at thermodynamic states characterized by long-range density fluctuations of the pure solvent because the dynamics of the latter is much slower. In addition the diffusion coefficient of the samples has been studied as a function of temperature and/or pressure in the range 24–374 °C. The experimental results collected so far are promising for applying DLS to probing colloidal behavior in hot water under high pressure including supercritical area.

Keywords Supercritical water · Colloid stability · Light scattering · High pressure · High temperature

Introduction

Water in near-critical and supercritical (SC) regions of the phase diagram is an interesting medium for colloid science since its properties, such as polarity and viscosity, can be easily varied by moderate changes in pressure and/or temperature [1]. For instance the density of SC water along an isotherm can be tuned in the whole interval from gaslike to liquidlike by an increase in pressure. Hence values of the dielectric constant ranging from nonpolar to polar ones are readily achieved and SC water presents a great opportunity to study the stability of colloids as a function of medium properties without any additives. Of particular interest is the intermediate density region, especially the one close to the critical point, where unusual solvating effects have been experimentally observed or predicted by computer simulations in the case of small solute molecules [2, 3, 4, 5]. The water compressibility in this area acquires very large values, leading to significant density inhomogeneities. In SC fluids the long-range density fluctuations are particularly pronounced close to the critical point, but recently it has been shown that they may extend well beyond it into the SC region [6, 7]. Working with inhomogeneous SC water is very difficult because small changes in the thermodynamic parameters provoke large changes of the medium properties and a precise control of pressure and temperature is needed. However, at the same time it is the most attractive one for studying colloids since the expected new solvating effects would influence greatly their stability.

In this paper we describe our attempts to apply dynamic light scattering (DLS) to probe the behavior of colloids in SC water. DLS turns out to be a convenient method for that purpose, since it provides direct information about the diffusion coefficients of the dispersed particles and at the same time it can give some insight into the interparticle interactions. Our study includes investigating the scattering of light by pure water and performing experiments with colloids, known to be very stable at normal conditions.

Experimental

DLS measurements were carried out using an experimental system designed to perform measurements under high pressure and temperature, which has already been described elsewhere [8]. It consists of a custom-made DLS-820 photometer (Otsuka Electronics Co., Osaka, Japan) equipped with a He–Ne (10 mW, 632.8 nm) and a solid-state laser (50 mW, 532 nm) and a high-temperature and high-pressure cell connected to a flow-type pressurizing system (AKICO Co., Tokyo, Japan). The design of the cell allows intensity and DLS measurements at 90° in the range 25–450 °C and 0.1–40 MPa, i.e. including a region above the critical point of water (374.0 °C, 22.064 MPa).

Prior to the experiments the entire system was thoroughly washed and filled with deionized water (Millipore, USA) or a colloidal dispersion. Then the pressure was increased up to a given value, the flow through the cell was terminated, and the heating was started. After equilibrating the cell at a given temperature and pressure, intensity or DLS measurements were performed. The pressure and the temperature of the cell were maintained with an accuracy of ± 0.025 MPa and ± 0.1 °C during the experiments.

In DLS measurements the intensity autocorrelation function (ACF), $g^{(2)}(\tau)$, was primarily recorded (homodyne regime) and the corresponding electric field ACF, $g^{(1)}(\tau)$, was calculated and used to determine the diffusion coefficient, D, of the colloidal particles by means of cummulant or exponential sampling method [10]. The Stokes–Einstein formula, $D = k_B T/(3\pi\eta d_H)$, where $k_B T$ is the thermal energy and η is the viscosity of water) was applied to calculate the particle hydrodynamic diameter, d_H. The water properties, such as viscosity, refractive index, and density, were calculated using the method in Ref. [9]. We present the results for a polystyrene sulfate latex (PSL), $d_H = 50$ nm (Dow Chemical Co.), and colloidal gold (CG), $d_H = 60$ nm (Polysciences). The latex sample was dialyzed against water to remove the initially contained NaN_3. Both samples were diluted with Millipore water prior to the experiments.

Results and discussion

Light scattering from pure water

The study of pure SC water was necessary because the long-range density fluctuations may have a significant contribution to the scattered light intensity as well as to its ACF recorded in DLS experiments. Firstly, we investigated the behavior of the average intensity of the light scattered by pure water, I_{SC}, as a function of pressure along a few isotherms. Up to near-critical region I_{SC} measured for liquid water is negligibly small, but it becomes more sensitive towards changes in pressure and temperature and starts to increase rapidly when the critical point is approached. This effect is related to the presence of significant density fluctuations over distances comparable to the wavelength of the visible light leading to the appearance of critical opalescence around the critical point [1]. In SC water I_{SC} demonstrates interesting behavior as is seen in Fig. 1 showing data recorded slightly above the critical point at 376.0 °C. The solid line presents the corresponding values of the water density, ρ [9]. Within the range investigated I_{SC} has a maximum at 22.7 MPa situated on the right side of the critical pressure line. The pressure dependence of the intensity is most pronounced in the intermediate density region and the location of the I_{SC} peak occurs approximately at the critical density ($\rho_{cr} = 0.322$ g/cm^3), i.e. at the inflection point of the density curve. Therefore in this area the long-range density fluctuations are still quite significant and we deal with a highly inhomogeneous fluid. The curves recorded at higher temperatures (not presented here) show that the maximum diminishes and shifts to higher values of pressure (e.g. at 380 °C, the maximum is located at 23.7 MPa), but it exists even at 405 °C. The values of pressure and temperature corresponding to the intensity maximum form a ridge along the extension of the liquid–gas coexistence curve into the SC area when plotted on the water phase diagram. It marks the middle of a region where the fluid density changes rapidly from gaslike to liquidlike and the water behaves as a highly inhomogeneous medium. A more detailed study of scattering from pure water is described elsewhere [11]. In general the results suggest that the contribution from the scattering of pure SC water should be taken into consideration when DLS measurements of colloidal dispersions are carried out at the conditions of the intensity peak.

To check the possible influence of the long-range density fluctuations on the intensity ACF, we performed DLS measurements with pure SC water at thermodynamic states with and without a noticeable contribution from density inhomogeneities. The results show that a correlation of the signal is obtained only when working at temperatures close to the critical point and at pressures around the maximum in I_{SC}. Two intensity ACFs measured for water at 376.0 °C are presented in Fig. 2.

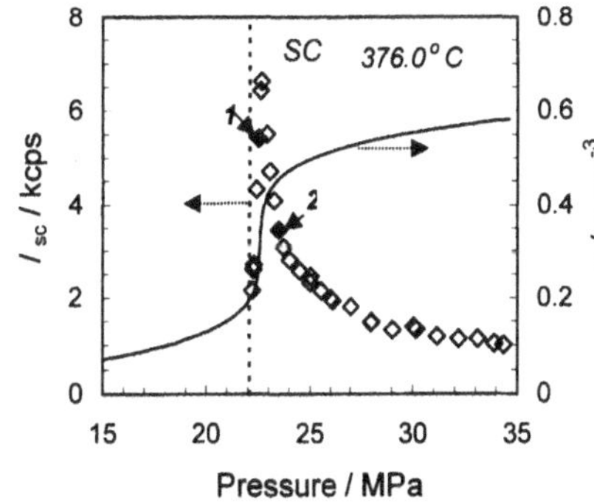

Fig. 1 Intensity of the light scattered by pure supercritical (*SC*) water, I_{sc}, (*symbols*) and water density, ρ, (*solid curve*) as a function of pressure at 376.0 °C. The *broken line* marks the critical pressure of 22.064 MPa. The *solid symbols 1* and *2* are the measurement conditions for curves 1 and 2 in Fig. 2

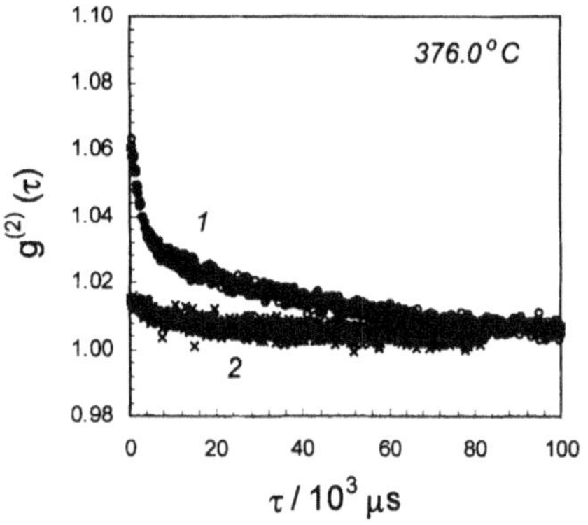

Fig. 2 Intensity autocorrelation function (ACF), $g^{(2)}(\tau)$, measured for pure SC water at 22.5 MPa (*curve 1*) and 23.5 MPa (*curve 2*). The temperature is 376.0 °C. Conditions for the two curves are the same as for points 1 and 2 in Fig. 1

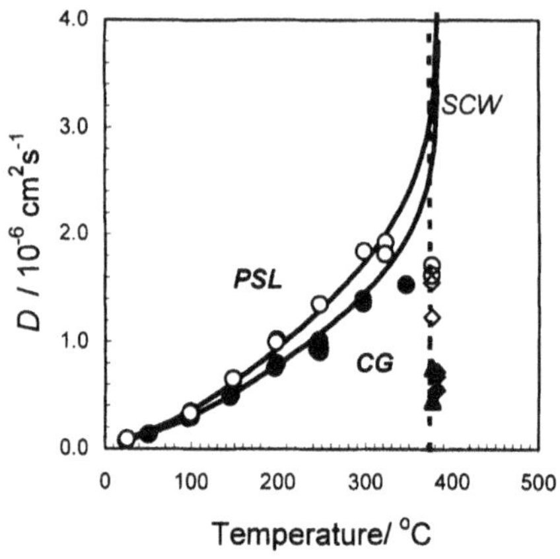

Fig. 3 Diffusion coefficient of a polystyrene sulfate latex (*PSL*) (*open symbols*) and colloidal gold (*CG*) (*solid symbols*) particles measured by dynamic light scattering as a function of temperature. The *solid lines* are theoretical curves for the diffusion coefficient of particles with constant size calculated at 25.1 MPa. The data marked by *circles* were measured at 25.1 MPa, the *solid triangles* correspond to 30.3 MPa, and the *diamonds* correspond to 35.0 MPa. The *broken line* presents the critical temperature (374.0 °C). The sample concentration is 3×10^{-4} g/cm^3 for PSL and 1×10^{-5} g/cm^3 for CG

Both curves show that the process contributing to the correlation of the signal has slow dynamics which is typical for systems close to criticality. Curve 1 was recorded close to the maximum of I_{SC} at 22.5 MPa and at a water density of 0.249 g/cm^3 $< \rho_{cr}$, where the long-range density fluctuations are well developed and the scattering of light by pure water is strong (Fig. 1, point 1). Two different parts of the intensity ACF can be distinguished: one decaying faster at small delay times and one with slower decay at larger τ. This shape of the ACF is probably determined by the contributions of single and multiple (double and higher-order) scattering coming from density fluctuations. Multiple scattering has the effect of introducing faster components to the decay process and therefore it should be responsible for the initial steeper part of $g^{(2)}(\tau)$. Curve 2 in Fig. 2 is obtained at higher density $\rho = 0.455$ g/cm^3 $> \rho_{cr}$ (23.5 MPa), but its conditions still fall into the area of the intensity peak (Fig. 1, point 2). It has a shape similar to that of curve 1, but it is noisier and the correlation is much less pronounced. When going farther from the intensity maximum by increasing the pressure or raising the temperature the correlation of the signal coming from pure SC water diminishes and disappears at a given point.

DLS experiments with colloidal particles

The colloidal behavior of PSL and CG particles was studied in the whole interval from room temperature to SC water. Figure 3 summarizes the experimental data for the diffusion coefficients measured as a function of temperature at a constant pressure of 25.1 MPa except for some points in the SC area measured at 30.3 and 35.0 MPa. Below the critical point the influence of pressure on the measured results was found to be negligibly small [8]. The solid curves present the temperature dependence of D calculated for particles with constant size assuming that the Stokes–Einstein formula is applicable in the whole range investigated. One sees that the experimental data are in good agreement with the theoretical predictions even at temperatures as high as 300 °C, which means that both dispersions remain colloidally stable (no significant changes in the particle size were observed during the measurement time) in a broad temperature range. However, in the range 300–360 °C D started to decrease with time when the sample was kept for a longer period, which indicated most probably the beginning of coagulation and resembled the behavior of a larger ($d_H = 206$ nm) polystyrene latex studied earlier [8]. At these conditions the dielectric constant of water has already reached a value of 20 and below and one could expect a considerable modification of the electrostatic repulsive interaction owing to the counterion condensation.

DLS measurements with colloidal particles above the critical point of water were carried out in the homogeneous and in the highly inhomogeneous region of the phase diagram. $g^{(2)}(\tau)$ measured for a PSL sample at 376.0 °C and 35.2 and 22.5 MPa is presented in Fig. 4a. The scattering of light by density fluctuations at 35.2 MPa is negligible and SC water behaves as a homogeneous fluid (the right-most point in Fig. 1) with liquidlike density of 0.585 g/cm^3.Curve 2 was recorded in the presence of pronounced density fluctuations at the same conditions as curve 1 in Fig. 2. In this case not only an increase in the background scattering (Fig. 1, point 1) but also a correlation of the intensity owing to the pure water has to be expected. However, the comparison between the two curves in Fig. 4a shows that the shape of

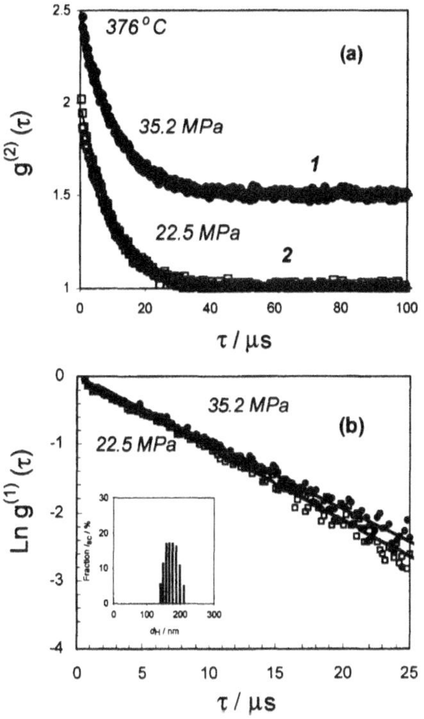

Fig. 4. a Intensity ACF, $g^{(2)}(\tau)$, measured for PSL particles at 35.2 and 22.5 MPa at 376.0 °C and **b** the log–normal plots of the corresponding $g^{(1)}(\tau)$. *Curve 2* was obtained at the same conditions as curve 1 in Fig. 2. *Curve 1* in **a** is shifted by a constant value of 0.5 from its original position. The *solid lines* present the theoretical fit of the ACF. The inset in **b** shows the distribution of the apparent hydrodynamic diameter, d_H, at 22.5 MPa calculated using the Stokes–Einstein formula

the measured ACF is pretty similar in both cases despite the differences in the homogeneity of the dispersion medium. Moreover it does not differ significantly from the one recorded at ambient conditions for the same PSL particles (not shown here) except for the increased noise and the shorter time scale of the process at 376.0 °C. At this temperature the water viscosity is between 0.033 and 0.068 cP within the pressure range studied. The average relaxation time of the diffusion at 376.0 °C was determined to be $\tau_D = 21$ µs (35.2 MPa) and $\tau_D = 19$ µs (22.5 MPa), which is more than 1 order of magnitude smaller than that at 24.2 °C ($\tau_D = 244$ µs). The linearity of the corresponding logarithmic plots of $g^{(1)}(\tau)$ presented in Fig. 4b suggests that the polydispersity of PSL particles dispersed in SC water is not very large, which is also confirmed by the particle size distribution at 22.5 MPa and 376.0 °C given in the inset. Consequently one may conclude that the intensity ACF at 22.5 MPa (curve 2) is determined mainly by the motion of the latex particles despite the pronounced density fluctuations. Indeed, when comparing $g^{(2)}(\tau)$ recorded in the case of pure water (Fig. 2, curve 1) and PSL (Fig. 4a, curve 2) one sees that the time scale of density fluctuations is much slower (the relaxation time of the process is of the order of milliseconds) than that of the particle diffusion. The latter takes place within a time period that is shorter than the time separating two correlation channels (two points of the ACF) in the case of pure water. On the basis of these observations one may expect that it would be possible to distinguish between the particle diffusion and the dynamics of water density fluctuations by using DLS.

All the diffusion coefficients measured in SC water are smaller (Fig. 3) than the values expected for particles with constant size, if we assume, of course, that the Stokes–Einstein formula can be still applied, which infers that the average size of the latex and gold particles in SC water is larger than the initial size. The observed decrease in the measured diffusion coefficient could be ascribed to several effects that are likely to occur in near-critical and SC water, such as particle clustering, swelling and solvation, local density augmentation around a dispersed particle, a preferential diffusion of the particles in water regions of higher local viscosity, and modifications of interparticle interactions. It is also necessary to have more clues about how the Brownian motion proceeds in conditions of density inhomogeneities and the possible role of the critical region effects in order to interpret correctly the diffusion coefficients measured by DLS in terms of particle size.

Conclusions

This paper presents the very first experimental data obtained by DLS for the diffusion coefficients of latex and CG particles suspended in SC water. The colloidal stability of the samples observed in a wide range from ambient to near-critical temperature is quite interesting and appealing for more detailed experimental and theoretical analysis to reveal the mechanism of stabilization at elevated temperatures and pressures. The results accumulated so far are promising for the future application of DLS to studying colloids dispersed in hot water under high pressure including SC area and eventually to probing the behavior of long-range density fluctuations in SC water in the presence of particles.

References

1. Clifford T (1999) Fundamentals of supercritical fluids. Oxford University Press, New York
2. Kajimoto O (1999) Chem Rev 99:355
3. Tucker SC (1999) Chem Rev 99:391
4. Petsche IB, Debenedetti PG (1991) J Phys Chem 95:386
5. Bennet GE, Johnston KP (1994) J Phys Chem 98:441
6. Nishikawa K, Morita T (2000) Chem Phys Lett 316:238

7. Morita T, Kusano K, Ochiai H, Saitow K-I, Nishikawa K (2000) J Chem Phys 112:4203
8. Alargova RG, Deguchi S, Tsujii K (2001) Colloids Surf A 183–185:303
9. Harvey AH, Peskin AP, Klein SA (2000) NIST standard reference database (NIST/ASME steam formulation for general and scientific use), version 2.2. US Department of Commerce, Maryland
10. Schmitz KS (1990) An introduction to dynamic Light scattering by macromolecules. Academic, San Diego
11. Alargova RG, Tsujii K (2004) (in preparation)

Progr Colloid Polym Sci (2004) 126: 139–145
DOI 10.1007/b93975

D. Kalnin
P. Quennesson
F. Artzner
O. Schafer
T. Narayanan
M. Ollivon

Monitoring both fat crystallization and self-assembly of sodium caseinate in model emulsions using synchrotron X-ray diffraction

D. Kalnin (✉) · M. Ollivon · F. Artzner
UMR CNRS 8612 Equipe Physico-Chimie des Systèmes Polyphasés, Châtenay-Malabry, France
e-mail: daniel.kalnin@web.de
Tel.: +33-1-46835465;
Fax: +33-1-46835312

D. Kalnin · P. Quennesson
Nestlé Product Technology Center
Beauvais, Z.I. no. 2, 60000 Beauvais, France

O. Schafer
Nestlé Research Center,
Vers-chez-les Blanc, 1000 Lausanne 26, Switzerland

T. Narayanan
European Synchrotron Radiation Facility, ESRF, BP 220, 38043 Grenoble Cedex, France

Abstract X-Ray diffraction analysis at small angles is proposed as a nondestructive method for characterization of emulsions. The paper focuses on determinations of both, the concentration of sodium caseinate (SC) used as a stabilizer and the organization of triglyceride crystals, located outside and within emulsion droplets, respectively. Measurements performed at very small angles using ID 02a at a sample-to-detector distance of 6.5 m made it possible to monitor the SC submicelle size variations at $q \sim 0.03$ Å^{-1} while exploring a SC/water binary system. Comparison with the scattering found in model emulsion shows congruent results. The characterization of the crystalline organizations of triacylglycerols was also performed simultaneously at $q \sim 0.1$ Å^{-1}. Phase transitions were directly observed within the emulsion droplets. Both characterizations allowed direct quantitative determinations of the submicellar SC concentration in the presence of emulsion and of the respective proportions of the various polymorphic forms of the fat within emulsion droplets. These experiments have shown that, knowing the primary concentrations of all ingredients, X-ray scattering at very small angles can be used, to monitor submicellar casein aggregation and fat polymorphism in a single analysis.

Keywords Triglyceride · Miceller formation · Sodium caseinate · SR · Polymorphism · Small-angle X-ray scattering · Emulsion · Interface

Introduction

Characterization of thermal and structural behavior of fat in emulsion droplets or more generally of any substance in finely dispersed systems represents a main challenge in many fields. In this respect, techniques based on scattering of light whatever its wavelength are well suited to study colloids of biological, pharmaceutical, medical or industrial interest. However, for instance, until recently, droplet size, concentration of fat in the sample and complexity of the systems prevented any direct identification of the crystalline varieties formed by triacylglycerols inside emulsion droplets [1]. Crystallization within droplets of a mean diameter of around 0.5 μm can now be monitored easily using synchrotron X-ray diffraction [1, 2, 3]. As the crystallization kinetics depend not only on the origins of the triglycerides (TAGs) and their composition but also on the size of the fat globules and the interface composition [3] it is important to handle the influence of both factors. Conversely, crystallization can be used as a reporter of the interface composition and organization [3], presuming the partition coefficients of the amphiphilic molecules between interface, bulk and dispersed phases are known.

Most dispersed multiphase systems are thermodynamically unstable per se and thus require stabilization. In the food industry, actually stabilization of emulsions is obtained by the addition of proteins such as sodium caseinate (SC). The addition of this protein has a double stabilizing effect since it provokes together with other added surfactants an increase of both viscosity of the continuous phase and stability of the interface [4, 5].

Crystallization behavior not only depends on the time/temperature history of the samples but also on the interface of the fat globule. Then, multicomponent systems – with any component being susceptible to influence the physical chemistry of the others – render investigation of emulsions very difficult. Numerous phases coexist in emulsions. Moreover, a supplementary phase is added to the system each time one of the components crystallizes or is at least partly crystalline. In this respect, the crystallization of the interfacial film or of the oily phase frequently with melting and/or crystallization points close to room temperature even further increases the complexity of the study of thermal properties – especially when TAGs are involved [1].

The use of destructive methods for the determination of the interfacial concentration of proteins or other lipid surfactants is really time-consuming, costly and not in any case realizable. Moreover, the dependence on temperature of the interface composition is difficult to assess with these techniques.

Differential scanning calorimetry (DSC) is mainly used to elucidate thermal and structural behavior of TAGs – the main constituents of fats – which exhibit complex monotropic polymorphism. The numerous DSC peaks observed whilst heating and cooling need further identification with the help of techniques that yield information about structures (e.g. X-ray or neutron diffraction, IR spectroscopy). The coupling of both small-angle X-ray scattering (SAXS) and wide-angle X-ray diffraction (WAXS) with DSC represents a powerful technique for the characterization of crystallization in emulsion and of the influence of system parameters such as droplet size and interface composition.

This paper complements a series of studies on TAG crystallization in emulsions by providing a nondestructive and convenient method using SAXS for both the characterization of SC in the aqueous phase together with the monitoring of the polymorphic transitions of the fat within the globules.

Material and methods

Binary SC/water system

SC was industrial food grade(Armor Protéines). The same SC was used for binary water systems and emulsion stabilization. Samples for the study of the binary SC/water system were prepared by dispersing weighed amounts of SC solubilized in Millipore water and afterwards weighing for concentration determination. Each sample, about 3 cm^3, was prepared without sodium azide. An X-ray diffraction study was carried out about 24 h after preparation to ensure equilibration of protein solutions.

Model oil-in-water emulsions

Ingredients of industrial food grade were weighed before emulsion preparation. The lipid emulsifier used was GMS mainly composed of glycerol monostearate provided by Danisco Coulter (Denmark). SC was first dissolved in preheated water and was so hydrated. Then, under mechanical mixing conditions, preheated fat was added at 60° for 20 min. Model emulsions with 40% fully hydroginated vegetal fat content were prepared using a pilot plant homogenizer at about 350 bars. Different emulsifier concentrations ranging from 0.5 to 1.5% were used. Emulsifier was added to the aqueous phase prior to homogenization. All emulsions were controlled for emulsion droplet mean size using a laser light diffraction apparatus (Malvern Mastersizer, Malvern, UK). One has to note that those emulsions are thermodynamical then stable for months. They display the same fat globule size distribution providing they are lightly agitated by hand shaking before the actual measurement.

X-ray diffraction

X-ray diffraction was performed using the high flux of the synchrotron radiation beam at beamline IDE02a of ESRF (Grenoble, France) using a wavelength of 1 Å (flux through sample 8x10^{12}ph s^{-1}/100 mA at 12.5 keV). An image-intensified FReLoN charged-coupled-device SAXS detector was used for data collection using sample-to-detector distances of 1.5 and 6.5 m. All X-ray diffraction patterns were recorded by transmission using glass capillaries (GLAS W. Müller, Berlin, Germany) (diameter 1.5 mm, wallthickness 0.01 mm). Samples were prepared by filling these glass capillaries with about 20 μl sample using a laboratory-made special syringe. The sample was tempered in a multiple sample holder (20 holes) designed in our laboratory for a temperature range from 10 to 100 °C. X-ray scattering was monitored at 20 °C for equilibrium and during temperature ramps. Temperature ramps were from 20 to 60 °C, then from 60 to 10 °C for SC solutions and from 60 to 10 °C with an isothermal regime at 60 °C prior to the actual measurement in order to eliminate all temperature/time history of the sample for emulsions, respectively.

Results and discussion

Scattering patterns of emulsions stabilized by SC present an intense central scattering [1]. This scattering is already detectable at a sample-to-detector distance of approximately 1.2–1.5 m, but the diffracted signal is not quantitatively exploitable in this region owing to the vicinity of the beamstop cut at about $q=0.025$ Å^{-1} [1]. Such observation led us to the investigation of this scattering at a sample-to-detector distance of 6.5 m. Increasing the distance leads to a higher resolution and the possibility to observe scattering at lower q values. In order to determine the origin of these scattering peaks, samples containing SC solutions were examined first, then the results were compared to those of the emulsions.

Scattering from SC solutions

At a sample-to-detector distance of 6.5m, SC solutions scatter X-ray as a function of protein concentration. At first glance and from rapid analysis, the scattered

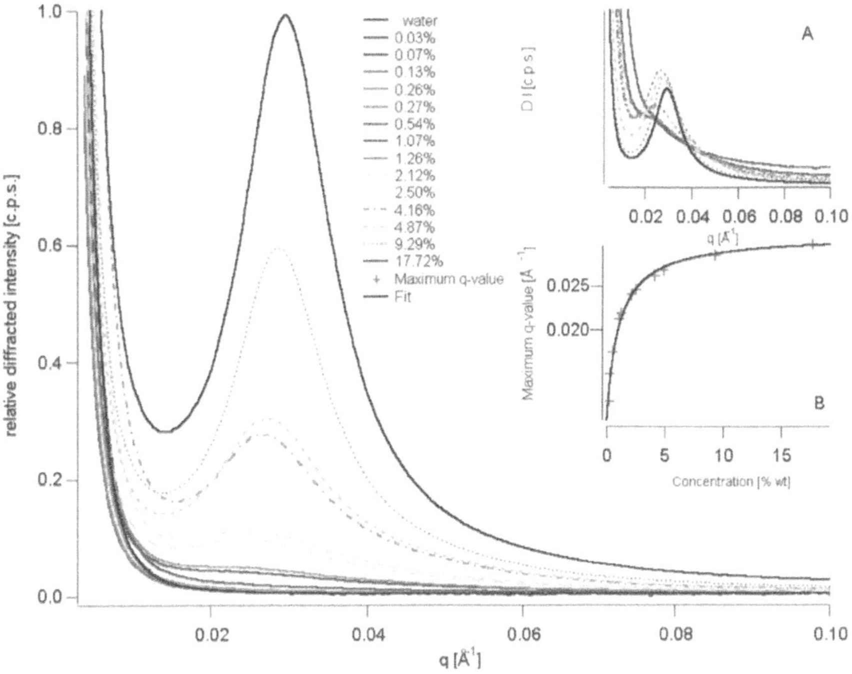

Fig. 1 X-ray scattering at very small angles (*SAXS*) patterns at a sample-to-detector distance of 6.5 m as a function of sodium caseinate (*SC*) concentration in aqueous solutions shows an increase of the scattering due to "micelle" formation. The SC concentrations examined are listed and attributed to the different line shapes *Insert A*: SAXS pattern of normalized intensity of aqueous SC solutions. Normalization is carried out by subtraction of pure water scattering as a function of the water content and by division through the SC concentration. *Insert B:* Evolution of the maximum-scattering peak position as a function of SC concentration determined from the normalized patterns using Gaussian fits

intensity is relative to the amount of SC engaged in the micelle structures. Conversely, solutions can be identified following their SC concentration very easily from scattering curves (Fig. 1).

The normalization of the scattering curves taking in account the scattering of pure water and the concentration of SC is shown Fig. 1, insert a. The plot was only normalized for SC concentration since the setup (multiple position sample holder) design did not allow rigorous scattering normalization and measurements of absolute intensities. Consequently, this plot cannot be directly compared for intensities since they were obtained from different capillaries (scattered intensities are not corrected for the different capillary sizes and sample absorption of X-ray; this mostly explains the intensity dispersion of the curves Fig. 1, insert a). However, the shift of the scattering maximum observed in this normalized plot shows that the aggregation state of the protein changes as a function of its concentration. (Fig. 1, insert a). Figure 1, insert b further evidences the variation of the scattering peak maximum as a function of SC concentration. Practically, scattering can be detected and well fitted on normalized diffraction patterns for an SC concentration of 0.26 wt% or greater. The variation of this scattering maximum, $q_{\max}$, is not linear and beyond a certain SC concentration the shift observed becomes weaker and weaker. Two regimes are observed, up to about 3 wt% and beyond 3 wt%. The latter regime is asymptotic, with an asymptote at about $q = 0.033$ Å^{-1} which correlates to a distance of about 18 nm. What are the exact origins of this scattering and of its dependence upon SC concentration?

Taking into account the domain of the scattering vector, q, in which scattering is observed, the direct scattering from the whole casein micelles was ruled out [7, 8, 9, 10]. Casein micelles which are aggregates known to exhibit a mean diameter in the range 0.2–0.3 μm [9] are too large for the scattering lengths at which scattering is observed. If not related to casein micelles, then the scattering patterns observed result from internal organization into casein submicelles [11]. Considering the complexity of the internal organization of the casein submicelles with possibly different types of casein submicelles, the distances corresponding to the minimum and the maximum of the scattering patterns are not necessarily directly correlated to some specific length within the casein micelle or submicelle. The submicelle shape factor and the organization of the SC molecules within this aggregate should be considered to find a model of the internal organization of casein micelles into subunits. This is out of the scope of this study. While the existence of the casein submicelle has been discussed for many years [7]. Kumosinski et al. [12] proposed recently from SAXS analysis a model for the organization of these particles. According to them, the submicellar particles consist of an inner, spherically symmetrical, hydrophobic and relatively electron-dense core surrounded by a hydrophilic and less electron-dense regions. The latter loose regions of adjacent micelles interact with each other [12]. This model was refined from the comparison with electron microscopy studies of casein submicelles [13]. As this submicelle model is in agreement with the data presented earlier for scattering lengths, we concluded that the scattering signal observed in caseinate

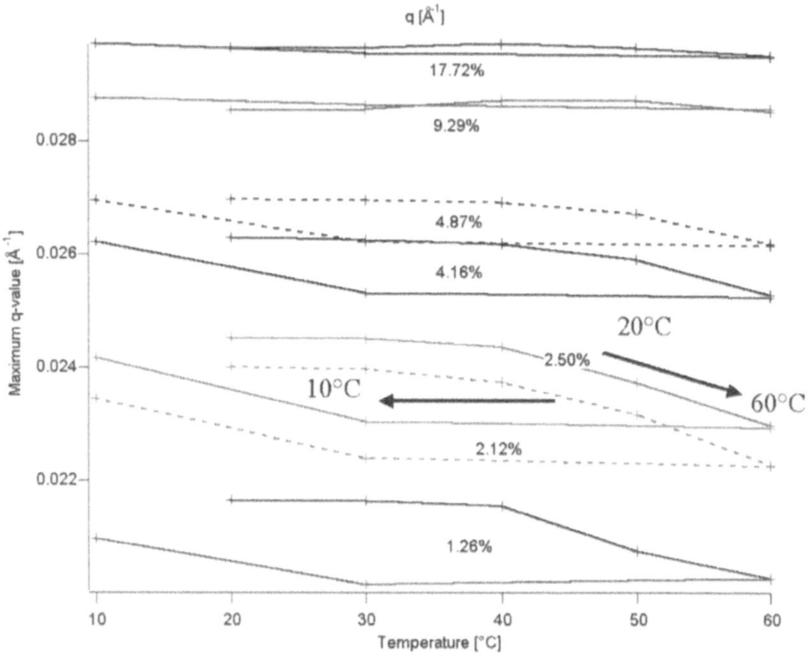

Fig. 2 Evolutions of maximum positions of SAXS peaks, determined from the normalized patterns in conditions as in Fig. 1, as a function of SC concentration and temperature. Each sample was first heated from 20 to 60 °C and then cooled to 10 °C, as indicated by *arrows*

solutions originated from a submicellar organization of SC.

Temperature dependence of the SC scattering

Solubility of SC in aqueous solutions is influenced by many factors, including temperature. The changes of the scattering maximum, q_{max}, as a function of temperature as recorded during heatings and subsequent coolings of SC solutions of various concentrations are shown in Fig. 2. As seen from Fig. 2, we were able to monitor even slight decreases of the q_{max} value thanks to peak-fitting. The q_{max} shifts indicate for all concentrations that a lower concentration of submicellar SC is observed at low temperature and this is consistent with the known higher solubility of casein at lower temperatures [9]. Such a solubility dependence is also reflected more directly from the variation of scattering intensity as a function of temperature from the same capillary, whatever SC concentration considered, a systematic decrease is observed at lower temperature (scattering intensity variations are not shown since, again, the scattering plots from different concentrations cannot be directly compared for intensity). The scattering intensity changes by about 8% at maximum, while the half height width does not compensate this temperature-induced variation. However, both intensity and q_{max} shifts indicate that fewer submicelles are formed at high temperature probably because of the increased solubility of the unaggregated casein [9].

Moreover, from the q_{max} variation versus temperature we deduced that these temperature-induced solubility changes are reversible (Fig. 2). The reversibility of this effect can also be monitored by X-ray diffraction. As shown in Fig. 2, hysteresis is observed whatever the concentration considered. This hysteresis has been attributed to slow equilibration kinetics within the submicelles. The exact time dependence of this equilibrium has not been measured. Further studies would be necessary to determine the rate of equilibration.

Scattering from SC solutions of oil-in-water fat emulsions

Emulsions stabilized with different concentrations of SC also show changes in the central scattering (Fig. 3). For SAXS, a q range from 0.03 to 0.4 Å is sufficient in order to monitor fat crystallization in these emulsions (data not shown) [2, 3]. However, X-ray recording at a sample-to-detector distance of 6.5 m (SAXS at very small angles) shows better evidence of the scattering phenomena owing to both dispersion of the fat in fat globules with a mean size of around 0.5 μm and self assembly of SC. The first-mentioned effect should be about the same for all emulsions measured because the fat concentration in the fat globule and the sizes are the same in the different emulsions. The only factor changing between the different emulsions is the SC concentration. The variations of the SAXS patterns as a function of surfactant and emulsifier content are shown in Fig. 3. In spite of the superimposed strong scattering due to the emulsion, the SC scattering of the solution is observed. This scattering can be compared to that obtained at different SC concentrations in the absence of oil droplets. Can we

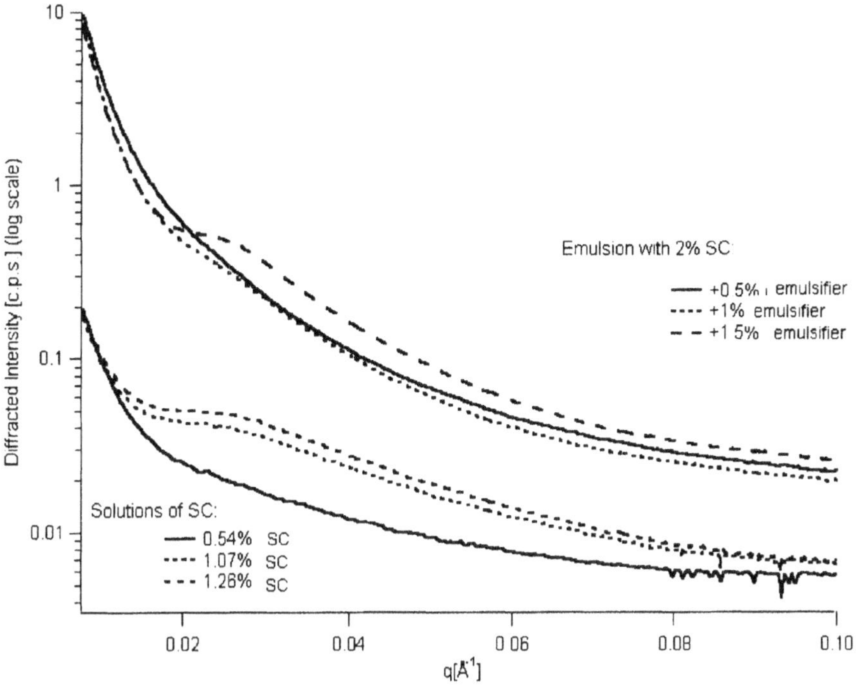

Fig. 3 SAXS patterns due to caseinate self-assembly in model emulsions (2% SC) at the different lipid emulsifier concentrations (*top*) compared for convenience to the scattering patterns (*bottom*) of three solutions of the binary SC/water system shown in Fig. 1. Both series of measurements were carried out at 20 °C

use this scattering to directly evaluate the submicellar SC content?

In an emulsified system, the "solubilization" of SC results in fact from the distribution of SC molecules at different locations and with different organizations. According to the literature, solubilization of SC is achieved by (1) SC molecules really solubilized as monomeric or oligomeric entities, and forming a true protein solution, (2) SC molecules organized in micelles and submicelles from which the scattering at very small angles originates (see earlier) and (3) that adsorbed at the fat globule interface. In this last location, depending on their nature, SC molecules can be either lying at the interface as a polymer with loop and train segments or in the form of submicelles [5, 6, 11, 12, 13]. In homogenized emulsions, the presence of casein submicelles at the globule interface can probably be ruled out. Although emulsions cannot be considered as systems at equilibrium, it can be assumed that the distribution of the SC molecules between the different locations obeys some kind of equilibrium. This can be deduced of the SC mobility and the existence of different locations without strong interactions. Then, molecules which are not in the aqueous phase can be supposed to be either at the fat globule interface or "solubilized" within the micelles and submicelles.

The comparison of SC scattering curves obtained before and after heating the emulsion from 20 to 60 °C allowed us to rule out any influence of the presence of lipid on the scattering. In this respect, it has been confirmed that a single heating did not affect the size distribution of the emulsions. Determining the maxima of the scattering peak thus makes it possible to evaluate the concentration of SC solubilized in the form of submicelles.

Diffraction from TAG crystallization within the fat globule of oil-in-water emulsions

The TAG crystallization within droplets of a SC model emulsion containing lipid surfactants has been monitored by time-resolved SAXS during sample cooling at 2 K min^{-1} from 60 to 10 °C [1, 2, 3]. Only the last frames of the time/temperature evolution are shown Fig. 4. The model emulsion (Fig. 4A) contains a high concentration of SC (5%) and 0.5% added monoglycerides. On the SAXS recordings, the large bump of scattering encountered at $q \sim 0.18$ Å^{-1} is due to the fact that at this temperature liquid fat is partially liquid-crystalline. During the cooling, crystallization in the emulsion which starts by the formation of an α (rotator) chain packing of TAG precedes to an α-to-β' transformation (WAXS data not available). The longitudinal stacking change from untilted to tilted chain position visible from SAXS which accompanies this lateral chain packing transition is shown Fig. 4B. The evolution of the SAXS during the same time temperature/history at the same surfactant concentration as emulsion concentration but at a lower concentration of SC (2%) is also shown in Fig. 4B. The rate of α-to-β' transformation is strongly affected by the change in the concentration of SC.

These results demonstrate that TAG crystallization can be quantitatively monitored within the emulsion

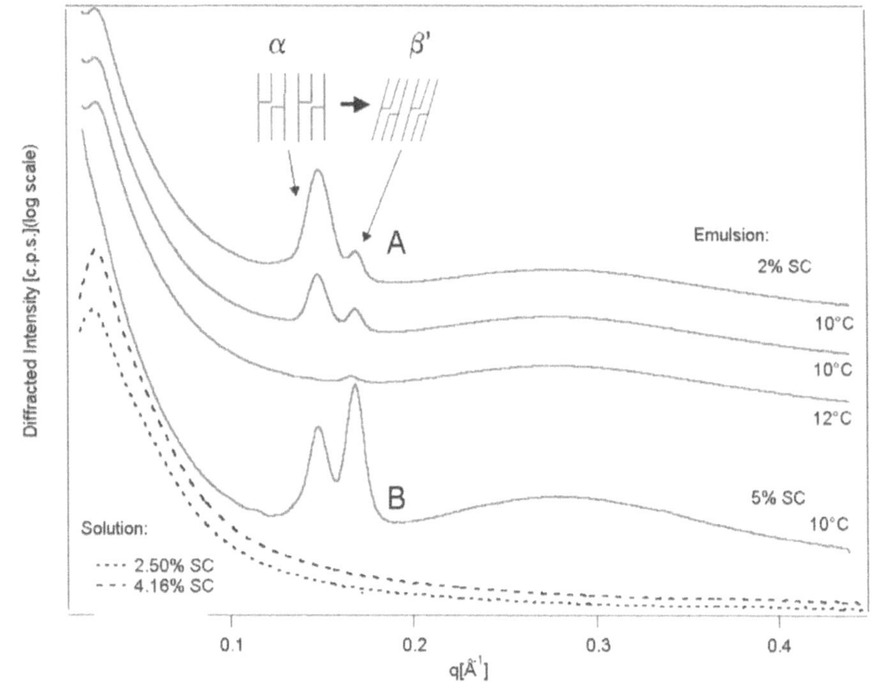

Fig. 4 SAXS monitoring at a sample-to-detector distance of 1.5 m of triglyceride crystallization within droplets of a SC model emulsion containing lipid surfactants. A Time/temperature evolution of an emulsion with a high concentration of SC and 0.5% added monoglycerides (only last frames of the time-resolved X-ray diffraction experiment are shown). *b* Same time temperature/history, same surfactant concentration as for the emulsion shown in *a* but at a lower concentration of SC (x %) (only the last pattern is drawn). Drawn *below* for comparison in *dashed lines* are the binary SC/water solutions, the concentrations of which are indicated

droplets, is affected by the concentration of SC and that SC is located partially at the interface. The data also show that the interface composition influences the TAG transition and crystallization. This point will be analyzed in more in detail elsewhere.

Competition between SC and other emulsifiers for the globule interface

The changes in the scattering patterns induced by the addition of between 0.5–1.5% monoglycerides as emulsifiers to the model emulsion are shown in Fig. 3. An increase of the scattered intensity observed at q_{max} is found upon addition of monoglycerides. How does the addition of monoglyceride affect the SC aggregation? Classical analysis is that the addition of monoglycerides has an effect on the fat globule interface, meaning that owing to its higher affinity for lipids it partially depletes the fat globules from their initial SC coating. The displacement of the protein from the interface to the solution should increase the number of submicelles and the scattering. Both influences of the addition of emulsifier on SC aggregation and on lipid crystallization will be examined in a further study.

The presence of monoglycerides in the emulsion also influences the crystallization behavior of the emulsified fats as seen Fig. 4. This means that the presence of monoglycerides in the aqueous phase has possibly a double impact on the fat globule interface: through the logical presence at the interface as a polar lipid that directly influences the crystallization behavior of the emulsified fats; through the modification of the partition coefficient of SC between the globule interface and the aqueous phase. However, it cannot be determined from this study if one of the two mechanism is preponderant.

Conclusion

SAXS permit the quantitative characterization of oil-in-water emulsions at two different levels: in the aqueous phase by the characterization of the protein self-assembly and within the fat globules by the determination of the TAG organization and polymorphism. Emulsions are complex systems in which the interface plays a critical role; however, this interface cannot be studied directly. We demonstrated here that the behavior or the structure of both phases studied is influenced by the composition and organization thanks to the location between the two phases examined. It is expected that most of the change in the interface is reflected by changes at least in one of them if not in both as observed with SC. In the latter case, then, a comparison of the influence of the two phases is very promising. However, taking into account the difficulties in the discrimination of the diverse scattering phenomenona, the influence on TAG diffraction looks simpler to analyze and to be more sensitive to the changes.

Acknowledgements The authors thank Nestlé (Beauvais, Verschez-les-Blanc) and ANRT (Paris, France) for supporting this research. EEC support for ESRF beamtime at ID02a is also acknowledged.

References

1. Lopez C, Lesieur P, Bourgaux C, Ollivon M (2000) J Colloid Interface Sci 229:62
2. Kalnin D, Arnaud G, Amenitsch H, Ollivon (2002) Food Res Int (in press)
3. Lopez C et al (2001) J Am Oil Chem Soc 78:1233
4. Cornec M, Wilde PJ, Gunning PA (1998) J Food Sci 63:39
5. Dickinson E (1999) Int Dairy J 9:305
6. Farrer D, Lips A (1999) Int Dairy J 9:281
7. Walstra P (1979) J Dairy Res 46:317
8. Berry GP, Creamer LK (1975) Biochemistry 14:3542
9. Gayot P, Lorient D (1998) Structures et technofonctions des protéins du lait. Tec et Doc Lavoisier
10. Storthart DG, Cebula DJ (1982) J Mol Biol 227:391
11. Walstra P (1999) Int Dairy J 9:189
12. Kumosinski TF, Pessen H, Farrell HM Jr, Brumberger H (1988) Arch Biochem Biophys 266:548
13. Kumosinski TF, Uknalis J, Cooke PH, Farrell HM Jr (1996) Lebensm Wiss Technol 29:326

Progr Colloid Polym Sci (2004) 126: 146–150
DOI 10.1007/b94009

Giacomo Martini
Sara Morandi
Simona Rossi
Sandra Ristori

Characterization of carriers for drug delivery in boron neutron capture therapy. Electron spin resonance study of nitroxide-containing liposomes

G. Martini (✉) · S. Morandi ·
S. Rossi · S. Ristori
Department of Chemistry, University of Florence, Polo Scientifico,
50019 Sesto F.no, Florence, Italy
e-mail: martini@csgi.unifi.it
Tel.: +39-573-4573042
Fax: +39-55-4573036

Abstract Electron spin resonance (ESR) spectroscopy of nitroxides was used in the study of structural and dynamic features of cationic liposomes built up with 1,2-dioleoyl-3-trimethylammoniumpropane and L-α-dioleoyl-phosphatidylethanolamine surfactants. Particular attention was devoted to the properties of the liposomes after loading with different amounts of boronated compounds such as [1,2-dicarba-*closo*-dodecarboran(12)-1-ylmethyl](β-D-galactopyranosyl)-(1→4)-β-D-glucopyranoside and 1,2-di-(β-D-glucopyranosyl-ox)methyl-1,2-dicarba-*closo*-dodecaborane(12), to be used in boron neutron capture therapy (BNCT). The ESR line shapes suggested that the structure and the internal dynamics of the liposomes did not change appreciably after small loading. In contrast, variations in both the mobility and the order degree of the nitroxides inside the bilayer were observed when carborane was loaded in a large amount. ESR spectroscopy was shown therefore to be a valuable tool for the study of liposome properties at a molecular level.

Keywords Electron spin resonance · Boron neutron capture therapy · Liposomes · Nitroxides · Borocompounds

Introduction

Boron neutron capture therapy (BNCT) is a targeted cancer treatment which is based on two steps: the selective deposition of ^{10}B nuclei into cancerous cells; the irradiation of ^{10}B nuclei with thermal neutrons coming from a nuclear reactor [1]. Many efforts in recent years have been directed toward the use of particle accelerators to obtain neutrons. The second step consists of the boron fission to heavy particles (typically ^{7}Li and ^{4}He). These particles have penetration depths in human tissues of the order of the cell size itself. Ionization-tracking and cellular damage then result. The lethal effect is limited only to the cells where boron nuclei are located or to those in the nearest proximity. In order to have good efficiency the delivery must be site-specific and about 15–20 μg boron is necessary per gram of tissue.

Because the selective uptake of borocompounds by malignant cells is a primary concern for the success of BNCT, liposomes are the preferred structures for drug delivery. Cationic liposomes have been loaded with the carboranyl compounds in Structures 1 and 2 which were

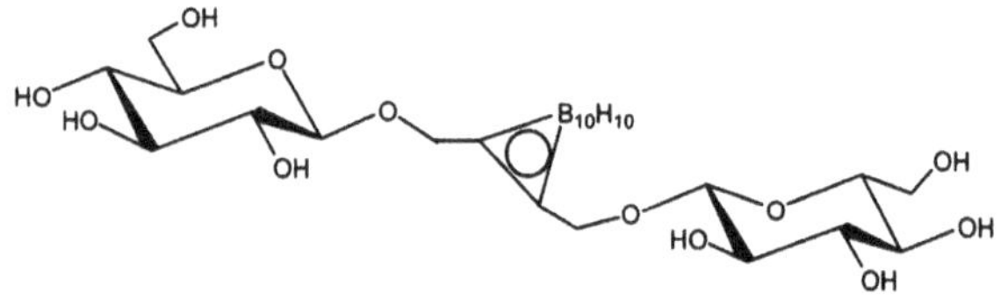

DGCOB

1,2-di-(β-D-gluco-pyranosyl-ox)methyl-1,2-dicarba-closododecaborane(12)

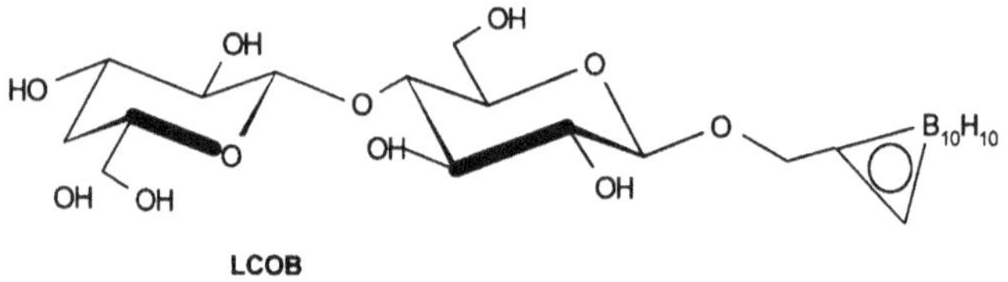

LCOB

[1,2-dicarba-closo-dodecarboran(12)-1-ylmethyl](β-D-galactopyranosyl-)-(1→4)-β-D-glucopyranoside synthesized by Giovenzana et al. [2]. These molecules have a boron cluster with ten boron atoms and two carbon atoms (*closo*-carborane unit)

Cationic liposomes are indeed known to penetrate the eucariote cell membranes and to cross nuclear membranes [3, 4, 5]. They are also currently used for gene therapy and other therapies based on drug delivery [6, 7]. We used cationic 1,2-dioleoyl-3-trimethylammonium-propane (DOTAP) and neutral L-α-dioleoyl-phosphatidylethanolamine (DOPE) as the components of the liposomes:

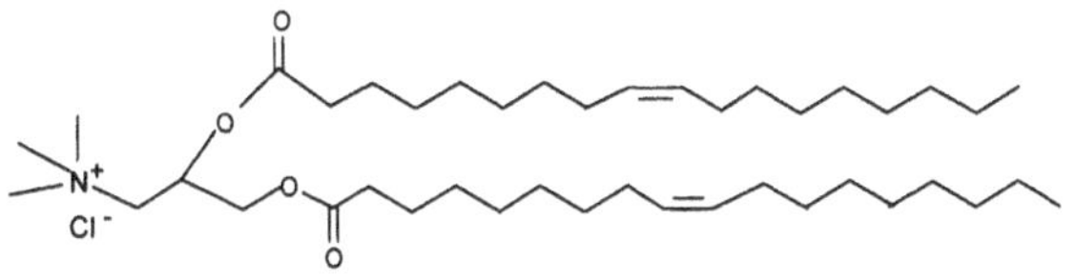

DOTAP

(1,2-dioleoyl-3-trimethylammonium-propane)

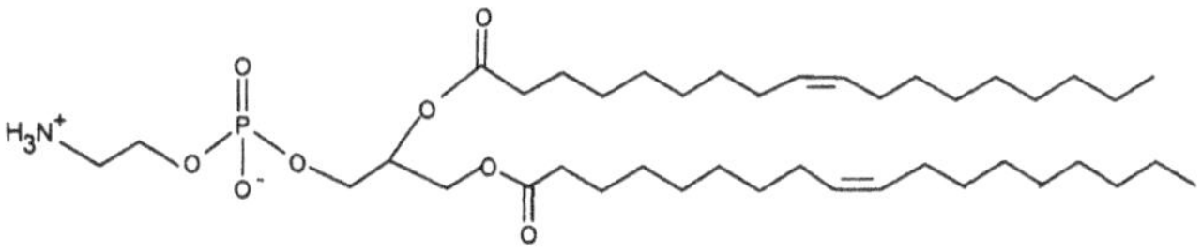

DOPE

(L-α-dioleoyl-phosphatidylethanolamine)

These systems were studied in our laboratory with experimental techniques as different as light scattering (LS), differential scanning calorimetry (DSC), NMR and electron spin resonance (ESR) [8, 9]. In this paper we report the use of ESR in the characterization of some structural and dynamic features of these systems. As well as other spectroscopic methods, ESR spectroscopy is able to give information at a molecular level that is not obtained with other techniques which are designed for massive properties [8, 10, 11, 12]. Appropriately tailored paramagnetic probes have been widely used in the last 30 years for the study of self-assembled systems because of the sensitivity that this technique exhibits toward dynamical events on time scales of 10^{-7}–10^{-8} s and below. The analysis of the ESR line shape was carried out in terms of the magnetic and motional parameters obtained from the fitting procedures reported by Freed and coworkers [13, 14]. The ESR spectra of nitroxides inserted into liposome membranes give valuable information on relevant features of the membranes themselves, such as mobility, polarity, and order degree [10, 11, 13]. Probe mobility in the lipophilic core may be inferred from the values of the motional correlation time, whereas hyperfine coupling constants report the polarity sensed by the probe.

Experimental

DOPE and DOTAP surfactants were purchased from Northern Lipids, Vancouver, Canada. Liposomes were extruded as aqueous solutions directly from a 1:1 w/w mixture of the individual components with the Lipofast apparatus (Avestin, Ottawa, Canada).

The total lipid content was 10 mg/ml, pH 7.4. The mean liposome diameter was 120 ± 10 nm as given by the manufacturer and checked by quasielastistic LS. The dispersions were typically stored at 277 K. Their stability is given to be 6 months at least.

The nitroxide radicals were purchased from Sigma Chemicals, Munich, Germany, and were used without further treatment.

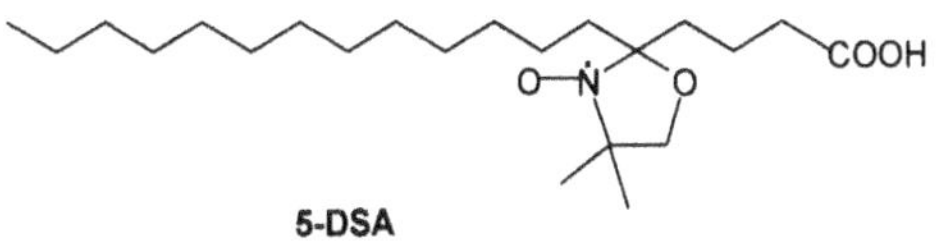

5-DSA

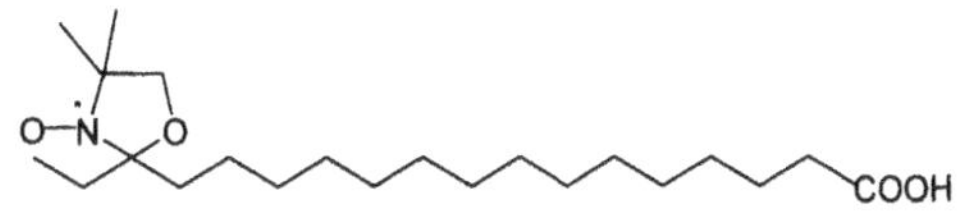

16-DSA

Phospate buffer (10^{-2} mol/l) in 0.15 mol/l NaCl solution was used. In dilute aqueous solution both 5-doxyl stearic acid (5-DSA) and 16-doxyl stearic acid (16-DSA) nitroxides give rise to the well-known three-line pattern owing to the interaction of the unpaired electron with the ^{14}N nucleus [8, 10, 13, 15]. We used 5-DSA and 16-DSA spin probes as they have been extensively used for the characterization of bilayer-based structures, mainly because they can be easily inserted in a large variety of lipid membranes, the orientation of the N–O bond is fixed and it is parallel to the lipid chain long axis (bilayer director), and different regions of the bilayer are monitored depending on the position of the doxyl unit in the stearic chain.

Carborane-containing DOTAP/DOPE liposomes were prepared by dissolving the appropriate amount of lipid and LCOB or DGCOB mixtures in chloroform. Details of the procedure can be found in Refs. [8, 9].

ESR spectra were taken with the aid of a Bruker model 200D spectrometer. Data acquisition and handling were performed with STELAR software. The temperature was varied with the VTB 3000 accessory (accuracy ± 0.5 K). Quartz tubes of inner diameter less than 2 mm were used for ESR runs.

ESR data treatment was done by using the procedures given by Freed and coworkers [13, 14]. These procedures are reported in many papers and reviews. The Brownian diffusional model for molecular reorientation was always assumed, and microscopic order/macroscopic disorder was applied when ESR probes were inserted into the liposomes. The values of the correlation times, of the motional anisotropy and of the order parameter S_{20} were obtained from these calculations.

Results and discussion

The carboranes we used in this work clearly resembled amphiphilic compounds, because of the contemporaneous presence of a lipophilic moiety (the *closo*-carborane unit) and of a hydrophilic moiety (the sugar moiety). When this kind of molecule was dissolved in aqueous buffer the first problem to be approached was the aggregation shown in fluid solution. ESR spectroscopy was helpful in solving this problem [15]. The 298 K ESR spectra of 5-DSA taken from dispersions in a physiological solution of LCOB at concentrations in the range from 10^{-3} to 10^{-2} mol/l are shown in Fig. 1. 5-DSA nitroxide localized near the external water/membrane

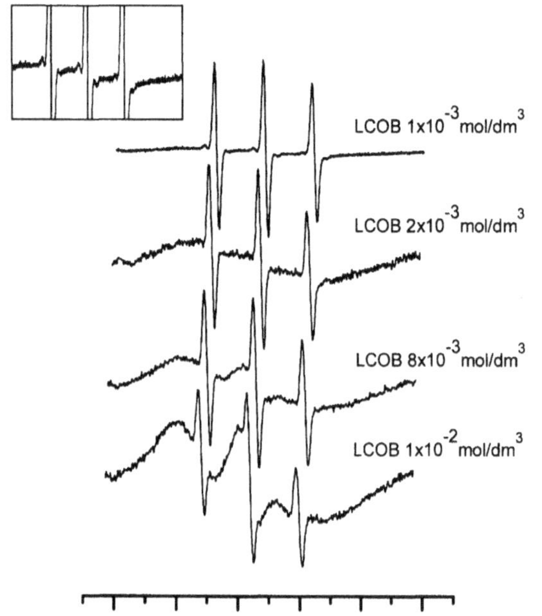

Fig. 1 Electron spin resonance (*ESR*) spectra at 298 K of 5-DSA added to phosphate buffered solutions of [1,2-dicarba-*closo*-dodecarboran(12)-1-ylmethyl](β-D-galactopyranosyl)-(1→4)-β-D-glucopyranoside (*LCOB*) at a concentration from 10^{-2} to 10^{-3} mol/l. The *inset* shows the enlargement of the top spectrum

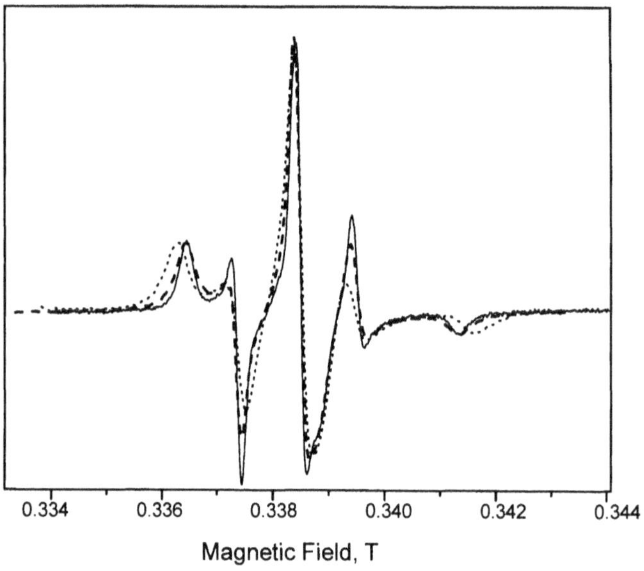

Fig. 2 ESR spectra at 298 K of 5-DSA in 1,2-dioleoyl-3-trimethylammoniumpropane (*DOTAP*)/L-α-dioleoyl-phosphatidylethanolamine (*DOPE*) (1/1 w/w) liposomes containing LCOB at molar fractions $x_{LCOB} = 0.066$ (*dashed line*) and $x_{LCOB} = 0.66$ (*dotted line*) compared with the spectrum of 5-DSA in carborane-free DOTAP/DOPE liposomes (*full line*)

interface and its pattern was largely dominated by the dynamical and structural events occurring in that region. From a very classical analysis of the ESR line shape we could assess that only at a carborane concentration lower than 10^{-3} mol/l might the occurrence of surfactant aggregates be ruled out. This arose from the fact that the top spectrum in Fig. 1 was the usual three-line absorption of 5-DSA in almost free motion owing to nitroxide molecules rapidly reorienting in an isotropic, nonviscous medium. A broad signal underlying the narrow triplet was observed only when the spectral intensity was enhanced, as shown in the inset. All the other spectra shown in Fig. 2 were complex spectra dominated by a large ESR absorption due to spin probes in restricted-motion conditions, as typically happens when paramagnetic probes are inserted in large, slowly moving aggregates. Quasielastic LS data agree with this behavior [9]. A similar trend was observed for DGCOB in aqueous solution.

The second point to be addressed was what kind of information ESR spectra can give on liposomes themselves. Figure 2 shows that the experimental ESR spectrum (full line) of nitroxide with the –NO unit near the polar head (5-DSA in our case) were typical of slowly moving species as happens after adsorption of nitroxide molecules in restricted-motion domains, where the **g** and **A** tensor components are not completely averaged. Many parameters were needed for the calculation and many dynamical parameters were obtained from the fitting, such as the perpendicular component of the motional rate of the >NO unit ($\tau_{\perp} = 7.2 \times 10^{-9}$ s) which reflected the mobility of the environment, the order parameter ($S_{20} = 0.49$), the anisotropy parameter ($n = 10$) for the motion.

When carborane molecules were inserted in the cationic liposomes the ESR line shape depended on the carborane concentration (Fig. 2, dashed and dotted lines). The order parameter increased and the mobility of the radical, i.e. the mobility of its environment, decreased at high carborane concentrations (dotted line: $x_{LCOB} = 0.66$; $\tau_{\perp} = 1.9 \times 10^{-9}$ s, $S_{20} = 0.63$). Only a small increase in the overall spectral width appeared at lower concentration (dashed line) and this was indicative of a less uniform environment sensed by the probe.

The dependence of the environment mobility on the concentration of the drug, at least as it could be measured from ESR parameters, is given in Fig. 3, where the separation of the extreme peaks is shown as a function of DGCOB concentration in the DOTAP/DOPE liposomes at two different temperatures. These data agreed for an almost linear decrease of the mobility with the increase of concentration. The same trend was observed with LCOB. The structure of the borocompound seemed therefore to be not relevant. This meant that the sugar moiety played only a minor role in the dynamic changes of the liposome bilayer; and the hydrophobic moiety, namely the boron cage, which was the same in both compounds, was responsible for the

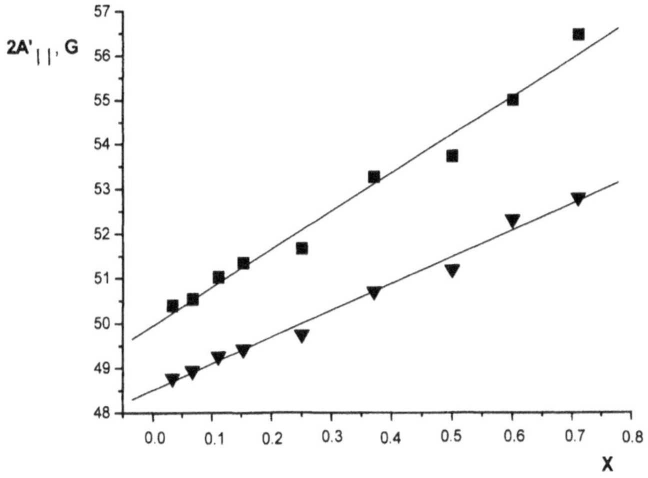

Fig. 3 Outer hyperfine splitting values measured from the extreme peaks in the ESR spectrum of 5-DSA inserted in DOTAP/DOPE liposomes as a function of the 1,2-di-(β-D-glucopyranosyl-ox)methyl-1,2-dicarba-*closo*-dodecaborane(12) loading. 298 K (*squares*), 310 K (*triangles*)

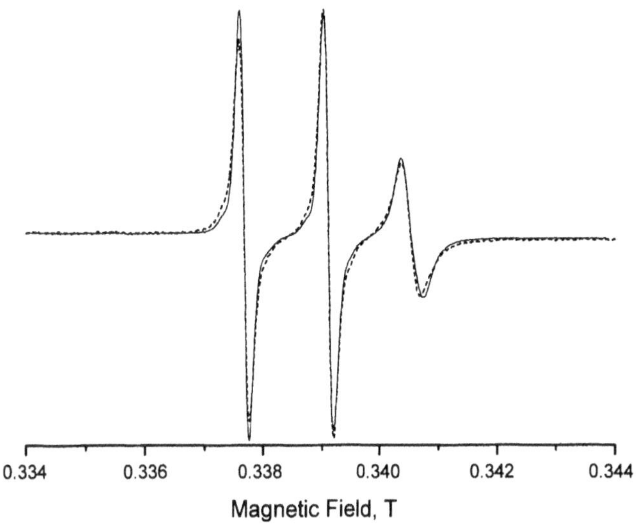

Fig. 4 ESR spectra of 16-DSA in carborane-free DOTAP/DOPE liposomes (*full line*) and in DOTAP/DOPE liposomes containing $x_{LCOB} = 0.5$ (dashed line)

insertion into the bilayer. This insertion was extensive as it results from ^{11}B and ^{1}H NMR [16].

ESR spectroscopy was therefore perfectly able to establish that original micelles or other kinds of aggregates were disrupted while carborane molecules entered the liposome membranes, without significantly altering the molecular bilayer properties to which ESR spectroscopy was sensitive. Considering these results, together with other results of the dependence of the correlation time on the temperature and with DSC data [8], we could assess that the presence of nitroxides and/or carboranes at relatively low concentrations did not greatly affect the behavior of DOTAP/DOPE liposomes and the results we got could be translated to the nitroxide-free systems.

We also used 16-DSA as the paramagnetic probe because it reflected the inner region of the liposome bilayer. No additional information was obtained since the spectral differences between carborane-free and carborane-loaded liposomes tended to vanish (Fig. 4) as expected if the system was considered in terms of the sizes of the carboranyl cage and of the full length of the hydrocarbon chains of the surfactants forming the liposomes. The spectra reported in Fig. 4 clearly showed that the perturbation induced by carborane insertion was mainly limited to the polar region of the lipid bilayer.

Conclusion

From the results reported the following conclusions can be drawn:

1. ESR spectroscopy of nitroxides was highly sensitive to changes of the molecular motion and of the molecular order where they occurred. In the systems described in this paper changes occurred only near the water/bilayer surface as a consequence of carborane insertion and only 5-DSA gave useful results.
2. ESR spectroscopy was able to report a small variation of the motional and dynamical parameters in the more hydrophobic regions of liposomes loaded with carboranes.

The ESR technique was therefore revealed to be one of the most powerful tools to study properties at a molecular level.

Acknowledgements The authors are in debt to the University of Florence and to the Consorzio per le Superfici e per le Grandi Interfasi for financial support. Thanks are due to L. Panza for the gift of the borocompounds used in this work.

References

1. Hawthorne MF (1998) Mol Med Today (April) 174
2. Giovenzana GB, Lay L, Monti D, Palmisano G, Panza L (1999) Tetrahedron 55:14123
3. Johnsson M, Bergstrand N, Edwards K (1999) J Liposome Res 9:53
4. Lasic DD, Templeton NSM (1996) Adv Drug Delivery Rev 20:221
5. Zuidam NJ, Barenholz Y (1998) Biochim Biophys Acta 1368:115
6. Meidan VM, Cohen JS, Amariglio N, Hirsch-Lerner D, Barenholz, Y (2000) Biochim Biophys Acta 1464:251
7. Heynes JA, Nicolescu-Duvaz D, Cooper RG, Springer CJ (2002) J Med Chem 45:99

8. Perissi I, Ristori S, Rossi S, Dei L, Martini G (2002) J Phys Chem B 106:10468
9. Morandi S, Ristori S, Berti D, Panza L Becciolini A, Martini (2004) Biochim Biophys Acta (in press)
10. Marsh D (1981) In: Grell E (ed) Membrane spectroscopy. Springer, Berlin Heidelberg New York, pp 51–142
11. Korstanje LJ, van Faassen EE, Levine YK (1989) Biochim Biophys Acta 982:196
12. Wikander G, Eriksson PO, Burnell EE, Lindblom G (1990) J Phys Chem 94:5964
13. Schneider DJ, Freed JH (1989) In: Berliner LJ, Reuben J (eds) Biological magnetic resonance Spin labeling theory and applications, vol 8. Plenum, New York, p 1
14. Budil DE, Lee S, Saxena S, Freed JH (1996) J Magn Reson 120:155
15. Baglioni P, Ottaviani MF, Martini G (1986) J Phys Chem 88:5107
16. Donati A, Ristori S, Bonechi C, Martini S, Panza L, Martini G, Rossi C (unpublished results)

Progr Colloid Polym Sci (2004) 126: 151–154
DOI 10.1007/b94020

J. M. López-López
A. Schmitt
J. Callejas-Fernández
R. Hidalgo-Álvarez

Time-resolved measurements of the cluster-size distributions arising in charge-heteroaggregation processes

Abstract Charge-heteroaggregation processes arising in 1:1 mixtures of positively and negatively charged particles were studied at low and high electrolyte concentration. Single-cluster light scattering was employed to monitor and resolve the detailed cluster size distribution up to hexamers. The results obtained show that long-range electrostatic interactions between oppositely charged particles favour aggregation at low electrolyte concentration. Specially, dimers are formed from oppositely charged monomers at an increased rate which is almost twice the value found for pure diffusion-limited aggregation.

Keywords Single-cluster light scattering · Colloidal heteroaggregation · Diffusion-limited cluster aggregation

J.M. López-López (✉) · A. Schmitt
J. Callejas-Fernández · R. Hidalgo-Álvarez
Biocolloid and Fluid Physics Group,
Department of Applied Physics,
University of Granada,
Campus de Fuentenueva,
18071 Granada, Spain
e-mail: jmlopezl@ugr.es
Tel.: +34-958-246175
Fax: +34-958-243214

Introduction

Heteroaggregation of colloidal materials having different composition, charge, shape or size is critical to many technological processes, such as mineral flotation [1], core–shell nanocomposites [2] and cell recovery [3], among others. The simpler case of homoaggregation, where only one type of particle is present, is well studied and documented in the literature [4]. In this case, two limiting regimes have been observed: (1) diffusion-limited cluster aggregation (DLCA), in which every collision between particles results in the formation of a bond [5], and (2) reaction-limited cluster aggregation, in which a large number of particle collisions is necessary before a stable bond is established [6]. However, when systems are composed of several species, the reaction kinetics is much more complex and no general kinetic theory is so far available [4, 7].

In this work, heteroaggregation processes arising in aqueous suspensions of colloidal particles with opposite charge were monitored experimentally by means of single-cluster light scattering (SCLS). So far, mainly static and dynamic light scattering experiments have been carried out [8]. Both techniques assess only some average cluster size. SCLS, however, allows us to follow the time evolution of the detailed cluster size distributions up to heptamers. The aim of this work is to study the influence of the electrostatic interactions on the kinetics of charge-heteroaggregation processes arising in 1:1 mixtures of oppositely charged model colloids. For this purpose, the range of the electrostatic interactions was modified by changing the electrolyte concentration of the dispersion medium. At high ion concentration, the particle charge is completely screened and the electrical double layer around the particle is confined to the particle surface. In this case, interactions of electrostatic origin are expected to have little, if any, influence on particle diffusion and aggregation. At low ionic concentration, however, the electrical double layer reaches out far into the solution and, thus, long-range attractive and repulsive forces due to double-layer overlap become important.

Materials and methods

Experimental systems

Two aqueous suspensions of polystyrene microspheres of identical size with oppositely charged surfaces were used as model colloidal systems. The cationic latex was supplied by Interfacial Dynamic Corporation (Portland, OR). It has a surface charge density of $+17.8\ \mu C\ cm^{-2}$ arising from amidine groups. The particle diameter was (525 ± 14) nm as determined by transmission electron microscopy and the polydispersity index was 1.002. The critical coagulation concentration of (0.17 ± 0.01) M KBr was determined by means of low-angle nephelometry. The anionic latex was synthesized in our laboratories according to the recipe described in Ref. [9] and had a surface charge density of $-8.7\ \mu C\ cm^{-2}$ owing to sulphate groups, a particle diameter of (524 ± 19) nm, a polydispersity index of 1.005 and a critical coagulation concentration of (0.34 ± 0.02) M KBr.

Sample preparation

Prior to aggregation, the samples were sonicated for 10 min (ultrasonic bath Ultrasons, Selecta, 150 W) in order to approach monomeric initial conditions. All experiments were started by mixing equal volumes of the corresponding reagents. For this purpose, the samples were poured thorough a Y-shaped mixing device. The initial monomer concentration, i.e. the concentration of individual colloids, in the reaction vessel was always $1\times10^8\ cm^{-3}$ so that single particle detection could be ensured during the complete processes. The temperature was stabilized at (20 ± 1) °C. Ultrapure water was obtained using Millipore equipment (Milli-Q Academic System, Millipore Corporation, MA). Under these conditions, the Bjerrum length is 0.7 nm. When electrolyte was added to the samples, its concentration was always chosen well above the critical coagulation concentration of both samples at 1.0 M KBr. In this case, the Debye length is approximately 0.3 nm. Since this value is smaller than the Bjerrum length, the electrical double layer may be considered to be completely suppressed. Without any added electrolyte, the Debye length is mainly determined by the pH of the medium and was estimated to be of the order of one particle radius. This means that, here, the electrical double layer reaches out far into the surrounding medium.

Single-cluster light scattering

The SCLS device used for this study is basically a counter and classifier of clusters. It allows time-resolved measurements of the cluster size distribution, $c_n(t)$, up to heptamers and also provides the total number of aggregates, $\Sigma c_n(t)$. The instrument separates the aggregates by means of hydrodynamic focusing, forcing them to flow one by one across a focused laser beam. As they pass, they scatter a pulse of light that is detected at low scattering angle. Under these conditions, the light pulse intensity is directly related to the cluster size and, therefore, the cluster size distribution can be obtained by simply counting the pulses as a function of the scattering intensity. Further details on the technique can be found elsewhere [10, 11].

Obtaining the aggregation rate constants

The dimer formation rate constant, k_{11}, can be obtained from the time evolution of the monomer concentration, $c_1(t)$, using the well-known constant kernel solution of the Smoluchowski rate equation which is given by [12]

$$c_1(t) = \frac{c_0}{\left(1 + \frac{c_0 k_{11}}{2} t\right)^2}, \tag{1}$$

where c_0 is the initial particle concentration. This equation can be transformed into

$$\frac{1}{\sqrt{c_1(t)}} = \frac{1}{\sqrt{c_0}}\left(1 + \frac{c_0 k_{11}}{2} t\right). \tag{2}$$

Hence, plotting the inverse square root of the monomer concentration versus time should give a straight line if this theory is valid. k_{11} may then be calculated from the slope once the initial particle concentration is known. The usefulness of this method has been shown for homoaggregation processes starting from monomeric initial conditions [12]. Nevertheless, when using Eq. (2), one should take into account that it is strictly valid only for the constant aggregation kernel. For real aggregation processes, however, the linear behaviour for the inverse square root of the monomer concentration is usually maintained only for relatively short times. At larger aggregation times, however, the curve tends to bend owing to deviations of the real aggregation kinetics from the constant kernel model. In charge-heteroaggregation experiments, the situation is even more complicated since here three different dimer formation rate constants have to be defined. These are the rate constants for dimers formed by two positive particles, two negative particles and the mixed case. Hence, the meaning of the fitted dimer formation rate constant, k_{11}, is not clear a priori. In this work, however, we performed only heteroaggregation experiments of stable systems where aggregation takes place only between particles of opposite charge. This implies that only mixed dimers are formed. Since two monomers of the same class do not aggregate, the value of k_{11} obtained can be identified as the dimer formation rate constant for pure heteroaggregation.

Results and discussion

Before performing heteroaggregation measurements of 1:1 mixtures of positively and negatively charged particles at low and high ionic strength, we checked that both samples were stable when no electrolyte was added. This means that long-range repulsive electrostatic interactions impede the particles approaching each other and stick together owing to attractive London–van der Waals type forces.

Then, we proved that the repulsive interactions between particles of the same type can be completely shielded by raising the electrolyte concentration to 1.0 M KBr. In this case, the samples are expected to aggregate in the purely diffusion-limited regime since no repulsive interactions restrict the diffusive motion of the particles. The experimentally obtained cluster size distributions exhibit the typical DLCA behaviour for homoaggregation and so confirm this hypothesis (the corresponding curves are not shown here owing to space restrictions.). The same is true for the dimer formation rate constants, k_{11}, which were obtained as described in the previous section. The results are $(4.7\pm0.5)\times10^{-12}$ and $(4.3\pm0.5)\times10^{-12}$ cm^3s^{-1} for cationic and anionic samples, respectively. Both values are well comprised within the interval of $(6\pm3)\times10^{-12}$ cm^3s^{-1} given by Sonntag and Strenge [13] as an average value for experimental DLCA measurements.

At high ionic strength (1.0 M KBr), not only the repulsive electrostatic interactions among equally charged particles are shielded but also the attractive forces between oppositely charged particles should be completely suppressed. This was confirmed by aggregating a 1:1 mixture of positively and negatively charged particles at high electrolyte concentration. Also here, the cluster-size distribution obtained shows the typical DLCA behaviour (Fig. 1) and the experimental dimer formation rate constant, $k_{11}=(4.3\pm0.5)\times10^{-12}$ cm^3s^{-1}, is quite close to the average experimental value given by Sonntag and Strenge for DLCA. Consequently, the system acts as if it were an unstable monocomponent colloid and neither attractive nor repulsive forces alter the diffusive motion of the particles.

In order to show the presence and importance of long-range attractive forces between the oppositely charged particles, we studied the time evolution of the cluster size distribution arising in a 1:1 mixture without any added electrolyte. The results obtained are shown in Fig. 2. Here, completely different behaviour was found which does not correspond to any of the known aggregation regimes. Of course, the monomer concentration decreases monotonously with time since monomers can only react among themselves or with any other cluster forming dimers or larger aggregates. The evolution of the concentration of clusters larger than monomers is a competition between formation from smaller aggregates and disappearance due to aggregation with any other cluster. Consequently, the corresponding curves exhibit a maximum. Only dimer formation is so fast that the initial increase to the maximum could not be resolved. At high electrolyte concentration, however, this maximum could be clearly distinguished. This implies that dimer formation is slower when the electrostatic interactions are shielded. Consequently, the electrostatic interactions between oppositely charged particles at low electrolyte concentration must be attractive and of long range. It should also be pointed out that the dimer concentration decays much faster than that of monomers and any other type of clusters. This behaviour may be explained taking into account that dimers have to be formed by two oppositely charged monomers. Hence, they should be almost electrically neutral or, at most, bear a relatively low net charge. This means that dimers can react in a diffusive manner with all types of charged and neutral aggregates and monomers. The corresponding dimer formation rate constant, $k_{11}=(8.2\pm1.3)\times10^{-12}$ cm^3s^{-1}, was determined from the time evolution of the monomer concentration as described in the Materials and methods section. This value is almost double that obtained for pure homoaggregation and so corroborates the high monomer reactivity due to additional long-range attractive interactions.

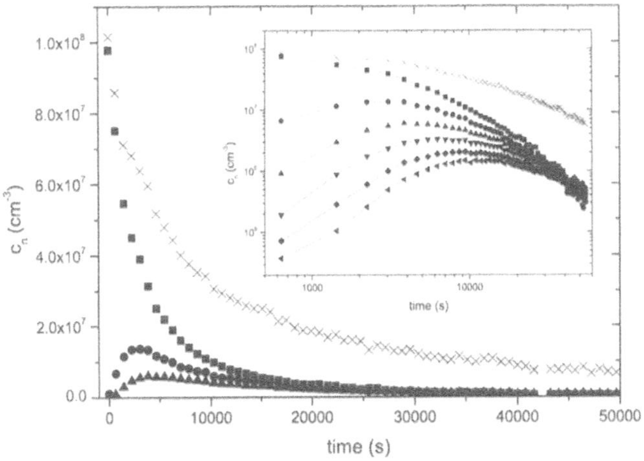

Fig. 1 Time evolution of the cluster size distribution arising in a 1:1 mixture of oppositely charged particles at high electrolyte concentration (1.0 M KBr). The data points correspond to the total number of aggregates (*crosses*), monomers (*squares*), dimers (*circles*), trimers (*triangles*), etc. The *inset* shows the same data on a double-logarithmic scale for monomers up to hexamers. The *lines* are drawn only as a guide for the eye

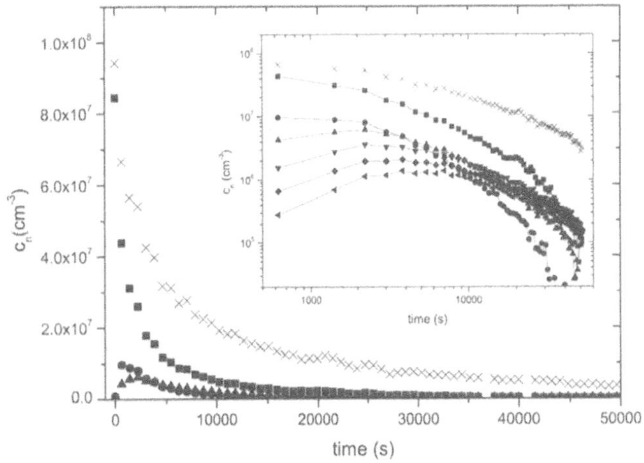

Fig. 2 Time evolution of the cluster size distribution arising in a 1:1 mixture of oppositely charged particles without any added electrolyte. The symbols have the same meaning as in Fig. 1

Conclusions

Charge-heteroaggregation processes arising in 1:1 mixtures of positively and negatively charged particles were studied at low and high electrolyte concentration. SCLS was employed to monitor and resolve the detailed cluster size distribution up to hexamers. The results obtained show that long-range electrostatic interactions between oppositely charged particles favour aggregation at low electrolyte concentration. Especially, dimers are formed from oppositely charged monomers at an increased rate which is almost twice the value found for pure diffusion-limited aggregation.

Acknowledgement This work was supported by the Ministerio de Ciencia y Tecnología (Plan Nacional de Investigación Científica, Desarrollo e Innovación Tecnológica (I + D + I) project MAT 2003-08356-c04-01).

References

1. Bandini P, Prestidge CA, Ralston J (2001) Miner Eng 14:487
2. Luna-Xavier JL, Guyot A, Bourgeat-Lami E (2002) J Colloid Interface Sci 250:82
3. Hayashi H, Nihei T, Ono M, Tsuneda S, Hirata A, Sasaki H (2001) J Colloid Interface Sci 243:109
4. Hidalgo-Álvarez R, Martín A, Fernández A, Bastos D, Martínez F, de las Nieves FJ (1996) Adv Colloid Interface Sci 67:1
5. Meakin P, Lin MY, Lindsay HM, Weitz DA, Klein R, Ball RC (1990) J Phys Condens Matter 2:3093
6. Meakin P, Lin MY, Lindsay HM, Weitz DA, Klein R, Ball RC (1990) Phys Rev A 41:2005
7. Islam AM, Chowdry BZ, Snowden MJ (1995) Adv Colloid Interface Sci 62:109
8. Puertas AM, Fernández-Barbero A, de las Nieves FJ (2001) J Chem Phys 114:591
9. Goodwin JW, Hearn J, Ho CC, Ottewill RH (1974) Colloid Polym Sci 252:464
10. Fernández-Barbero A, Cabrerizo-Vílchez MA, Martínez-García R, Hidalgo-Álvarez R (1996) Phys Rev E 53:4981
11. Fernández-Barbero A, Schmitt A, Cabrerizo-Vílchez MA, Martínez-García R (1996) Physica A 230:53
12. Holthoff H, Schmitt A, Fernández-Barbero A, Borkovec M, Cabrerizo-Vílchez MA, Schurtenberger P, Hidalgo-Álvarez R (1997) J Colloid Interface Sci 192:463
13. Sonntag H, Strenge K (1987) Coagulation kinetics and structure formation. VEB Deutscher Verlag der Wissenschaften, Berlin

Progr Colloid Polym Sci (2004) 126: 155–158
DOI 10.1007/b94013

F. Royer
D. Jamon
J.J. Rousseau
D. Zins
V. Cabuil
S. Neveu
H. Roux

Magneto-optical properties of $CoFe_2O_4$ ferrofluids. Influence of the nanoparticle size distribution

F. Royer (✉) · D. Jamon, J.J. Rousseau
Dispositifs et Instrumentation en Opto-électronique et Micro-onde, Université Jean Monnet, 23, rue Michelon, 42023 St. Etienne, France
e-mail: francois.royer@univ-st-etienne.fr
Tel.: +33-4-77485076
Fax: +33-4-77485039

D. Zins · V. Cabuil · S. Neveu
Liquides Ioniques et Interfaces Chargées, Université Pierre et Marie Curie, 4, place Jussieu, 75252 Paris, France

H. Roux
Département de Physique des Matériaux, Université Claude Bernard, 43, Bd. du 11 Novembre, 69622 Villeurbanne, France

Abstract This paper reports an experimental investigation on size-sorted cobalt ferrite ($CoFe_2O_4$) ferrofluids. The sample preparation and the size-sorting process are exposed. Faraday rotation measurements, performed with a null ellipsometer, and transmittance spectra are discussed using the size of the nanoparticles and the Co-to-Fe ratio. Comparison with the properties of γ-Fe_2O_3 particles is made. Finally the study shows that the ferrofluid factor of merit is enhanced by small particles.

Keywords Ferrofluid · Particles size sorting · Magneto-optical effects · Transmittance spectrum

Introduction

Composite materials such as magnetic nanoparticles embedded in a silica gel matrix are promising candidates for future magneto-optical applications [1]. They exhibit large remanent or field-induced magneto-optical effects and good transparency [1, 2]. These effects are dependent on the particle size. Thus, in order to improve the properties of magneto-optical devices, a systematic study was carried out, with the liquid phase used for the sol–gel preparation, i.e., ferrofluids.

Magnetic fluids, or ferrofluids, are colloidal dispersions of nanosized magnetic particles in a liquid carrier. Each particle being a magnetic monodomain, the particle orientation in the direction of an applied magnetic field results in magnetization of the solution [3, 4]. Moreover under the influence of a longitudinal magnetic field (magnetic direction parallel to light beam), ferrofluids induce a magneto-optical effect known as Faraday rotation (FR). The magnitude of this effect is proportional to the particle volume fraction, ϕ, and to the cell length and depends on the field intensity [5, 6].

In previous work [7], the particle size dependence of FR in aqueous colloidal dispersions of maghemite (γ-Fe_2O_3) was studied; it was shown that using small particles provides a large increase of the ferrofluid factor of merit, i.e., the effective FR divided by the optical absorption coefficient, α.

Cobalt ferrite ($CoFe_2O_4$) ferrofluids exhibit a higher magnitude of FR than maghemite ones. A systematic study on size-sorted cobalt ferrite ferrofluids was thus performed in order to obtain a higher-quality material.

This paper is organized as follows. The synthesis of the ferrofluid samples and their characteristics are described in the following section 1. Optical results are presented and discussed in the subsequent section.

Experimental

Synthesis of particles

The cobalt ferrite particles were synthesized according to the method in Ref. [8] and were modified in order to obtain water-dispersible particles of controlled size. Sodium hydroxide (160 ml; 10 mol l^{-1}) was abruptly added under vigorous stirring to a solution of cobalt and ferric nitrate (350 ml, sum of iron and cobalt concentrations of 0.5 mol l^{-1}; ratio of iron-to-cobalt concentrations of 2). A dispersion of a nonmagnetic amorphous precipitate was obtained. It was stirred under heating (100 °C) for 2 h. The magnetic precipitate was isolated and washed three times with distilled water, then dispersed in a solution of nitric acid (200 ml; 2 ml l^{-1}) for 5 min. It was again isolated, and was stirred in a trisodium citrate solution (1 l, 0.1 mol l^{-1}) at 90 °C, for 30 min.

The particles obtained through this procedure had an average diameter around 9 nm, with a rather wide polydispersity that is well described using a log–normal distribution of diameters [3]:

$$P(D) = \frac{1}{\sqrt{2\pi}sD}\exp\left\{-\frac{1}{2(s)^2}\left[\ln\left(\frac{D}{D_0}\right)\right]^2\right\}.$$

After such a synthesis process, s is around 0.35.

The particles obtained by this procedure can be dispersed in water to obtain a stable aqueous colloidal dispersion, the pH of which is 7.

A size-sorting process was performed in order to reduce the polydispersity and to get particles of different diameters. This process is based on a controlled destabilization of the colloidal dispersion through the addition of salt [9]. It allowed us to fractionate the initial population of particles in three samples of average diameters 8, 10 and 12.5 nm. For all these samples, the chemical composition of the particles was found such that the ratio of cobalt-to-iron concentrations was around 0.45.

In order to get smaller particles, tartrate anions were added during the first step of the synthesis process (coprecipitation) according to the method in Ref. [10]. This allowed us to obtain particles with an average diameter around 6 nm, but the chemical composition of these particles had a smaller cobalt-to-iron concentration ratio (0.3). They were dispersed in water using the same kind of procedure as that already described.

The four dispersions obtained were adjusted to the same particle volume fraction, $\phi = 0.2\%$.

Characterization of the samples

All the samples were characterized by chemical titration of iron and of cobalt in order to establish the composition of the particles and to determine the volume fraction of the particles in the colloidal dispersion. The size distribution was deduced from transmission electron microscopy pictures (Table 1, Fig. 1).

Magnetization curves were plotted for each colloidal dispersion using a vibrating magnetometer (Foner apparatus), between 0 and 8×10^5 A m^{-1}.

FR measurements were performed with a high-precision null ellipsometer [11], using a 1-mm-thick cell, free of residual birefringence. Several light wavelengths were used: 633, 820, 980, 1150, 1320, 1550 nm. The electromagnet associated with the ellipsometer can reach a magnetic field strength of 6.4×10^5 A m^{-1}.

Sample transmittance was determined using a PerkinElmer spectrophotometer (Lambda 900) in the range 550–1600 nm, with a 1-mm-thick cell.

Table 1 Sample characteristics. The median diameter, D_0, and standard deviation s of the diameter distributions are deduced from transmission electron microscopy pictures. Each particle volume fraction, ϕ, was adjusted to 0.2% by water dilution. [Co] and [Fe] are the results of a chemical titration

D_0 (nm)	s	ϕ (%)	[Co]/[Fe]
12.5	0.2	0.2	0.48
10	0.25	0.2	0.5
8	0.25	0.2	0.42
6	0.25	0.2	0.27

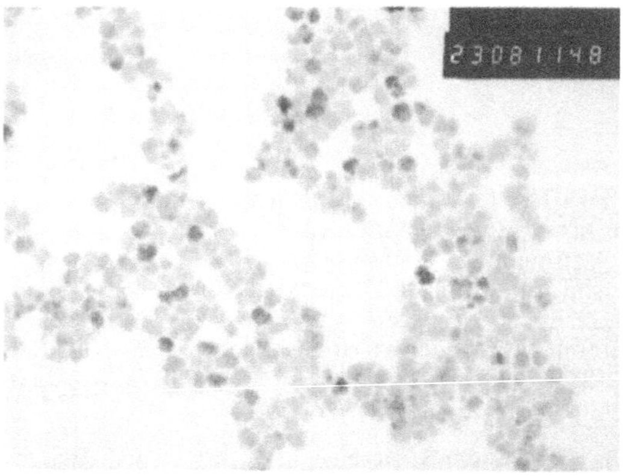

Fig. 1 Transmission electron microscopy picture of the largest particles

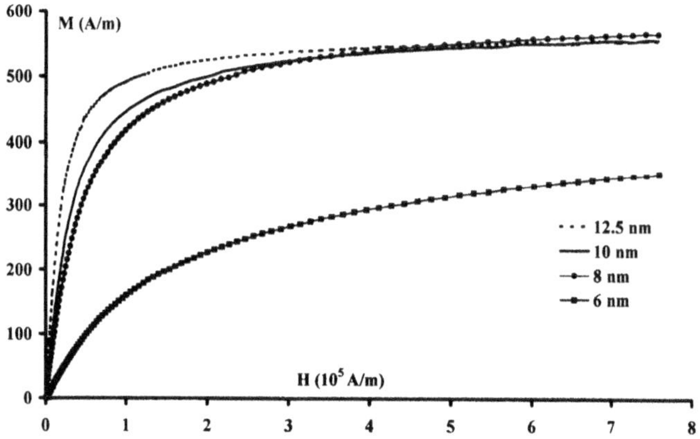

Fig. 2 Magnetization curves plotted for each colloidal dispersion using a vibrating magnetometer (Foner apparatus)

Results and discussion

Magnetisation curves are reported in Fig. 2. All the samples have the same particles volume fraction and, indeed, should have the same magnetization saturation level, but different behavior owing to their own size distribution. That is the case except for the 6-nm sample, which has a lower saturation owing to the low cobalt-to-iron concentration ratio.

FR results are reported in Fig. 3. The wavelength dependence is identical for all the samples and is characteristic of cobalt ferrite particles [11]: this spectrum evidences three main magneto-optically-active transitions around 633, 820 and 1550 nm. The first and the last are assigned to Co^{2+} ions in tetrahedral coordination [12, 13, 14] whereas the 820-nm one is assigned to Fe^{3+} ions in octahedral coordination [14]. No significant difference can be deduced between the 8-, 10- and 12.5-nm samples, although there is a 50% difference between 6- and 10-nm maghemite particles [7]. This result indicates that, in this range of diameters, there is no size effect on FR in $CoFe_2O_4$ ferrofluids.

The FR is plotted as a function of the cobalt-to-iron concentration ratio for 633 and 820 nm in Fig. 4. The quite linear variation shows that cobalt ferrite FR is very sensitive to this ratio and that the low FR of the 6-nm sample (Figs. 3, 4) seems to be much more due to its different ratio than its different mean size diameter.

In order to study the particle size effect on transmittance, the curves reported in Fig. 5 are referenced to a cell full of water. The wavelength dependence is characteristic of cobalt ferrite particles [11] and it points out a noticeable extinction size effect: small particles provide better transmittance than big ones. The same kind of effect has been evidenced for maghemite particles [7]. The particle diameter being small compared with the wavelength, the ferrofluid extinction is dominated by absorption [15]. Anyway, scattering losses are too weak to explain such transmittance size effects, which may be related to the size dependence of the refractive index of the particles [16].

Our experiments on size-sorted $CoFe_2O_4$ samples underline a size effect on transmittance but no significant size effect on FR. The factor of merit of the samples appears thus as the relevant parameter:

$$F = \frac{\gamma(^\circ m^{-1})}{\alpha(m^{-1})},$$

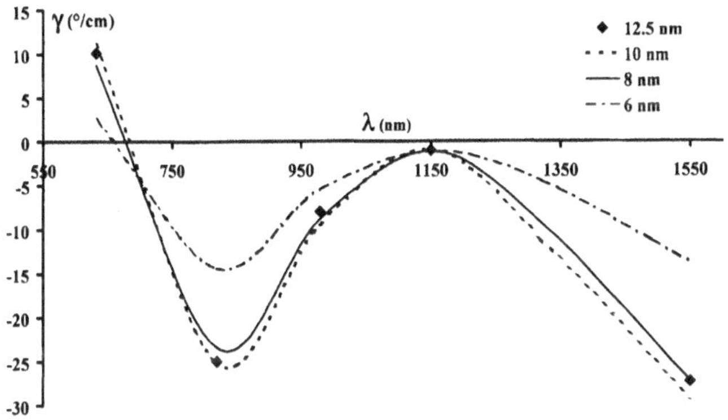

Fig. 3 Faraday rotation (*FR*) as a function of wavelength. γ is the effective FR due to the nanoparticles (water FR has been subtracted) in a magnetic field of 6.4×10^5 A m^{-1}

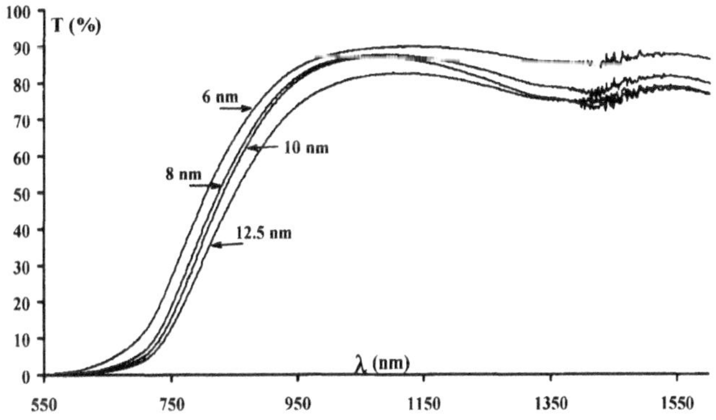

Fig. 5 Sample transmittance, *T*, versus wavelength, λ. Measurements were made with a 1-mm-thick cell and are referenced to a cell full of water. The noisy part of the curves near 1,450 nm is due to a water absorption band which gives a very low value to the reference

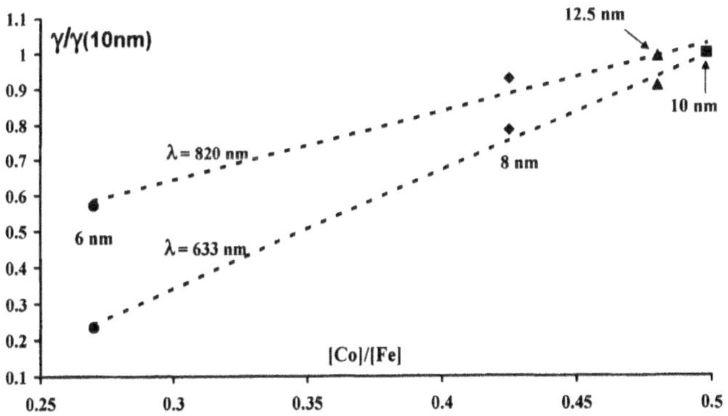

Fig. 4 FR as a function of [Co]/[Fe] in the case of light of wavelength 820 and 633 nm. [Co] and [Fe] are deduced from chemical titration. γ/γ (10 nm) is the FR rotation of the current sample normalized to the 10-nm sample, which has the best [Co]/[Fe], i.e. 0.5

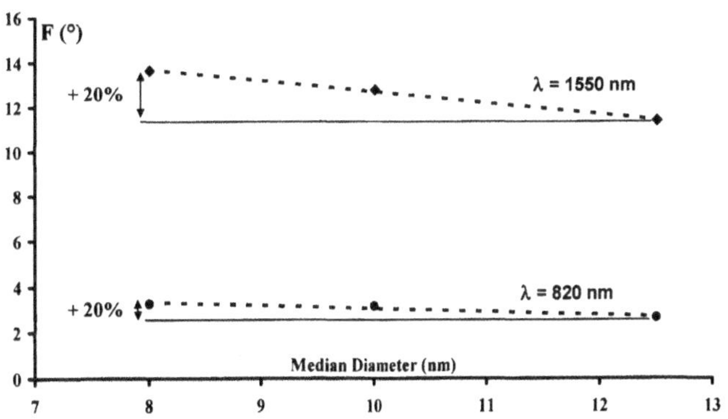

Fig. 6 Ferrofluid factor of merit as a function of median diameter for light of wavelength 820 and 1,550 nm. Three samples values are reported: 8, 10 and 12.5 nm. The 6-nm value is not relevant because of a too low [Co]/[Fe]

where γ is the effective FR and α is the absorption coefficient.

F is reported in Fig. 6 for wavelengths 820 and 1550 nm. This figure shows that for both wavelength the use of 8-nm-diameter particles provides a 20% increase in F compared with 12.5-nm particles. This result where cobalt ferrite properties should increase with decreasing particle diameter is of great interest in the course of magneto-optical applications.

Conclusion

This experimental investigation on size-sorted cobalt ferrite ferrofluids evidences that FR is almost size independent and is very sensitive to the cobalt-to-iron concentration ratio. But size effects revealed from the transmittance lead to an important increase of the factor of merit of the material when using small particles (8 nm compared with 12.5 nm). This is an interesting result for magneto-optical applications based on cobalt ferrite particles and trying to reduce more the particles diameter with a controlled cobalt-to-iron concentration ratio is very attractive.

Acknowledgement We thank D. Talbot for experimental help.

References

1. Bentivegna F, Nyvlt M, Ferre J, Jamet JP, Brun A, Visnovsky S, Urban R (1999) J Appl Phys 85:2270
2. Jamon D, Robert S, Donatini F, Rousseau JJ, Bovier C, Roux H, Serrughetti J, Cabuil V, Zins D (2001) IEEE Trans Magn 37:3803
3. Bacri JC, Perzynski R, Salin D, Cabuil V, Massart R (1986) J Magn Magn Mater 62:36
4. Langevin P (1905) J Phys 678
5. Davies HW, Leweellyn JP (1980) J Phys 13:2327
6. N.A. Yusuf, I. Abu-Alarayesh, Rousan AA, El-Ghanem HM (1990) IEEE Trans Magn 26:2852
7. F. Royer, Jamon D, Rousseau JJ, Cabuil V, Zins D, Roux H (2002) Trans Magn Soc Jpn 2(4)
8. Chantrell RW, Popplewell J, Charles SW (1978) IEEE Trans Magn Mag 14:975
9. Cousin F, Dubois E, Cabuil V (22001) J Chem Phys 115:6051
10. Neveu S, Bee A, Robineau M, Talbot D (2002) J Colloid Interface Sci (to be published)
11. Donatini F, Jamon D, Monin J, Neveu S (1999) IEEE Trans Magn 35:4311
12. Fontijn WFJ, van de Zaag PJ (1998) J Appl Phys 83:6765
13. Martens JWD, Peeters WL, van Noort HM, Erman M (1985) J Phys Chem Sol 46:411
14. Stichauer L (1994) PhD thesis. Université de Nancy
15. Lacoste D, Donatini F, Neveu S, Serughetti JA, Van Tiggelen BA (2000) Phys Rev E 62:3934
16. Lenglet J (1996) PhD thesis. Université Denis Diderot, Paris

Progr Colloid Polym Sci (2004) 126: 159–162
DOI 10.1007/b93972

E. Belamie
M.M. Giraud-Guille

Liquid-crystalline behavior in aqueous suspensions of elongated chitin microcrystals

Abstract Chitin elongated microcrystals are rigid rodlike particles known to form chiral nematic liquid-crystalline phases when suspended in acidic aqueous solutions. These phases remarkably mimic the cholesteric geometry observed in biological tissues such as crustacean exoskeletons where chitin constitutes the fibrous component of a composite material, in association with a protein matrix and a mineral phase. The aim of the present work is to investigate the phase diagram of chitin colloidal dispersions and in particular the chirality of the nematic phase and its dependence on the composition of the suspension. Both the pH and the ionic strength affect the concentrations in the isotropic, c_I, and chiral nematic, c_N, phases at coexistence. It turns out that c_N does not vary monotonically with HCl and goes through a minimum of $5.36 \pm 0.29\%$ (in weight) at a HCl concentration of around 5×10^{-4} M, with $c_I = 2.98 \pm 0.21\%$. When the ionic strength of the solution reaches about 10^{-2} M the system does not bulk phase-separate anymore; anisotropic droplets form throughout the sample but do not coalesce or sediment. The cholesteric pitch, P_0, in the biphasic domain of the diagram ranges from 27.5 ± 3.5 to $140 + 17$ μm depending on the composition of the suspension medium. The variations of P_0 follow a power law as a function of the concentration of crystalline rods in the coexistence domain $P_0 \sim c^{-\nu}$, with $0.3 < \nu < 1.2$ as a function of ionic strength. This set of results sheds light on the self-assembly mechanisms that lead to the formation of a complex and ordered biological composite like the cuticle of insects and crustaceans. A better description of the phase diagram of these colloidal liquid crystals is a useful basis for their application in designing, for instance, biomimetic composites and ordered mesoporous materials.

Keywords Chitin · Colloidal liquid crystal · Cholesteric · Phase separation · Chirality

E. Belamie (✉) · M.M. Giraud-Guille
Laboratoire de Chimie de la Matière Condensée, UMR 7574,
Université Pierre & Marie Curie,
CNRS, Ecole Pratique des Hautes Etudes,
12 rue Cuvier, 75005 Paris, France
E-mail: ebelamie@snv.jussieu.fr
Tel.: +33-1-44276552,
Fax: +33-1-44276539

Introduction

The use of colloidal liquid crystals as templates for the growth of metal oxide networks has recently gained attention. Such organic–inorganic composites may find applications as, for example, ordering quantum dots [1] or forming organized mesoporous materials. Much attention was devoted to the study of surfactant mesophases and their ability to accommodate, for instance, silica networks produced by in situ condensation of molecular precursors [2]. Many biological systems, mostly extracellular structures, are formed following similar strategies although the template molecules are proteins or polysaccharides. Connective tissues of vertebrates and arthropod exoskeletons are examples of biological materials in which a fibrous and organized organic matrix serves as a mould for the deposition of a reinforcing mineral phase. Collagen fibers in bones and chitin fibers in crustacean shells are both associated with calcium phosphate and/or calcium carbonate that diffuse and precipitate after the fibrous component has been excreted and stabilized. In these tissues, the supporting organic component is made of preformed nanometer- to micron-sized particles that arrange into supramolecular structures with geometries analogous to those of some liquid crystals [3, 4]. It was shown that, in most cases, the fibers (collagen, chitin, cellulose) can be extracted and dispersed in aqueous media to form colloidal suspensions [5, 6]. At appropriate concentrations, liquid-crystalline phases can be identified, indicating that in these systems, the rodlike particles tend to align with each other. Unlike surfactant systems, the shape of the particles is independent of the phase diagram, which can make it easier to control the behavior of the suspensions as a function of concentration, pH, ionic strength and mixing with other chemical species. Revol and coworkers [5, 7] established without doubt that it was possible to obtain cholesteric phases with chitin crystals suspensions. The aim of the present work was to define better the conditions that allow the formation of a stable liquid crystal. In the first part of the paper, we report data indicating that the system bulk phase separates in a large range of pH but at rather moderate ionic strength. We then investigate the chirality of the cholesteric phase by studying the evolution of the pitch as the composition of the mixtures changes.

Experimental results

Extraction and purification processes

Chitin extracted from shrimp shells was kindly provided by France Chitin (http://www.france-chitine.com). The particles were prepared as follows [5]. Chitin flakes (1 g) were submitted to hydrolysis in 20 ml 4N HCl at boiling for 90 min. The resulting suspension was thoroughly dialyzed against deionized water and a dispersion of the particles was achieved by sonication (Branson sonifier). Large particles and aggregates were eliminated by low-speed centrifugation. The supernatants were then spun with an ultracentrifuge at 50,000*g* and the pelleted particles were redispersed in the working medium. Stock suspensions were brought to a concentration of about 10% in weight, measured precisely by evaporation and weighing.

Phase diagrams

A series of dilutions were prepared by weighing in test tubes (6 mm in diameter) and inserted into round capillary tubes of 600-μm diameter for microscopic investigations. Measurements were made after the samples had reached equilibrium, which took typically a few days, and did not evolve further over a period of several months. Inspections of the samples were performed between crossed polarizers either with the naked eye or under a microscope.

As the concentration of chitin increases, the samples go from isotropic to cholesteric (Fig. 1) through a biphasic domain where the two phases coexist with concentrations c_I and c_N, respectively. The onset of the phase separation, i.e. c_I, does not change significantly as the HCL concentration increases, until it reaches 5×10^{-4} M (Fig. 2). c_N follows a similar trend except for a slight decrease in the lowest part of the diagram, at higher pH and lower ionic strength. For HCl concentrations ranging from 5×10^{-4} to 5×10^{-3} M, both c_I and c_N increase. Samples prepared with 10^{-2} M HCl and beyond did not exhibit bulk phase separation. Instead, the birefringence seemed to increase continuously from dark to very bright samples when viewed between cross polarizers. Samples were also prepared with a background HCl concentration of 10^{-4} M to which various amounts of NaCl were added. The total ionic strength, I_{tot}, refers to mobile ions added to the suspensions and is thus the sum of HCl and NaCl concentrations. As I_{tot} increases (Fig. 3), c_I and c_N follow the same trend as with HCl only beyond 10^{-4} M.

When the volume fraction of the nematic phase in biphasic samples is plotted against the total chitin concentration (Fig. 4), assuming a linear dependence, extrapolations to 0 and 1, respectively, give the concentrations in the isotropic and nematic phases. The presence of birefringent droplets suspended in the isotropic phase is easily identified and such samples were not taken into account in the linear regressions. The extrapolated coexistence concentrations measured in capillary tubes are reported in Fig. 5 together with the boundaries observed visually. The values for c_I are in good agreement and can be fitted with those predicted by the classical Onsager model [8] adapted to polydispersed samples [9] (continuous lines). In contrast, the concentrations at the biphasic/nematic boundary seem to depend on the tube diameters.

Cholesteric pitch

One important parameter to describe the cholesteric phase is its pitch, P_0, i.e. the distance that separates rods having equivalent orientations after 360 ° rotation. The evolution of P_0 as a function of HCl concentration is plotted in Fig. 6 and appears to be nonmonotonic. Each data point is the average of all the values measured in a dilution series for a given condition (HCl concentration). The inset in Fig. 6 shows the variation of P_0 as a function of chitin concentration and the corresponding log–log plots give straight lines in all conditions. Their slopes give the scaling exponent, ν, assuming an exponential relation of the form $P_0 \sim c^{-\nu}$ as suggested by theory for threaded rods [10, 11]. The values of ν obtained for different HCl concentrations are reported in Fig. 7.

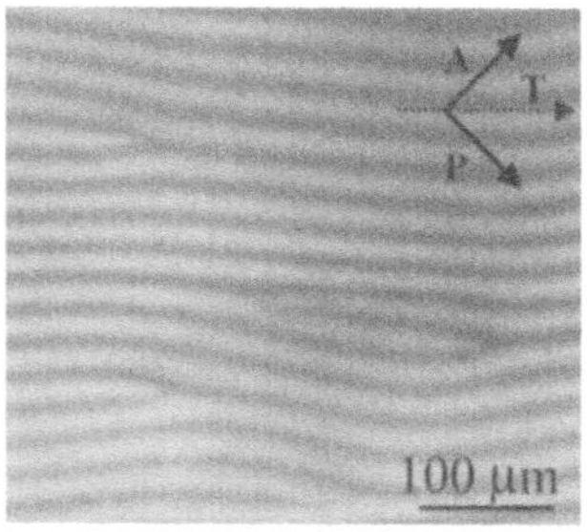

Fig. 1 Micrograph showing the fingerprint pattern characteristic of the cholesteric phase as observed between polarizers. For this sample, the HCl concentration is 2.5×10^{-3} M, the chitin concentration is $4.52 \pm 0.13\%$ (in weight) and $P_0 = 53.6 \pm 3.1$ μm. The long axis of the tube (T) is indicated as a *horizontal dotted arrow* and those of the polarizers are represented as *straight arrows* making an angle of 45 ° with respect to the tube axis

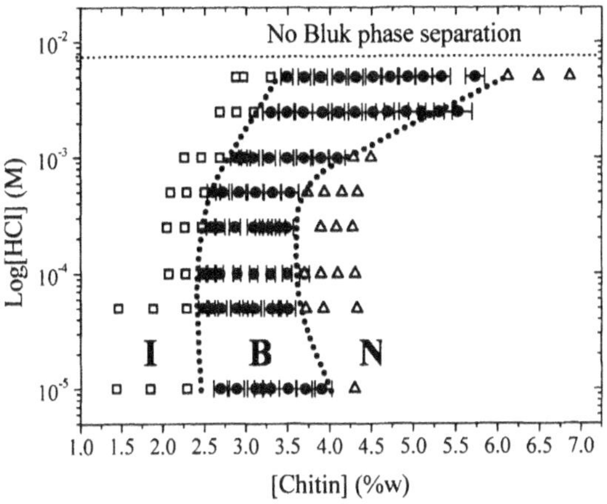

Fig. 2 Phase diagram obtained for chitin elongated particles dispersed in different HCl solutions. The diagram was mapped out by visual inspections of samples with different chitin concentrations between crossed polarizers. No bulk phase separation was observed for samples at a HCl concentration of 10^{-2} M and beyond. Isotropic (*squares*), biphasic (*circles*), chiral nematic (*triangles*)

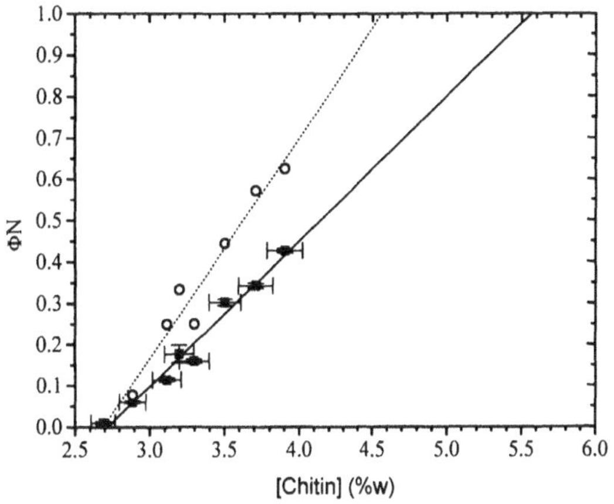

Fig. 4 Evolution of the nematic volume fraction ΦN as a function of chitin concentration in biphasic samples with a HCl concentration of 10^{-5} M. The tube diameter is either 6 mm (*circles*) or 600 μm (*squares*). All measurements were made after the tubes had been left standing for 2 weeks at room temperature. The *straight* and *dotted lines* are linear regressions of the data from which extrapolated values of c_I and c_N were extracted

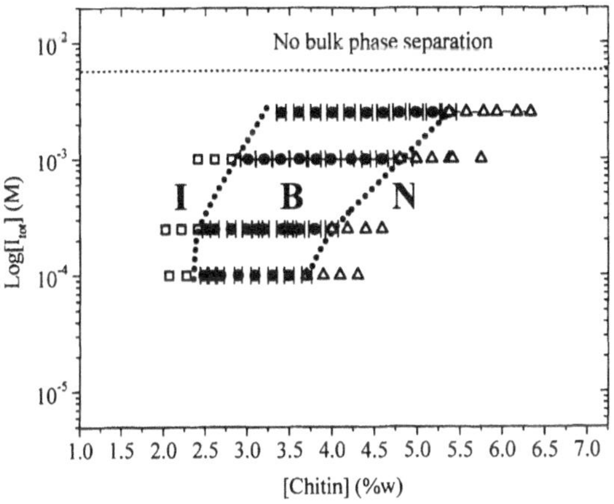

Fig. 3 Phase diagram similar to Fig. 2 but obtained in 10^{-4} M HCl to which various amounts of NaCl were added. I_{tot} refers to mobile ions added to the suspensions, and is thus the sum of HCl and NaCl concentrations. For samples prepared with $I_{tot} > 5\times10^{-3}$ M, no bulk phase separation was observed. See legend of Fig. 2 for details.

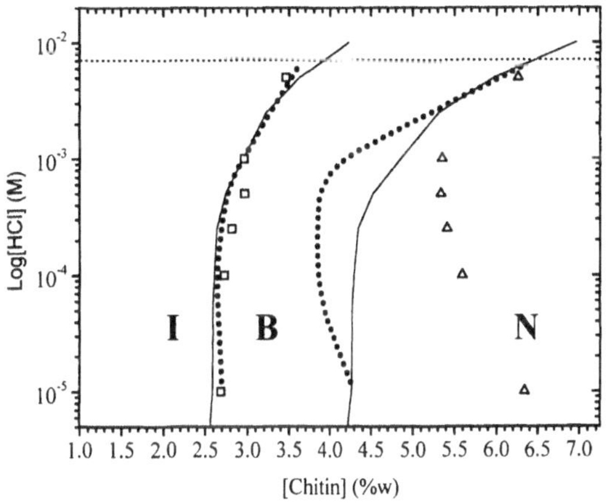

Fig. 5 Variation of the biphasic domain boundaries (c_I and c_N) as a function of HCl concentration as determined by visual inspection between crossed polarizers (*dotted lines*, see Fig. 2); extrapolation of the linear regressions (as in Fig. 4), c_I (*squares*) and c_N (*triangles*); and predictions given by a model adapted from Onsager (*straight lines*)

Discussion and conclusions

The phase diagrams obtained for chitin microcrystals suspended in HCl solutions with or without added salt can, for the most part, be explained with an Onsager-type model [8]. Chitin suspensions are stabilized by charges arising from the protonation of amino groups present on the surface of the crystallites. The clouds of counterions and coions surrounding the rods are responsible for an increase in diameter of the order of a few Debye lengths, κ^{-1}. The volume from which one rod is excluded owing to the presence of its neighbor is thus proportional to an effective diameter, D_{eff}, which is affected by the ionic strength. The predicted coexistence concentrations in the isotropic phase are in good agreement with the experimental data, taking into account the twisting effect introduced by electrostatic repulsions [12], and introducing polydispersity to the length distribution of the rods [9]. The contribution of the crystallites to the total ionic strength was included and the surface charge density was varied considering the pK of the amino group to be 6.3 [13]. We do not explain the discrepancy between values of c_N determined in tubes of different diameters but it can hardly be attributed to polydispersity. In addition, the nonmonotonic evolution of c_N as a function of HCl concentration is not expected, even taking into account changes in surface charge density, the increase of which is

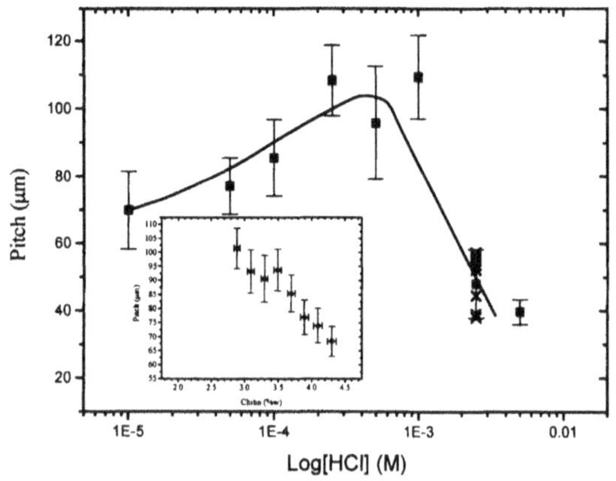

Fig. 6 Variation of the average cholesteric pitch in biphasic samples prepared at different HCl concentrations. Each data point is the average value of data taken along a dilution series in the biphasic domain. The entire set of pitch values is shown for a HCl concentration of 2.5×10^{-3} M (*crosses*). The *line* is a guide for the eye. The *inset* shows the evolution of the pitch as a function of chitin concentration for the case where the HCl concentration is 10^{-4} M. Every pitch value reported in the graph is an average taken over 10–50 measurements made along the capillary tube axis

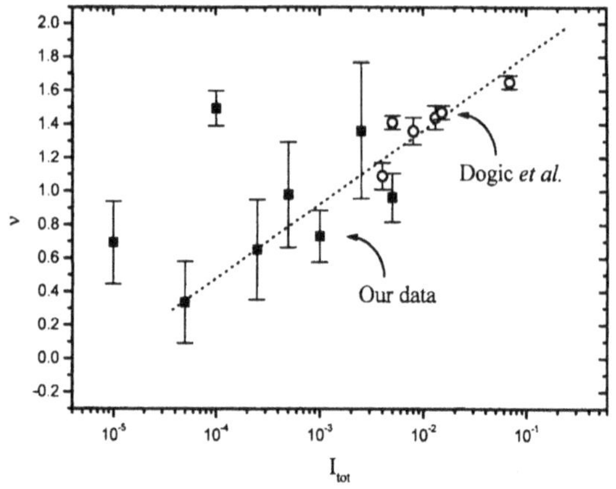

Fig. 7 Values of the scaling exponent ν (see text) at different strengths obtained in this study for chitin microcrystals (*squares*) and for the fd virus in tris(hydroxymethyl)aminomethane–HCl buffers by Dogic and Fraden [15] (*circles*). The dotted line is a guide for the eye

overcome by the consequent larger contribution of the crystallites to the ionic strength.

The origin of chirality in cholesterics is still a much debated question. The most commonly accepted picture is that for threaded rods interacting with electrostatic repulsions, the most favorable situation is when the axis of one rod fits into the groove of the neighboring one. According to some theories [10, 11], the exponent ν that relates P_0 with the concentration of rods should be 1 for hard chiral rods and 5/3 for semiflexible chiral chains, whereas Pelcovits [14] predicts a value of 2, independent of the chain flexibility. We find that the exponent ranges from about 0.3 to 1.2, depending on the ionic strength, as shown in Fig. 7. These values taken at low ionic strength fall on the same master curve as those measured by Dogic and Fraden for the fd virus [15] (open circles in Fig. 7) in tris(hydroxymethyl)aminomethane–HCl buffers, at higher salt concentrations. Extreme situations have been reported for double-stranded DNA with no variation of the pitch as a function of concentration [16] and for poly(*g*-benzyl-L-glutamate) in dioxane with $\nu=2$ [17]; and a number of other different values are available throughout the literature for similar systems. Being able to tune the chirality of colloidal nematic phases is a key technical and theoretical hurdle that needs to be further addressed in order to open the way to the preparation of nanohybrids with controlled structures.

References

1. Lee S-W, Mao C, Flynn CE, Belcher AM (2002) Science 296:892
2. Sanchez C, Soller-Illia GJ de AA, Ribot F, Lalot T, Mayer CR, Cabuil V (2001) Chem Mater 13: 3061
3. Bouligand Y (1972), Tissue Cell 4:189
4. Neville AC (1970) Insect ultrastructure. Blackwell, Oxford
5. Revol J-F, Marchessault RH (1993) Int J Biol Macromol 15:329
6. Giraud-Guille M-M (1992) J Mol Biol 224:861
7. Li J, Revol J-F, Naranjo E, Marchessault RH (1996) Int J Biol Macromol 18:177
8. Onsager L (1949) Ann NY Acad Sci 51:627
9. Lekkerkerker HNW, Coulon P, Van der Haegen R, Deblieck R (1984) J Chem Phys 80:3427
10. Straley JP (1976) Phys Rev A 14:1835
11. Odijk T (1987) J Phys Chem 91:6060
12. Stroobants A, Lekkerkerker HNW, Odijk T (1986) Macromolecules 19:2232
13. Li J, Revol J-F, Marchessault RH (1996) J Colloid Interface Sci 183:365
14. Pelcovits R (1996) Liq Cryst 21:361
15. Dogic Z, Fraden S (2000) Langmuir 16:7820
16. Van Winkle DH, Davidson MW, Chen WX, Rill RL (1990) Macromolecules 23:4140
17. Robinson C (1961) Tetrahedron 13:219

Progr Colloid Polym Sci (2004) 126: 163–168
DOI 10.1007/b93987

Gabriella Caminati
Elisa Peroni
Anna Maria Papini
Piero Baglioni

Photophysical investigation of lipopeptides of myelin basic protein in phospholipid vesicles

G. Caminati (✉) · P. Baglioni
Department of Chemistry, University of Florence, Via della Lastruccia 3, 50019 Sesto F.no, Florence, Italy
e-mail: caminati@unifi.it

E. Peroni · A.M. Papini
Department of Organic Chemistry "Ugo Schiff, University of Florence, Via della Lastruccia 13, 50019 Sesto F.no, Florence, Italy

Abstract We studied lipopeptide analogs of the human myelin basic protein, one of the putative autoantigens of multiple sclerosis disease, in the attempt to determine whether a lipid chain in myelin basic protein peptides may facilitate the insertion of the protein into the membrane bilayer. We used biomimetic models of the cellular membrane of lymphocytes; in particular, we examined how the presence of a lipid moiety in selected positions of the peptide sequence affects the incorporation of the peptide in vesicles of dipalmitoylphophatidylcholine. The resulting systems were characterized by means of UV–vis and fluorescence spectroscopy. The study of the solvatochromic properties of 7-amino-4-methylcoumarin-3-acetic acid provided information on the distribution of the peptides in the phospholipid vesicle and on the local dielectric properties experienced by the label. Further information on the mechanism of interaction of the lipopeptides with the phospholipid vesicles was obtained from fluorescence quenching experiments. The results showed the existence of different localization sites of the lipopeptide in the bilayer characterized by different accessibilities to the quencher. In the case of lipid-bound quenchers we obtained direct access to the position of the fluorescent label in the phospholipid vesicles for the various lipopeptides examined. These findings were interpreted in terms of molecular interactions and lipopeptide conformation and were correlated to clinical trials of the same substances.

Keywords Phospholipid · Unilamellar vesicles · Myelin basic protein · Fluorescence quenching · Fluorescence resonance energy transfer

Introduction

Myelin basic protein (MBP) is a major protein of the myelin membrane in the central nervous system and is directly involved in autoimmune demyelinating diseases, such as Multiple Sclerosis (MS) and in its animal model, i.e. experimental allergic encephalomyelitic (EAE). Vertebrate nervous axons are electrically insulated by the myelin sheath wrapped around them. During MS relapses, patches of white matter in the central nervous system that normally contain tracts of myelinated neurons become inflamed and lose their myelin. Destruction of the myelin sheath interferes with the conduction of nerve impulses along the axons, and results in diminished eyesight, motor coordination, and sensory perception [1].

The long-term goal of the research is the identification of a possible vaccine for EAE and MS [2]; in fact, recent work established that lipid modification of antigenic peptides enhances in humans their immunogenity [3].This result suggests that the lipophilic residue may favor vehiculation of the peptide through the cellular membrane.

In an attempt to clarify the molecular basis of this behavior we studied the interactions between lipopeptide of the human MBP (hMBP) with phospholipid bilayers. We synthesized different peptides, i.e. [Lys^{75}(AMCA)]GpMBP(74–85), Palm-[Lys^{75}(AMCA)]GpMBP(74–85) and AMCA-ωAud-GpMBP(74–85), introducing a palmitoyl moiety and/or a cumarin fluorophore (7-amino-4-methylcoumarin-3-acetic acid, AMCA) in selected positions of the sequence of a well-known MBP immunodominant epitope, i.e. a peptide able to elicit the immune response in EAE.

Vesicles of dipalmitoylphophatidylcholine containing the lipopeptides were prepared both by coextrusion and by incubation as a function of lipopeptide concentration and incubation time. The study of the solvatochromic features of the UV–vis spectra of AMCA provided information on the distribution of the peptides in the phospholipid vesicle and on the local dielectric properties experienced by the label. Photophysical methods such as fluorescence labeling and fluorescence quenching were used to investigate the incorporation of the peptides into phospholipid vesicles. Information on the mechanism of interaction of the lipopeptides with the phospholipid vesicles was obtained from fluorescence quenching experiments using water-soluble quenchers (KI) as well as lipid-bound quenchers (BODIPY). The results showed the existence of different localization sites of the lipopeptide in the bilayer characterized by different quenching mechanisms In particular, fluorescence resonance energy transfer (FRET) experiments allowed us to obtain direct access to the distance between the palmitoyl chain and the vesicle interface for the various lipopeptides examined.

Experimental

Materials

The fluorescent molecules AMCA and 11-[(7-amino-4-methylcoumarin-3-acetic)amino]undecanoic acid (AMCA-ωAud-OH) were synthesized according to a method previously described by the authors [4]. Peptide I, [Lys^{75}(AMCA)]GpMBP(74–85), peptide II, Palm-[Lys^{75}(AMCA)]GpMBP(74–85), and peptide III, AMCA-ωAud-GpMBP(74–85), were synthesized in the solid phase on a semiautomatic NovaSyn Gem apparatus and purified by reverse-phase high-performance liquid chromatography using a Beckman System Gold apparatus, as described previously [4]. Their chemical structures are reported in Fig. 1.

1,2-Dipalmitoyl-*sn* glycero-3-phosphatidylcholine (DPPC) was purchased from Northern Lipids (Vancouver, Canada) and was used without further purification. 2-(4,4-Difluoro-5,7-dimethyl-4-bora-3*a*,4*a*-diaza-*s*-indacene-3-dodecanoyl)-1-hexadecanoyl-*sn*-glycero-3-phosphocholine (BODIPY-DPPC) was purchased from Molecular Probes (Eugene, OR, USA) (see Fig. 1). Potassium chloride was purchased from Fluka. All experiments in an aqueous environment were performed using tris(hydroxymethyl)aminomethane (Tris) buffer (pH 7.4) purchased from Sigma (St. Louis, MO, USA). All the solvents used were of spectroscopic or pro analysis quality, and were carefully checked for the absence of fluorescent impurities.

Methods

Solutions of the peptides were prepared by dissolving the peptide in 0.2% dimethyl sulfoxide and by adding the proper amount of buffer solution. Vesicles of DPPC containing the lipopeptides were prepared in Tris buffer by two different procedures:

1. Procedure 1. Incubation of the lipopeptide in DPPC vesicles as a function of lipopeptide concentration and incubation time. Vesicles were prepared by dissolving DPPC in 3% of methanol and 7% of chloroform (2 mM). The solvent was removed under vacuum, for 1 night, leaving a thin film of lipid on the walls of the flask. The dried lipid was hydrated with an appropriate volume of Tris buffer and the suspension was vortexed to generate multilamellar vesicles (MLV). Unilamellar vesicles were prepared from MLV by extrusion at 65 °C with an extruder from Lipex Biomembranes (Vancouver, Canada). We obtained vesicles of size 60 ± 10 nm. The results reported in this work were obtained by dissolving the peptide in Tris (pH 7.4) and adding the solution to the liposome suspension (peptide 5×10^{-5}–2.5×10^{-6} M, DPPC 1×10^{-3} M). The system was incubated for 24 h at 37 °C.
2. Procedure 2. Coextrusion of lipopeptides and DPPC as a function of lipid-peptide molar ratio. In the present work, covesicles were made by dissolving the peptide and DPPC (peptide-to-DPPC ratio 1:20, 1:400) in 3% of methanol and 7% of chloroform. Fluorescence-labeled covesicles were made by dissolving the peptide, DPPC and the BODIPY-labeled phospholipids (100:5:1) in 3% of methanol and 7% of chloroform. Unilamellar vesicles were obtained using the procedure already described.

Emission spectra were recorded with a PerkinElmer LS50B spectrofluorometer. Absorption spectra were measured with a PerkinElmer Lambda 900 spectrophotometer. All spectra were recorded at 37 °C.

Results and discussion

Solvatochromism of AMCA-labeled peptides in phospholipid vesicles

It is well known that the wavelength of the absorption and emission maxima of the AMCA fluorophore shifts as a function of the solvent dielectric constant [5]; at the same time the intensity of the absorption and emission maxima of free AMCA increases with increasing polarity of the solvent [5]. Studies on solvatochromism of AMCA-labeled peptides in phospholipid vesicles [6] provided information on the distribution of the peptides in the phospholipid vesicle and on the local dielectric properties experienced by the label. We found that, differently from the free fluorophore, AMCA in the labeled peptides experiences a polar environment also in a solvent with low dielectric constant. Moreover, in the case of AMCA-labeled peptides in phospholipid vesicles, we also observed a decrease of fluorescence intensity along the series

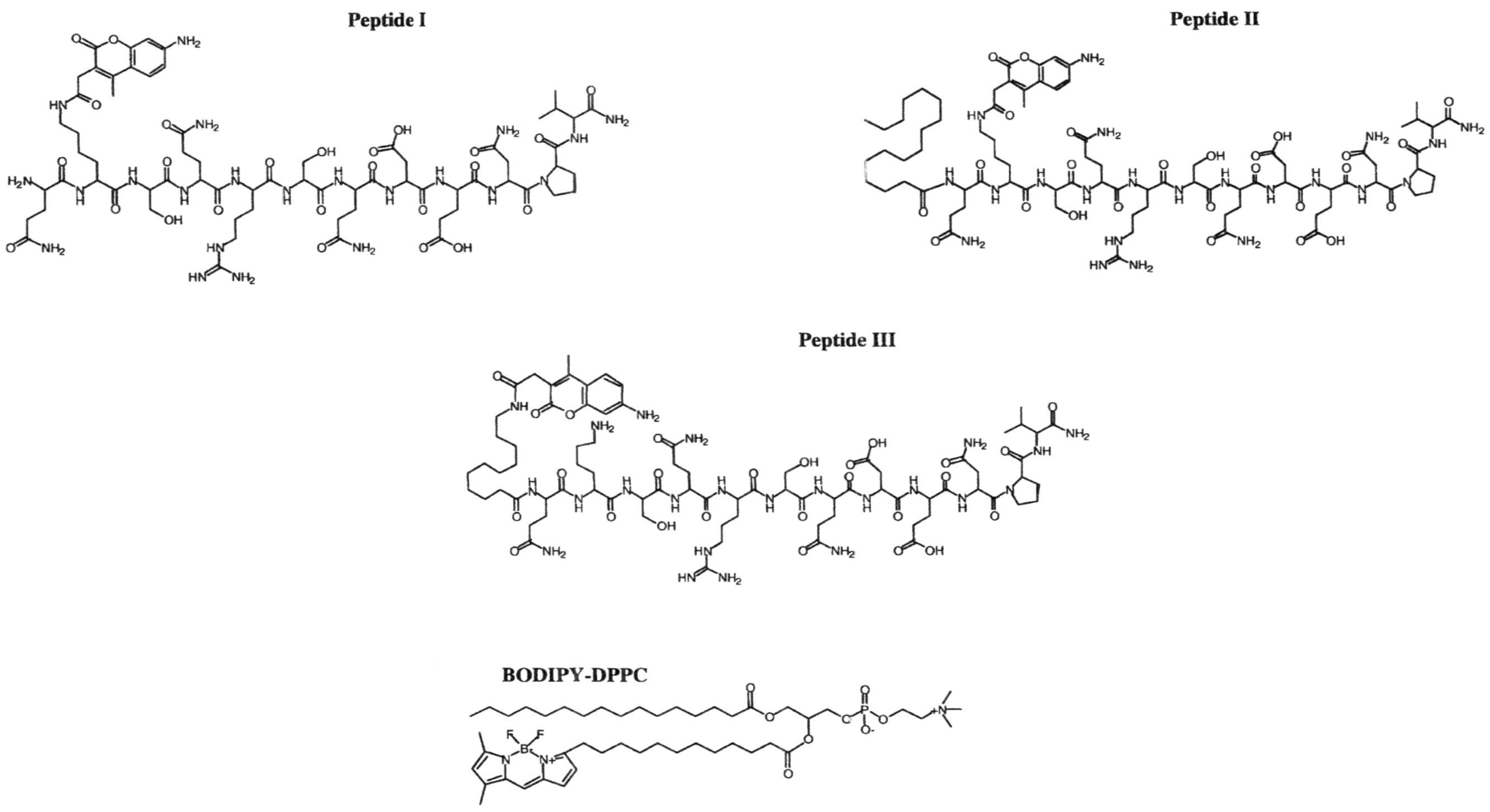

Fig. 1 Chemical structures of peptides I, II, III and 2-(4,4-Difluoro-5,7-dimethyl-4-bora-3*a*,4*a*-diaza-*s*-indacene-3-dodecanoyl)-1-hexadecanoyl-*sn*-glycero-3-phosphocholine (*BODIPY-DPPC*)

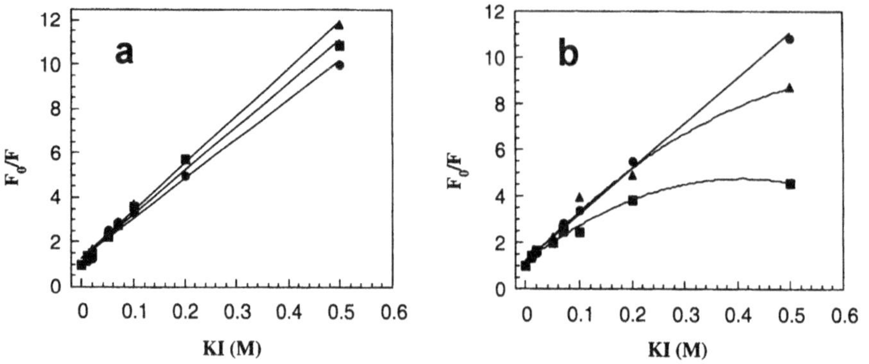

Fig. 2 a Fluorescence intensity as a function of potassium iodide concentration for peptide I. Peptide in buffer solution (*circles*), peptide incubated in DPPC vesicles in buffer solution (*triangles*), mixed peptide/DPPC covesicles in buffer solution (*squares*). **b** Fluorescence intensity as a function of potassium iodide concentration for peptide III. Peptide in buffer solution (*circles*), peptide incubated in DPPC vesicles in buffer solution (*triangles*), mixed peptide/DPPC covesicles in buffer solution (*squares*)

I > II > III. This result was found to hold for procedures 1 and 2. The solvatochromism exhibited by AMCA-labeled peptides in vesicles suggested that the lipophilic modification induces penetration of the peptide into the phospholipid bilayer, whereas the native peptide is located closer to the polar group region.

Further insight into the mechanism of interaction of the AMCA lipopeptides with the phospholipid vesicles was obtained from fluorescence quenching experiments. Quenching may occur by several mechanisms: collisional or dynamic quenching, static quenching, quenching by energy transfer and by charge-transfer reactions [6]. We selected two specific cases: collisional quenching of a water soluble quenchers (iodide ions) and energy transfer between the AMCA donor and an acceptor (BODIPY) covalently bound to the N-terminal position of the phospholipid alkyl chain. In the former case, a decrease of fluorescence intensity is expected only if the AMCA fluorophore in the vesicle is facing the outer aqueous phase. On the other hand, FRET experiments may be used as a molecular ruler to estimate the distance between the fluorophore and quencher.

Collisional quenching with potassium iodide

Fluorescence intensity was measured as a function of iodide concentration in the outer aqueous phase of phospholipid vesicles; we examined unilamellar DPPC vesicles obtained both with procedure 1 (incubated vesicles) and with procedure 2 (covesicles). Preliminary quenching experiments were also run for the peptides in Tris buffer solution. Typical results are reported in Fig. 2a and b for peptide I and peptide III, respectively.

In the case of peptide I, the ratio of the fluorescence intensity in the absence, F_0, and in the presence, F, of the quencher increases linearly with iodide concentration according to the Stern–Volmer equation [7]:

$$F_0/F = 1 + k_q \tau_0 [\mathrm{I}^-] = 1 + K_{SV}[\mathrm{I}^-], \tag{1}$$

where k_q is the rate constant, τ_0 is the lifetime of the AMCA fluorophore in quencher-free systems and K_{SV} is the Stern–Volmer constant.

A linear relationship is found for all the systems containing peptide I, although with different slopes, suggesting that the AMCA fluorophore bound to the lysine residue resides on the outer surface of the vesicle exposed towards the water phase.

A different behavior was observed for peptides II and III. Whereas for the simple buffer solution system the Stern–Volmer plot is linear as in the case of peptide I, for unilamellar vesicles we always observed a leveling off of the Stern–Volmer plot at higher iodide concentrations as shown in Fig. 2b.

This behavior is generally ascribed to the presence of two different fluorophore populations [7]: one accessible to the quencher, population a, and the other buried deeper in the phospholipid bilayer, population b. The fluorescence intensity can be written as

$$F_0 = F_{0,a} + F_{0,b}. \tag{2}$$

In the presence of quencher, the intensity of the accessible fluorophores decreases according to the Stern–Volmer equation and Eq. (2) becomes

$$F = F_{0,a}/(1 + K_{SV}[\mathrm{I}^-]) + F_{0,b}. \tag{3}$$

Rearrangement of Eq. (3) gives:

$$F_0/(F_0 - F) = F_0/ = (f_a K_{SV}[\mathrm{I}^-])^{-1} + (f_a)^{-1}, \tag{4}$$

where f_a is the fraction of accessible fluorophores at infinite quencher concentration.

Plots of $F_0/\Delta F$ versus $1/[\mathrm{I}^-]$.for peptides II and III in unilamellar vesicles were successfully fitted with Eq. (4). The f_a fitting parameters and the K_{SV}^* values are reported in Table 1 together with the K_{SV} values obtained by fitting the experimental data of peptide I with Eq. (1).

The accessible fluorophores experience a decrease in fluorescence upon collision with the iodide quencher; accessibility depends on exposure to the quencher as well as the lifetime of the fluorophore. Accessibility of AMCA is directly reflected in the Stern–Volmer constant: low values of K_{SV} indicate sites of low exposure. For peptide I, the Stern–Volmer constant increases slightly when the peptide interacts with the DPPC vesicle, indicating that the AMCA fluorophore is more exposed to the water phase compared to the peptide in homogeneous solution. Similar conclusions can also be drawn for the accessible population of peptides II and III; interestingly the fraction of such populations in the case of peptides II and III is lower when the peptide is inserted in DPPC vesicles by means of procedure 2 (covesicles). This suggests that in this case the distribution of the peptide in the bilayer is forced by the preparation procedure. We also recall that for peptides II and III we observed a decrease in fluorescence emission compared to water, indicating a relatively nonpolar environment [6] for the labeled peptides II and III.

Quenching by FRET

FRET is a distance-dependent interaction between the electronic excited states of two dye molecules in which excitation is transferred from a donor molecule to an acceptor molecule without emission of a photon [7].

Table 1 Stern–Volmer fitting parameters

	Peptide I	Peptide II			Peptide III		
	K_{SV} (M^{-1})	K_{SV} (M^{-1})	f_a	K_{SV}^* (M^{-1})	K_{SV} (M^{-1})	f_a	K_{SV}^* (M^{-1})
Solution	17.9	18.7			19.6		
Vesicles	19.6		0.95	19.2		0.95	18.9
Covesicles	21.5		0.69	18.9		0.89	20.7

FRET between the fluorescent AMCA donor and BODIPY was detected by measuring the emission of the acceptor fluorophore when excited at the wavelength of absorption of the donor. This wavelength would normally produce an emission from the acceptor only if energy transfer is involved. The energy-transfer process is also accompanied by a decrease of donor emission. The FRET was studied for unilamellar vesicles using DPPC molecules labeled with BODIPY either in the alkyl chain or in the polar head group region using the procedures reported in the Experimental section.

Typical results obtained for covesicles containing peptides II and III are reported in Fig. 3a and b, respectively. The results obtained for the other systems are reported elsewhere [6]; interestingly negligible quenching was detected for peptides II and III when the donor was covalently attached to the polar head group of the phospholipid.

In the case of the BODIPY donor bound to the phospholipid alkyl chain we observed a strong decrease in AMCA emission intensity and the appearance of the emission band of BODIPY: energy is transferred between the fluorescent labels when the acceptor is inside the phospholipid bilayer. The extent of quenching is higher for peptide III as shown by the efficiency values reported in Table 2.

The FRET efficiency, E, is obtained by measuring the fluorescence intensities of the donor with acceptor, F_{da}, and without acceptor, F_d:

Table 2 Fluorescence resonance energy transfer efficiency, *E*, and average 7-amino-4-methylcoumarin-3-acetic acid–4,4-difluoro-5,7-dimethyl-4-bora-3*a*,4*a*-diaza-*s*-indacene (*BODIPY*) distance for 1,2-dipalmitoyl-*sn* glycero-3-phosphatidylcholine (*DPPC*)/peptide/BODIPY-DPPC (100:1:5) vesicles

Peptide	*E*	D–A distance (Å)
I	0.24	60
II	0.63	46
III	0.76	41

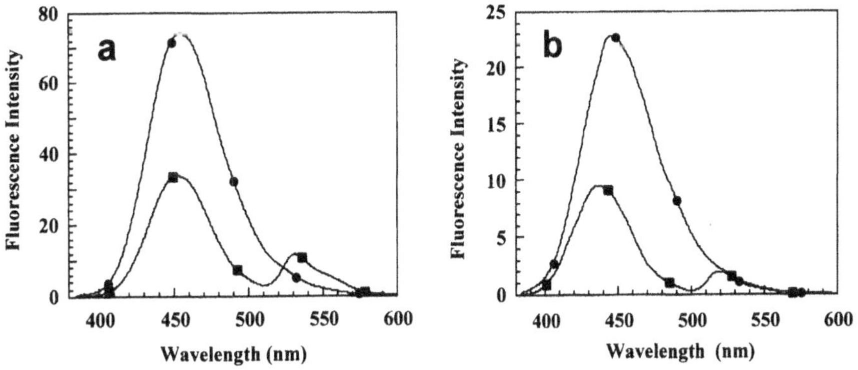

Fig. 3 a Fluorescence emission of DPPC/peptide II/BODIPY-DPPC (100:1:5) vesicles. Excitation wavelength 350 nm. **b** Fluorescence emission of DPPC/peptide III/BODIPY-DPPC (100:1:5) vesicles. Excitation wavelength 350 nm

$$E = 1 - F_{da}/F_d. \quad (5)$$

These results confirm the higher tendency of the alkyl peptides to insert their hydrophobic portion inside the phospholipid bilayer anchoring the peptide to the vesicles.

Measurements of FRET efficiency allow us to estimate the average distance between the fluorophores, AMCA donor and the BODIPY acceptor, from the following relationship:

$$\text{Donor} - \text{acceptor distance} = R = R_0(1/E - 1)^{1/6}, \quad (6)$$

where R_0 is the Förster distance. At R_0, there is an equal probability for resonance energy transfer and the radiative emission of a photon. The magnitude of R_0 is dependent on the spectral properties of the donor and acceptor molecules:

$$R_0 = \left(8.8 \times 10^{23} K^2 n^{-4} Q_d J\right)^{1/6}, \quad (7)$$

where K^2 is the dipole orientation factor, Q_d is the fluorescence quantum yield of the donor in the absence of acceptor, n is the refractive index of the medium and J is the spectral overlap integral, which represents the degree of overlap between the donor fluorescence spectrum and the acceptor absorption spectrum. The literature value for R_0 was used in the calculations [9], i.e. $R_0 = 50 \pm 2$ Å. The donor–aceptor distances reported in Table 2 show that the average donor–acceptor distance is rather large for peptide I, supporting the conclusions drawn from collisional quenching experiments: the AMCA fluorophore is located close to the polar head group region. For peptides II and III we observed smaller values for the average donor–acceptor distance; moreover, we found that this value decreases with increasing concentration of the acceptor in the vesicle [6]. As a rationale for these findings we suggest that both donor and acceptor are located in the inner portion of the bilayer.

Conclusions

The solvatochromism exhibited by AMCA-labeled peptides in vesicles showed that the lipophilic modification induces penetration into the phospholipid bilayer, whereas the native peptide is located close to the polar group region.

The collisional quenching results showed the existence of different localization sites of the lipopeptide in the bilayer: a population of peptide probably adsorbed at the interface where the fluorophore is directly exposed to the water phase and a population, not accessible to the quencher, incorporated deeper into the bilayer.

In the case of lipid-bound quenchers, we obtained direct access to the position of the AMCA label in the bilayer for the various lipopeptides examined by means of FRET. We observed FRET quenching only for peptides II and III, in particular for peptide III we observed higher FRET efficiency and a smaller average distance between AMCA and the quencher inserted in the hydrophobic portion of the vesicle bilayer.

Our results indicate that hMBP peptides modified with a lipophilic moiety at the N-terminal position interact with models of the cellular membrane with greater affinity than their unmodified analogues. These results may explain the behavior of immunological tests run with the same peptides used in this work [10]. In such studies an increased proliferative T cell response was obtained with the modified lipopeptide.

Acknowledgements The Ministero dell' Istruzione, dell'Universita' e della Ricerca (PRIN-2001), the Consorzio Sistemi a Grande Interfase and the Consiglio Nazionale delle Ricerche (Progetto Finalizzatto-MSTAII) are gratefully acknowledged for financial support.

References

1. (a) Johns TG, Kerlero de Rosbo N, Menon KK, Abo S, Gonzales MF, Bernard CCA (1995) J Immunol 154: 5536; (b) Cashman N, Martin C, Eizenbaum JF, Gegos JD, Bach MA (1982) J Clin Invest 70:387
2. (a) Deres K, Schild H, Wiesmuller KH, Jung G, Rammensee HG (1989) Nature 342:561; (b) Schild H, Deres K, Wiesmuller KH, Falk K, Jung G, Rammensee HG (1991) Eur J Immunol 2:2649; (c) Deprez B, Sauzet JP, Boutillon C, Martinon F, Tartar A, Sergheraert C, Guillet JG, Gomard E, Grass-Masse H (1996) Vaccine 14:375
3. Broecke S (1996) Nature 379:343
4. Peroni E, Caminati G, Baglioni P, Nuti F, Chelli M, Papini AM (2002) Bioorg Med Chem Lett 12:1731
5. Seixas de Melo J, Becker RS, Macanita AL (1994) J Phys Chem 98:6054
6. Peroni E, Papini AM, Baglioni P, Caminati G (in preparation)
7. (a) Lackowicz JR (1983) Principles of fluorescence spectroscopy. Plenum, New York; (b) Metzger JW, Sawyer WH, Willie B, Biesert L, Bessler WG, Jung G (1993) Biochem Biophys Acta 1149:29
8. Van Der Meer BW, Coker G III, Chen S-Y (1994) Resonance energy transfer theory and data. VCH, New York
9. (a) Keller RCA, ten Berg D, Nouwen N, Snel MME, Tommansen J, Marsh D, de Kruijff B (1996) Biochemistry 35:3063; (b) Keller, Silvius JR, de Kruijff B (1995) Biochem Biophys Res Comm 207:508; (c) Karolin J, Johansson LBA, Strandberg L, Ny T (1994) J Am Chem Soc 116:7801
10. Papini AM, Mazzanti B, Nardi E, Traggiai E, Ballerini C, Biagioli T, Kalbacher H, Beck H, Deeg M, Chelli M, Ginanneschi M, Massacesi L, Vergelli M (2001) J Med Chem 44: 3504

Progr Colloid Polym Sci (2004) 126: 169–172
DOI 10.1007/b94011

J.P. Lecomte
O. Soula
R. Meyrueix
A. Constancis
S. Angot
G. Soula

Adsorption and desorption of interferon onto colloidal hydrophobic surfaces

J.P. Lecomte (✉) · O. Soula · R. Meyrueix
A. Constancis · S. Angot · G. Soula
Flamel Technologies,
69693 Vénissieux, France

Abstract Medusa technology for protein delivery is based on the self assembly of amphiphilic polyaminoacids and proteins. The well controlled synthesis of random poly(1-leucin-b-1-glutamate) leads to nanoparticles which adsorb interferon alpha 2b in water. Preclinical results in dogs demonstrate a sustained release of the non denaturated protein over more than 5 days.

Keywords Interferon · Protein delivery · Polyaminoacids · Nanoparticles

Introduction

Interferon α2b is a protein naturally secreted in response to a viral attack. Recombinant human interferon α2b is being administered in case of hepatitis C or certain cancers. The clinical efficiency of this cytokine is limited by its rather short half-life (approximately 8 hours [1]). To improve the pharmaco-dynamics of interferon, and to reduce its side-effects, the half-life of circulation of interferon α2b has been extended by fusion with serum albumin [2] or by covalent attachment of poly(ethylene glycol) moieties to the protein [3, 4]. A drawback of this last reaction is that these derivatised interferons show diminished in vitro activities [5, 6]. Nonetheless, they have been commercialised, and show improved pharmaco-dynamics [7, 8]. A very interesting way of improving the kinetics and pharmaco-dynamics of interferon without modifying it and without the corresponding loss of activity would be to release the native protein in a controlled fashion. Such a controlled release of interferons (α, β or γ) has been sought for a long time, for example, by encapsulation in liposomes or by entrapment in poly(lactic–glycolic) films [9], but without success.

We present a sustained release of interferon α2b from Medusa self-assembled nanoparticles of random poly(L-leucine–L-glutamate) [10].

Polymer synthesis

The polymers were prepared by anionic ring-opening copolymerisation of *N*-carboxyanhydrides of L-leucine and L-glutamate (Fig. 1). The low polydispersity measured shows that a living polymerization mode was achieved.

Random copolymers were obtained by polymerising the mixed monomers. As the reactivity ratio of the two monomers differ ($k_{NCA\text{-}Glu}/k_{NCA\text{-}Leu} \approx 2$), "gradient-random" macromolecular chains with a leucine-rich end were obtained. The overall hydrophilic/lipophilic balance of the polymer was controlled by the glutamate-to-leucine ratio. The chain length was controlled by the initiator-to-monomers ratio.

This process allows the preparation of a large family of compounds, named rLE, with high reproducibility (Fig. 2).

Structure of the nanoparticles in water

These amphiphilic rLE are self-associating copoly(amino acid)s, which form nanoparticles dispersed in water (Table 1).

The rLE A nanoparticles can be envisioned as fuzzy structures of 10–20-nm diameter (Fig. 3) with a leucine-

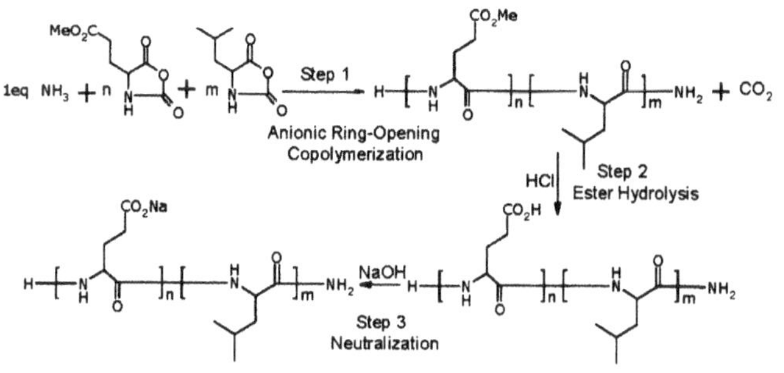

Fig. 1 Scheme of the rLE synthesis

Table 1 Elementary characteristics of self-assembled rLE nanoparticles in 0.15 M NaCl

Hydrodynamic diameter (quasielastic light scattering in NaCl 0.15 M, 2 g/l)	10–20 nm
M_w (flow FFF/SLS in 0.15 M NaCl)	2.3×10^5 g/mol
Mean number of chains per nanoparticle	30

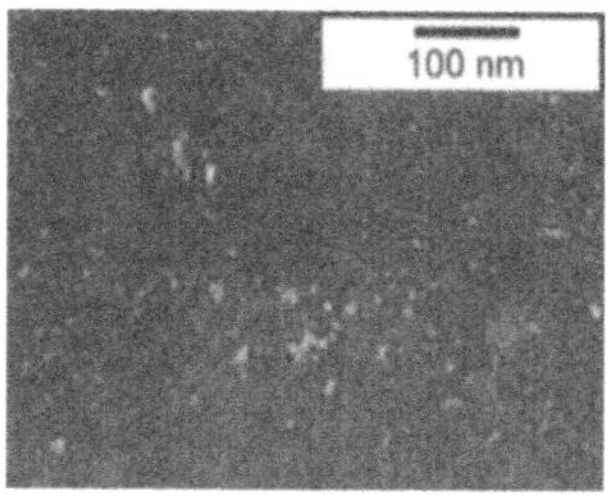

rLE : 10 to 20 nm

Fig. 3 Transmission electron microscopy micrograph of rLE A nanoparticles

rich core surrounded by loops and tails of poly(sodium glutamate)-rich polyelectrolyte segments preventing them from flocculating. The rLE A dispersion in water is completely transparent and clear, it can be sterilised by 0.2-μm filtration. The nanoparticle structure and properties are determined by the copolymer structure. rLE form a large and versatile class of nanoparticles made of entirely biodegradable and biocompatible components.

Association and dissociation of interferon α2b from Medusa nanoparticles

The protein loading was performed in very mild conditions, by spontaneous association of interferon α2b (IFNα2b, Figs. 4, 5) with the nanoparticles in water at pH 7.4 and 25 °C, upon simple gentle mixing: no solvent, no shear. The rLE A nanoparticles showed a high IFNα2b loading capacity: up to 25 wt/wt %.

The association isotherm is of mixed type (Fig. 6): 20% of association capacity with high affinity, 80% with low affinity for IFNα2b. The association and dissociation characteristics of IFNα2b onto rLE can be modified by changing the polymer structure and the formulation process (time, temperature, concentration, salinity, etc.). The complete dissociation of IFNα2b by sodium dodecyl sulfate indicates that the protein is not covalently bound to the polymer.

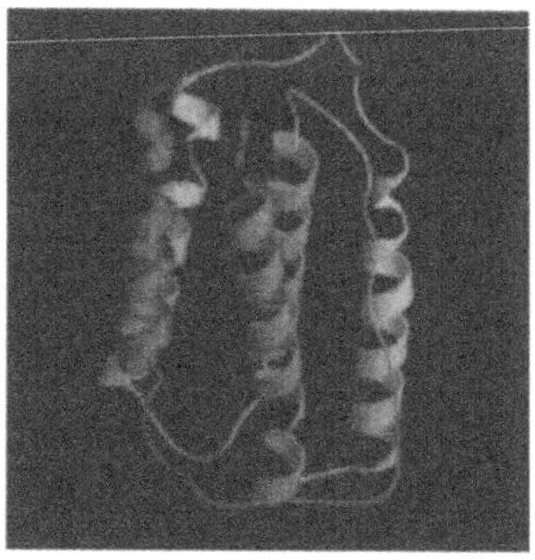

Fig. 4 Interferon α2b (IFNα2b): 165 AA 20 kDa, pI = 5.8

Medusa: in vivo proof of the concept of long-acting IFNα2b

An rLE–IFNα2b formulation was administered subcutaneously to dogs, and the plasmatic IFNα2b concentration was followed in blood samplings (Fig. 7, Table 2).

Fig. 2 *Left:* Size-exclusion chromatography (*SEC*) chromatogram of rLE A polymer, UV detection (214 nm). *Right:* Characteristics of rLE type A copoly(amino acid)

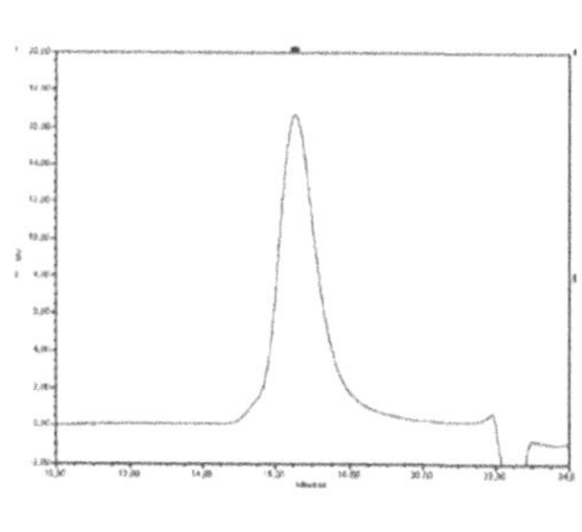

Polymer	rLE A
Type	Random
Mw (G/mole)	8000
Polydispersity index	1.1
Glutamate/Leucine molar ratio	0.7

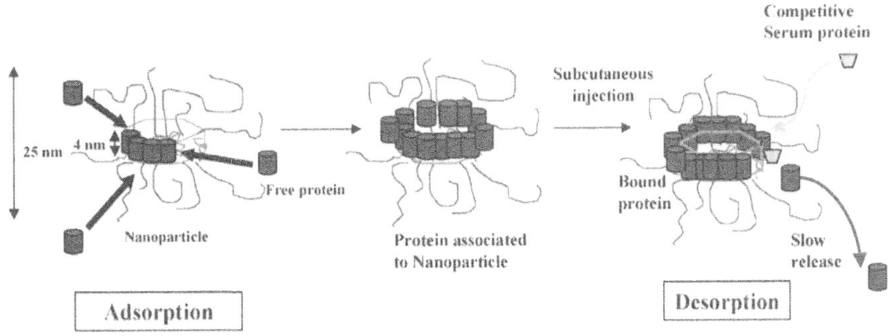

Fig. 5 Schematic association/dissociation of protein onto/from Medusa particles

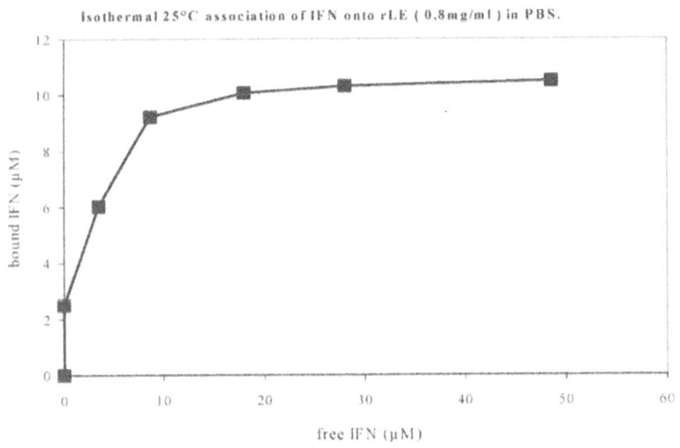

Fig. 6 Association isotherm of IFNα2b onto rLE nanoparticles (10 mg/ml) in phosphate buffer solution, measured by SEC

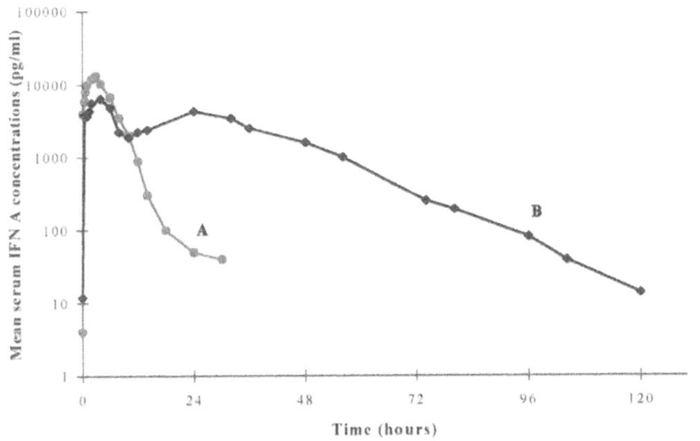

Fig. 7 Plasmatic IFNα concentration in dogs. **a** IFNα 3 MIU/kg, data from Ref. [11]. **b** rLE–IFNα 15 MIU/kg. Animal model male Beagle dogs 10.5–13.6 kg ($n=6$). IFNα2b dose 60 μg/kg, injected subcutaneously. IFNα2b concentration in plasma measured by ELISA (Amersham RPN hIFN-α kit)

Table 2 Comparison of the pharmacokinetic parameters for interferon α2b (IFNα2b) and Medusa rLE–IFNα2b. IFNα2b data from Ref. [2]

Formulation	Dose (MUI/kg)	t_{max} (h)	c_{max} (ng/ml)	Area under the curve (ng h/ml)	$T>10$ pg /ml (h)
IFNα	3	3	13.1	79	48
rLE–IFNα	15	4 and 24	6.44	184	>125

The association of IFNα2b onto rLE A nanoparticles (rLE–IFNα2b) leads to

- An increase of the duration of action up to 1 week: increased therapeutic efficacy and improved compliance.
- A reduction of the concentration at peak by a factor 2: reduced toxicity.

Conclusions

Random poly(leucine–sodium glutamate) forms in aqueous media a colloidal dispersion of nanoparticles that is stable, transparent and that can be sterilised by 0.2-μm filtration. IFNα2b associates to these particles by a spontaneous, nondenaturing and reversible process. Upon subcutaneous injection in dogs, the plasmatic concentration profile of IFNα2b is extended over more than 125 h while c_{max} is decreased. This technology is currently used for the clinical development of a sustained release dosage form of IFNα2b.

References

1. Schalm SW et al (1999) Gastroenterology 117:408–413
2. Osborne BL (2002) J Pharmacol Exp Ther 303: 540
3. Gilbert CW, Park-Cho MO (1999) US Patent 6,042,822, and references therein
4. Koslowski A et al (2001) BioDrugs 15:419–429
5. Youngster S et al (2002) Curr Pharm Des 8:2139–2157
6. Grace M et al (2001) J Interferon 21:1103
7. Gupta SK et al (2002) J Clin Pharmacol 42:1109–1115, and references therein
8. (a) Zeuzem S et al (2000) New England J Med 343:1666–1672, and references therein; (b) Heathcote EJ et al (2000) New England J Med 343:1673–1680
9. Eppstein DA et al (1986) NATO ASI Ser Ser A 125:277–283
10. Flamel Technologies (1996) US Patent 5,904,936
11. Hoffmann-La Roche Inc (1985) J Interferon Res 5:403–408

Progr Colloid Polym Sci (2004) 126: 173–177
DOI 10.1007/b93990

Anna Stradner
George Thurston
Vladimir Lobaskin
Peter Schurtenberger

Structure and interactions of lens proteins in dilute and concentrated solutions

A. Stradner (✉) · V. Lobaskin ·
P. Schurtenberger
Department of Physics,
University of Fribourg,
1700 Fribourg, Switzerland
e-mail: anna.stradner@unifr.ch
Tel: +41-26-3009120;
Fax: +41-26-3009747

G. Thurston
Rochester Institute of Technology,
College of Science,
85 Lomb Memorial Drive,
Rochester, NY 14623-5603,
USA

Present address: V. Lobaskin,
MPI for Polymer Research,
Ackermannweg 10, 55128
Mainz, Germany

Abstract We present results from a study of the structural properties of concentrated lens protein solutions using small-angle neutron and X-ray scattering. The lens proteins are particularly interesting and important, as they are ideally suited for an attempt to use well-defined and monodisperse proteins as model systems for colloids as well as because of their biological and medical relevance for cataract formation. We studied the original proteins (α- and γB-crystallin) as well as the structure of concentrated mixtures of proteins. In addition, we studied low concentrations of the pure solutions to determine the particle form factors needed and to study the interparticle potentials. The experimental investigations were combined with Monte Carlo computer simulations. Our aim is to quantitatively characterize and understand the molecular factors that determine the free energy of aqueous crystallin mixtures, at high protein concentrations comparable to those found in the ocular lens.

Keywords Eye lens proteins · Small-angle neutron scattering · Small-angle X-ray scattering · Monte Carlo computer simulations · Structure factor

Introduction

The study of the phase behaviour of eye lens proteins has been inspired by their role in cataract disease. A cataract is a gradual clouding of the eye lens as a result of increased scattering of light, and is the leading cause of blindness, worldwide. The eye lens cells contain mixtures of what are named crystallin proteins, at concentrations ranging from approximately 200 up to 400 mg ml^{-1}. Close, uniform packing of these crystallins at high concentrations underlies normal lens transparency, by creating a uniform index of refraction. Aggregation of the crystallin proteins and microphase separation are possible sources of increased light scattering in the predominant forms of cataracts [1].

There are three principal classes of eye lens crystallins, named α, β, and γ. The relative proportions and overall concentrations of the crystallins, and their subtypes, differ between mammalian species and vary with position inside a given lens.

The γ-crystallins comprise one of the most studied classes of crystallins to date, and are globular proteins of molecular mass near 21 kDa. γ-Crystallins exhibit attractive interprotein interactions that result in a rich phase behaviour that includes two phase transitions, liquid–liquid phase separation and crystallization [2, 3, 4, 5]. High concentration solutions of γ-crystallins generally undergo liquid–liquid phase separation upon lowering the temperature below a critical temperature, T_c [1, 4, 5, 6]. The value of T_c varies with the particular γ-crystallin, but is, in general, between 0 and 40 °C. In addition, the liquid–liquid coexistence curve is metastable with respect to the solid–liquid phase boundary. Theoretical analysis indicates that γ-crystallins have short-range attractive interactions that depend on temperature [7, 8, 9, 10, 11, 12, 13]. This is similar to observations in other colloidal systems [14] and is consistent with very recent developments in the theoretical description of the statistical mechanics of colloidal particles interacting via a potential consisting of a hard-core repulsion and a short-range attractive part [14].

α-Crystallin is a multisubunit protein of about 800 kDa, made up of approximately 40 globular subunits of types A and B, each 20 kDa. Previous investigations have shown that α-crystallins are globular polydisperse proteins and that their interactions can be described to a good approximation with a simple hard-sphere model [15, 16, 17].

There have been extensive phase boundary, light scattering, X-ray scattering, and theoretical studies of relatively high concentration solutions of α- and γ-proteins [2, 3, 4, 5, 6, 7, 8, 9, 16]; however, the concentrations investigated were still below the highest concentrations in the eye lens. Moreover, a comparable study of high concentration mixtures of these proteins is in its early stages [21, 22]. How different crystallins interact to determine the amount of light that is scattered in their mixtures, the stability of crystallin mixtures with respect to phase separation and aggregation, and the Gibbs free energy of such mixtures is largely unknown. Therefore we started an investigation of the pure α- and γB-crystallins up to concentrations close to those found in the eye lens as well as of highly concentrated mixtures of the proteins in order to perform a detailed investigation of the structure factor of such solutions. These studies are also very interesting in view of the recent developments in the theory of strongly interacting and highly concentrated colloidal suspensions [14]. In addition to small-angle X-ray scattering (SAXS) we are using small-angle neutron scattering (SANS) to probe an extended q range and combine our experimental investigations with Monte Carlo (MC) computer simulations. Our aim is to quantitatively characterize and understand the molecular factors that determine the free energy of aqueous crystallin mixtures, at high protein concentrations comparable to those found in the ocular lens, in a protein species-specific manner.

Sample preparation and methods

α-Crystallins and γ-crystallins were isolated from third trimester prenatal to 6-week-old calf lenses using size-exclusion chromatography of cortical fractions on Sepharose CL-6B, using 50 mM sodium phosphate buffer, pH 7, containing 2.5 mM dithiothreitol, 1.0 mM ethylenediaminetetraacetate, 0.2 M NaCl (α-crystallin) or of nuclear fractions on Sephacryl S-200 (γ-crystallin), using 0.275 M sodium acetate buffer, pH 4.5, followed for γB-crystallin by cation-exchange chromatography on sulfopropyl Sephadex C-50, using 0.275 M sodium acetate buffer, pH 4.8, and a 0–0.3 M NaCl gradient [2, 22]. All buffers contained sodium azide (0.02%). As solvent 0.1 M phosphate buffer in D_2O including dithiothreitol (20 mM) and sodium azide (0.02%) was used. To reach concentrations of the order of 300–350 mg ml^{-1}, solutions of known proportions were concentrated by ultrafiltration using disposable centrifugal filtration units. Protein concentrations were determined by UV absorption using published extinction coefficients. The highest concentrations used in these experiments were 30 mg ml^{-1} for the α-crystallins and 330 mg ml^{-1} for the γB-crystallins.

The SANS experiments were performed at the SANS I instrument of the Swiss spallation source SINQ at the Paul-Scherrer-Institut, PSI, Switzerland. A q range of 2×10^{-3} Å$^{-1}\leq q\leq 0.3$ Å^{-1} was covered by three sets of sample-to-detector distances (d=2, 12 and 18 m) and two different wavelengths of λ=8 and 12.67 Å, respectively. The samples were kept in stoppered quartz cells (Hellma, Germany) with path lengths of 1 and 2 mm. The neutron spectra of water were measured in a 1-mm path length quartz cell. The raw spectra were corrected for background from the solvent (D_2O buffer), sample cell, and electronic noise by conventional procedures. Furthermore, the two-dimensional isotropic scattering spectra were corrected for detector efficiency by dividing by the incoherent scattering spectra of pure water and were azimuthally averaged.

The SAXS experiments were performed with an Anton Paar SAXSess slit camera, which combines focusing mirrors with a block collimator and a two-dimensional imaging plate detection system. A q range of 8×10^{-3} Å$^{-1}\leq q\leq 0.6$ Å^{-1} was covered. All scattering experiments were performed at 25 °C.

The data analysis (model fitting and indirect Fourier transformation, IFT) was performed including the instrumental resolution functions [23].

Results and discussion

A comparison between experimental SANS and SAXS data on an arbitrary scale from a 49 mg ml^{-1} γB-crystallin solution is shown in Fig. 1a. At these concentrations the influence of interactions at low q values is already measurable and needs to be taken into account in a quantitative analysis of the data. SAXS provides access to higher q values, and allows the observation of the first minimum of the form factor, which seems to be located at lower q values than the form factor minimum in the neutron data. Such a difference has previously been found for other proteins and was interpreted with the different influences of a hydration shell on the contrast profile for neutron and X-ray scattering [24].

The SANS and SAXS data for 20 mg ml^{-1} α-crystallin are plotted in Fig. 1b. A detailed analysis of the data (model fitting and IFT) again shows a measurable influence of the interaction effects and that the structure factor suppresses the intensity even at low q. The data from our SANS experiments on an α-crystallin concentration series (concentrations from 2 up to 255 mg ml^{-1}) are shown in Fig. 2a. With increasing protein concentration, one observes a suppression of the scattered intensity at low q values followed by a clear structure factor peak as expected for repulsive hard-spheres interaction and as already observed [16, 19]. We

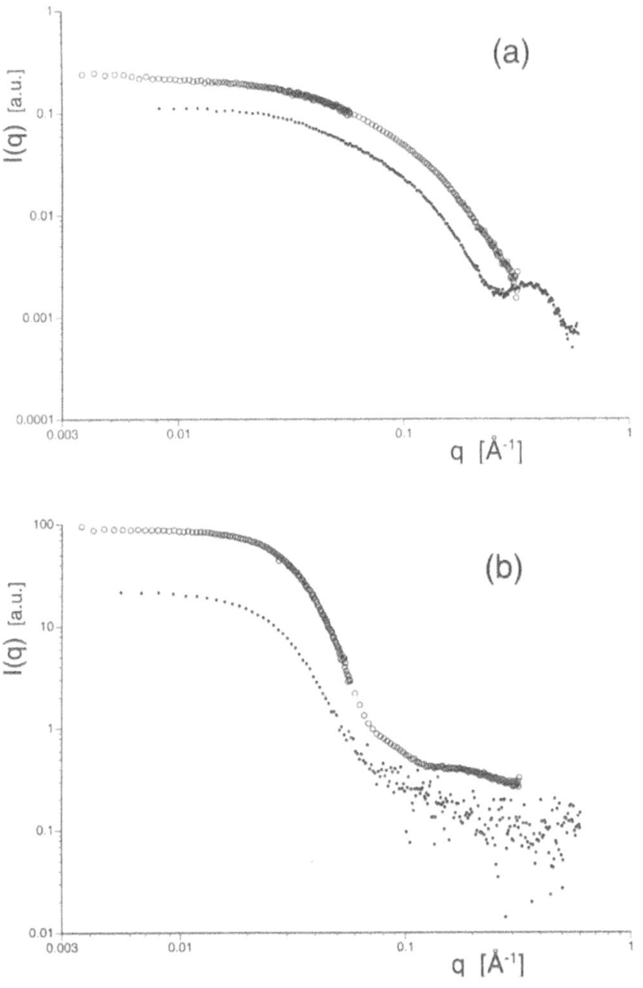

Fig. 1 Small-angle neutron scattering (*SANS*) (*open circles*) and small-angle X-ray scattering (*SAXS*) (*small dots*) data from **a** γB-crystallin at 49 mg ml^{-1} and **b** α-crystallin at 20 mg ml^{-1}

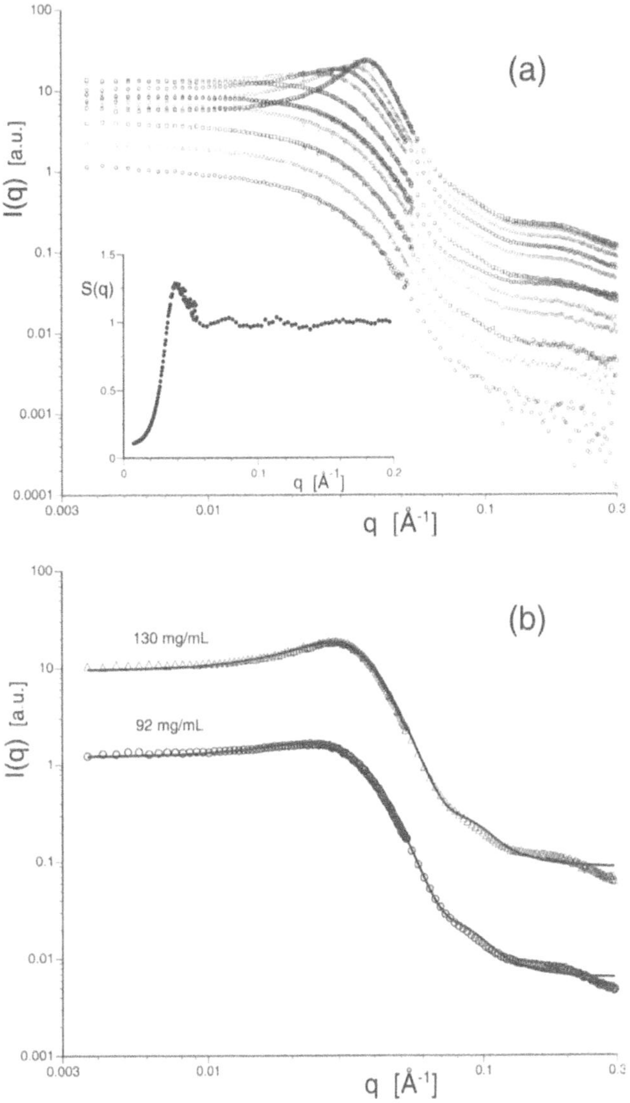

Fig. 2 SANS data of α-crystallin solutions: **a** α-crystallin concentration series (2 mg ml^{-1}$\leq c \leq$255 mg ml^{-1}); *inset* experimentally determined structure factor, $S(q)$, for 177 mg ml^{-1}. **b** α-Crystallin solutions (92 and 130 mg ml^{-1}) together with a fit to the data applying the Perkus–Yevick closure relation for polydisperse hard spheres (see text for details). The 92 mg ml^{-1} data are offset by a scale factor of 0.1

determined the experimental structure factors, $S(q)$, for the highly concentrated samples by dividing the respective scattering intensity, $I(q)/c$, by that from the 4 mg ml^{-1} data. As an example $S(q)$ for a protein concentration of 177 mg ml^{-1} is shown as an inset. We performed a detailed analysis of the α-crystallin data applying the Perkus–Yevick closure relation for polydisperse hard spheres. The results of the fits are very similar for all concentrations, giving a number-averaged radius $R = 60 \pm 2$ Å and a polydispersity $\sigma = 18 \pm 2\%$. As an example, in Fig. 2b the neutron scattering curves for 92 and 130 mg ml^{-1} α-crystallin are plotted together with the Perkus–Yevick fit. At high q values we find indications for an internal structure within the proteins and the data show slight deviations from the fit. At that resolution they cannot be represented by the simple assumption of homogeneous particles any more.

In addition to measurements of the pure components we also investigated mixtures made of a 230 mg ml^{-1} α-crystallin solution and a 260 mg ml^{-1} γB-crystallin solution. The spectra of the pure solutions together with three different mixtures are plotted in Fig. 3. Adding α-crystallin to the γB-solution first only slightly increases the forward scattering (12.5% α-crystallin). At a composition of 25% α-crystallin and 75% γB-crystallin, the intensity at low q reaches a maximum and at the same time the hard-sphere interaction peak starts to evolve. Already at 50% α–crystallin solution the forward scattering is highly suppressed owing to the repulsive interactions arising from the α-crystallins, and we find a very well pronounced structure factor peak.

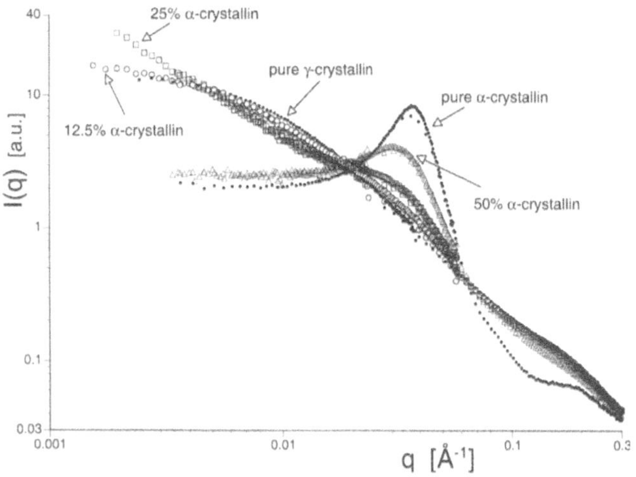

Fig. 3 SANS data of mixtures of α-crystallin (230 mg ml^{-1}) and γB-crystallin (260 mg $ml^{-1} = c_c$) solutions together with the pure protein solutions: pure α-crystallin (*small dots*); 50% α-crystallin/50% γB-crystallin (*open triangles*); 25% α-crystallin/75% γB-crystallin (*open squares*); 12.5% α-crystallin/87.5% γB-crystallin (*open circles*); pure γB-crystallin (*small filled squares*)

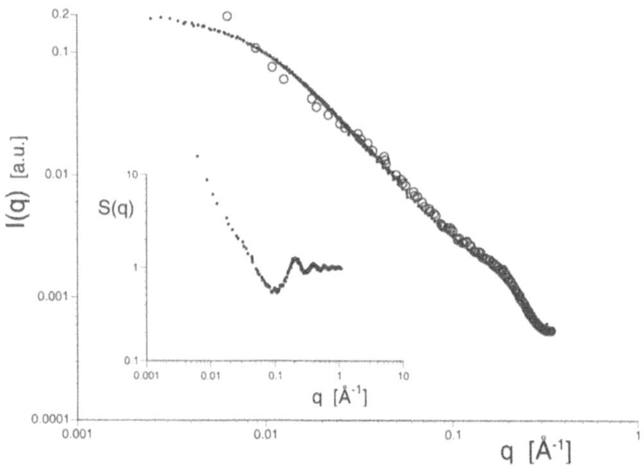

Fig. 4 SANS data of γB-crystallin (*small black dots*) at the critical concentration $c_c = 260$ mg ml^{-1} and 25 °C together with Monte Carlo simulation results at the same concentration and temperature (*open circles*). The Monte Carlo simulation reproduces the enhanced forward scattering as well as the structure factor peak at very high concentrations. *Inset* Structure factor, *S(q)*, at c_c from Monte Carlo simulations

In parallel to the experiments we started MC simulations of α-crystallin, γB-crystallin and mixtures of the proteins. An essential part of the simulation of protein solutions and mixtures is the choice of the appropriate interaction potentials for α–α, γB–γB, and α–γB interactions. Here we chose a hard-core potential for α-crystallins and for α–γB interactions, and a combination of a hard-core potential and an attractive Yukawa potential for the interactions between γB-crystallins. The choice of these potentials is verified by a comparison between the SANS data and the corresponding MC simulation. This is illustrated for γB-crystallins at the critical concentration, where the influence of the attractive interactions is particularly pronounced. The SANS at the critical concentration $c_c = 260$ mg ml^{-1} and 25 °C is shown in Fig. 4 together with MC results at the same concentration. In the MC simulations the strength of the attraction was set such as to reproduce the critical temperature. The hard-core radius was set to $R = 16$ Å and the interaction decay length to $0.5R$. $S(q)$ from MC simulations is shown as an inset. To calculate the scattering curve of the 260 mg ml^{-1} γB-crystallin solution we multiplied $S(q)$ obtained from MC simulations with an effective γB-crystallin form factor obtained from an IFT analysis of low concentration γB-crystallin SANS data which includes the appropriate convolution with the instrumental resolution function. The simulations reproduce the important features, which consist of a forward scattering that is enhanced by several orders of magnitude owing to the attractive potential and a structure factor peak at high q values owing to the density build-up at the protein contact distance at this very high concentration.

Conclusions

We have shown that crystallin solutions and mixtures exhibit a wealth of interesting structural features that range from simple hard-sphere behaviour for pure α-crystallins to phase separation and critical behaviour for pure γB-crystallins. It is clear that this has important implications both for the properties of eye lenses as well as for the use of these proteins as ideal model colloids. The main goals of our investigations are thus both related to biophysical/physiological as well as general condensed matter problems. By elucidating the relationships between the phase transitions for concentrated protein solutions and mixtures, and using the results from the computer simulations and the scattering experiments, insight will be gained into possible molecular origins of light scattering in cataracts. Moreover, we shall use these proteins in order to study the interplay and competition between various phase transitions found in concentrated suspensions of attractive colloidal particles such as phase separation, glass formation and crystallization and their dependence on the exact nature of the interparticle interaction potential.

As a first step towards this, it is obviously necessary to fully understand the pure systems (α–crystallin and γB-crystallin) at low (form factor) and high concentrations (structure factor). In this paper we could support previous findings about the hard-sphere behaviour of α-crystallin solutions as well as the short-range attractive interactions in γB-crystallin solutions up to very high concentrations. Moreover, we have been able to extend the q range substantially by combining SANS and

SAXS, and the combination of our scattering data with MC simulations for γB-crystallin yields encouraging agreement between experiment and simulation. First measurements of high concentration mixtures of the crystallins showed a combination of the same features already found in the pure systems. The next step in our long-term study will be the performance of further SANS measurements of mixtures as a function of concentration and temperature combined again with MC simulations.

Acknowledgements We gratefully acknowledge financial support from the Swiss National Science Foundation, COST (Action P1) and the Swiss Federal Office for Education and Research. We thank the PSI, Villigen, Switzerland, for providing the neutron research facilities and we gratefully acknowledge the expert help of our local contact Joachim Kohlbrecher.

References

1. Benedek GB, Pande J, Thurston GM, Clark JI (1999) Prog Retinal Eye Res 18:391
2. Thomson JA, Schurtenberger P, Thurston GM, Benedek GB (1987) Proc Natl Acad Sci USA 84:7079
3. Schurtenberger P, Chamberlin RA, Thurston GM, Thomson JA, Benedek GB (1989) Phys Rev Lett 63:2064
4. Broide ML, Berland CR, Pande J, Ogun O, Benedek GB (1991) Proc Natl Acad Sci USA 88:5660
5. Berland CR, Thurston GM, Kondo M, Broide ML, Pande J, Ogun O, Benedek GB (1992) Proc Natl Acad Sci USA 89:1214
6. Liu C-W, Lomakin A, Thurston GM, Hayden DL, Pande A, Pande J, Ogun O, Asherie N, Benedek GB (1995) J Phys Chem 99:451
7. Malfois M, Bonneté F, Belloni L, Tardieu A (1996) J Chem Phys 105:3290
8. Petitt P, Edwards ME, Forciniti D (1997) Eur J Biochem 243:415
9. Bonneté F, Malfois M, Finet S, Tardieu A, Lafont S and Veesler S (1997) Acta Crystallogr Sect D53:438
10. Koretz JF, Cook CA and Kaufman PL (1997) Invest Ophthalmol Vis Sci 38
11. Lomakin A, Asherie N, Benedek GB (1996) J Chem Phys 104:1646
12. Asherie A, Lomakin A, Benedek GB (1996) Phys Rev Lett 77:4832
13. Lomakin A, Asherie N, Benedek GB (1999) Proc Natl Acad Sci USA 96:9465
14. Foffi G, McCullagh GD, Lawlor A, Zaccarelli E, Dawson KA, Sciortino F, Tartaglia P, Pini D, Stell G (2002) Phys Rev E 65:031407
15. Tardieu A (1998) Int J Biol Macromol 22:211
16. Finet S, Tardieu A (2001) J Cryst Growth 232:40
17. Frenkel D (2002) Science 296:65
18. Tardieu A, Veretout F, Krop B, Slingsby C (1992) Eur Biophys J Biophys Lett 21:1
19. Veretout F, Delaye M, Tardieu A (1989) J Mol Biol 205:713
20. Vanhoudt F, Abgar S, Aerts T, Clauwaert J (2000) Biochemistry 39:4483
21. Thurston GM, Lurio L. Mochrie S.(2001) Invest Ophthalmol Vis Sci 42
22. Thurston GM, Pande J, Ogun O, Benedek, GB (1999) Invest Ophthalmol Vis Sci 40.299
23. Pedersen JS, Posselt D, Mortensen KJ (1990) J Appl Crystallogr 23:321
24. Svergun DI, Richards S, Koch MH, Sayers Z, Kuprin S, Zaccai G (1998) Proc Natl Acad Sci USA 95:2267

Progr Colloid Polym Sci (2004) 126: 178–183
DOI 10.1007/b93971

L. Dahbi
C. Bourgaux
M. Ollivon

Phase transitions of dilauroylphosphatidylcholine in excess water: a coupled small-angle X-ray scattering–wide-angle X-ray scattering study

L. Dahbi · C. Bourgaux (✉)
Laboratoire pour l'Utilisation du Rayonnement Electromagnétique, 91898 Orsay, France
e-mail: claudie.bourgaux@lure.u-psud.fr

M. Ollivon
Equipe Physico-Chimie des Systèmes Polyphasés, 92296 Châtenay-Malabry, France

Abstract Phase transitions or thermotropic evolution of dilauroylphosphatidylcholine (DLPC) are studied at very low scan rates by a new technique coupling small-angle X-ray scattering or wide-angle X-ray scattering and differential scanning calorimetry (DSC) in a single apparatus with the same sample. Fully hydrated saturated phosphatidylcholine molecules are known to undergo a "main" transition between the gel phase $P_{\beta'}$ and the liquid crystalline phase L_α . In contrast to the other saturated diacylphosphatidylcholines which exhibit a single main transition between $P_{\beta'}$ and L_α , DLPC shows by DSC two phase transitions between the $P_{\beta'}$ and L_α phases: a sharp peak corresponding to the main transition at −2 °C and a broad peak at 5 °C. This characteristics of DLPC were originally observed by Mabrey and Sturtevant by DSC and highlighted by X-ray diffraction (XRD) studies by Hatta et al. For the first time, the coupling of these two techniques (DSC-XRD) has been used to confirm the existence of an intermediate phase between the gel ripple phase $P_{\beta'}$ and the liquid-crystalline phase L_α,with a scanning rate as slow as 0.1 °C min^{-1}. Both wide angle and small angle diffraction profiles show discontinuities, highlighting, respectively, changes in the lamella spacing and in the hydrocarbon interchain "short" spacing.

Keywords X-ray diffraction · Differential scanning calorimetry · Dilauroyl phosphatidylcholine

Introduction

Phospholipids are amphiphilic molecules that can form flat bilayers in water. Unilamellar or multilamellar vesicles made by closed lipid bilayers, so-called liposomes, have been widely studied since they are efficient tools for cell membrane modelisation.

Lipid bilayers made from diacylphosphatidylcholines (C*n*PCs) are the most studied. These phosphatidylcholines are zwitterionic molecules that displays both lyotropic and thermotropic behavior [1, 2, 3, 4]. In excess water, saturated phosphatidylcholines with saturated linear chains of $n \geq 13$ carbons are known to undergo a "main" transition corresponding to the chain melting, at T_m between the liquid-crystalline phase L_α and the gel phase $P_{\beta'}$. In the L_α. phase, at sufficiently high temperature, the hydrocarbon chains have a liquidlike, highly disordered conformation; below the "chain-melting" transition, the chains are fully extended and tilted with respect to the plane of the lipid bilayers and the lamellae are distorted by a periodic ripple. At lower temperatures, a pretransition between $P_{\beta'}$ and the lamellar phase $L_{\beta'}$ is also observed at T_p. In specific conditions, a crystalline phase L_c may form at even lower temperature and it transforms into $L_{\beta'}$ upon heating at T_c. The classical transitions of C*n*PCs are [3]

$$L_c \xrightarrow{T_c} L_\beta \xrightarrow{T_p} P'_\beta \xrightarrow{T_m} L_\alpha \quad (1)$$

The temperatures of the transitions depend on the chain length.

In contrast, dilauroylphosphatidylcholine (DLPC, $n=12$) exhibits, by differential scanning calorimetry (DSC) a sharp peak at $T=-1.8$ °C and a broad peak spreading between −2 and 5 °C. This was originally observed by Mabrey and Sturtevant [5] and further confirmed by Finegold et al.[6]. On the basis of both small-angle and wide-angle X-ray diffraction (XRD), Hatta et al. [7] suggested the existence of an intermediate phase L_i between $P_{\beta'}$ observed below the sharp peak, and L_α, at temperatures above the broad peak. L_i would be a nontilted, partially disordered liquid-crystalline-phase [7]. The results of Raman spectroscopy [8], which is sensitive to chain conformational (gauche/trans bond ratios) and packing disorder, are consistent with a two-step process for the transition $P_{\beta'} \rightarrow L_\alpha$. On the other hand, after very specific prolonged incubation at $T<-6$ °C or after repeated temperature cycling, two peaks are also observed; both involve chain melting. NMR studies show the coexistence of a gel-like phase and a liquid-crystalline phase over a wide temperature range, throughout the transition region [6].

In this study, time-resolved XRD (wide-angle and small-angle scattering, WAXS and SAXS) as a function of temperature and high-sensitivity DSC were carried out simultaneously to correlate structural and thermodynamic data. Simultaneous recording of X-ray and DSC patterns warrants against any temperature lag or difference in thermal history of the sample. The low scanning rate (0.1 °C min^{-1}) allows us to be close to equilibrium and to evidence possible intermediate states. Indeed, polymorphic changes often depend on the kinetics. For instance, in dipalmitoylphosphatidylcholine (DPPC), either a stable or a metastable $P_{\beta'}$ phase is formed, depending on the rate of cooling from L_α [9, 10, 11, 12].

Materials and methods

The DLPC was purchased from Avanti Polar Lipids (Birmingham, AL, USA) and was used without further purification. A former study [6] showed that the purity of the phospholipid was 99%. All preparations were done in a 145 mM NaCl and 10 mM *N*-(2-hydroxyethyl)piperazine-*N*′-ethanesulfonic acid (Sigma, L'Isle d'Abeau, France) buffer (pH 7.4).

The samples were prepared by hydration of the dry powder of DLPC, until we obtained a buffer weight ratio of 80% w/w, which satisfied the condition of excess water. This preparation was then mixed by a vortex for 3 min at 5 Hz, then allowed to equilibrate and then mixed by a vortex again. Since the highest transition temperature in DLPC is at 5 °C, we are assured that at room temperature we are at least 10 °C above the gel phase to liquid-crystal transition temperature. A slight quantity of sample was then loaded (25–30 μl) in a Lindemann glass capillary of 1.4-mm (±0.1-mm) diameter (GLAS, Muller, Berlin, Germany).

The experiments were performed thanks to a technique allowing simultaneous time-resolved XRD as a function of temperature (XRDT) and high-sensitivity DSC to be carried out on the same sample in the same apparatus [13]. This device is available on the D24 bench of the synchrotron Laboratoire pour l'utilisation du rayonnement électromagnétique (LURE), Orsay, France, operating at $\lambda=1.5498$ Å. The reference used to calibrate the detector was the β form of tristearin, characterized by 44.96-Å long spacings and by short spacings of 4.59, 3.85 and 3.70 at room temperature [14]. No X-ray-induced degradation of the sample is expected at the brightness of the D24 beam (10^9 photons mm^{-2} s^{-1}) for the duration of the experiments even at the slowest scanning rates employed (0.1 C min^{-1}).

Results and discussion

Thermotropism of fully hydrated DLPC upon heating and cooling is illustrated in Fig. 1. The thermal scans were recorded at a very slow scanning rate (0.2 or 0.1 °C min^{-1}) in order to allow sample equilibration. Two peaks are observed both on cooling and on heating, a sharp peak at low temperature and a broad transition spreading above, which is peculiar to DLPC. Other C*n*PC (with $n>12$) in a fully hydrated state exhibit a sharp main transition. Such behavior is quite unusual in "pure" compounds (C*n*PC mixtures in excess water are expected to behave as pure compounds in the two-phase domain corresponding to excess water) [4, 5].

The onset temperature of the sharp peak on heating is −1.8 °C, whereas the onset temperature on cooling is −2.9 °C. It is noteworthy that the onset of the exotherm observed on cooling is very close to the onset of the

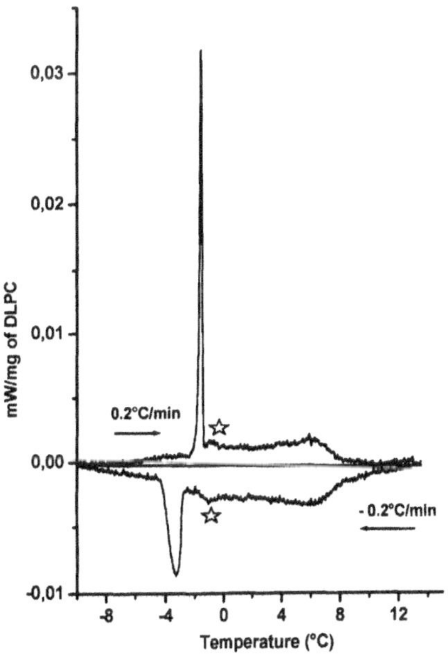

Fig. 1 Differential scanning calorimetry (*DSC*) recordings of dilauroylphosphatidylcholine (*DLPC*) in excess water (80% water w/w) during cooling and heating at 0.2 C min^{-1} between 15 and −10 °C (sample 18.0 mg, DSC7-Pyris). *Arrows* indicate the scan direction; *stars* show an additional possible transition (see text)

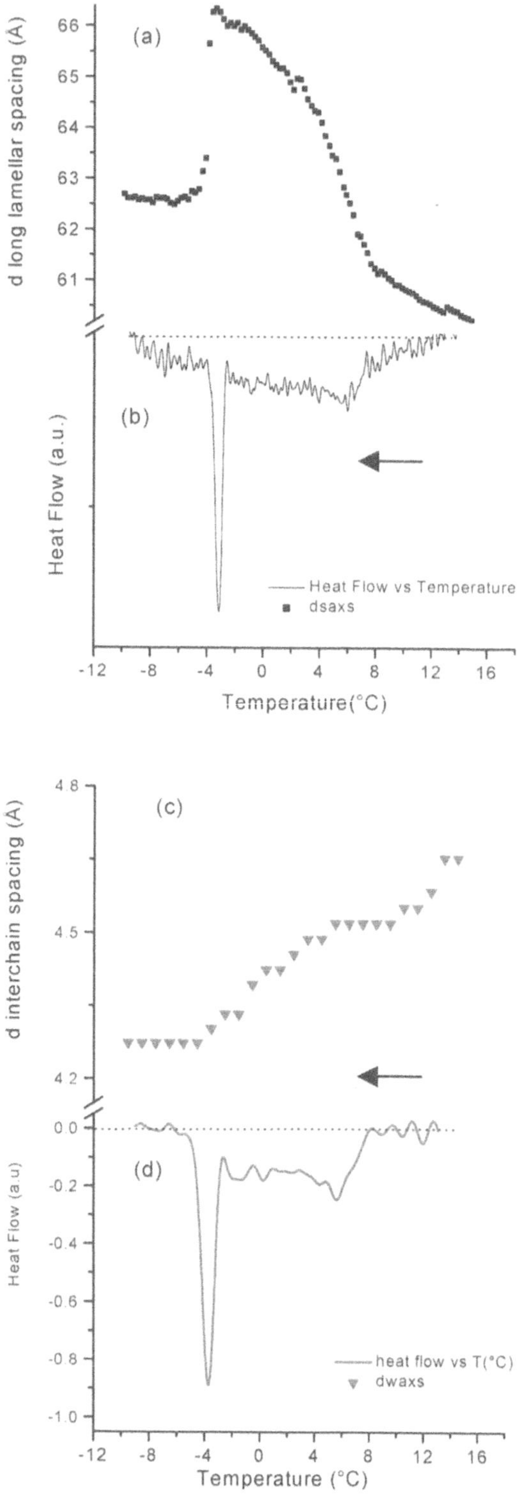

Fig. 2a–d Simultaneous structural and thermal evolutions of DLPC in excess water as recorded by coupled DSC-XRDT during a cooling scan at 0.2 °C min^{-1} between 15 and −10 °C. **a** SAXS period evolution and **b** corresponding DSC scan. **c** WAXS period evolution and **d** corresponding DSC scan

heating endotherm (quasi absence of undercooling suggests that no metastable state is present).

The enthalpy associated with the sharp peak is 1.51 kcal mol^{-1} (not taking into account the area between the peak aisles and the baseline) and is slightly less than previous values of 1.69 and 1.8 kcal mol^{-1} reported, respectively, by Mabrey and Sturtevant [5] and Finegold et al. [6]. The total area of the broad endothermic curve between −10 and 10 °C corresponds to 4.4 kcal mol^{-1}.

There are discrepancies between data previously reported for the total enthalpy: they range from 3.8 [6] to 6.75 kcal mol^{-1} [15]. Vervegaert et al. [16] found 4.3 kcal mol^{-1} [16] and Mabrey and Sturtevant 4.77 kcal mol^{-1} [5]. Quadratic extrapolation of the main transition temperature versus n from dimyristoylphosphatidylcholine (DMPC) ($n=14$) to distearoylphosphatidylcholine ($n=18$) gives a transition temperature of 1.2 °C for DLPC [3]. This correlation gives support for thinking that melting of the chains occurs throughout both endotherms. On the other hand, the enthalpy of chain melting for DLPC, expected from a simple extrapolation from DMPC and DPPC, ranges between 2 and 3 kcal mol^{-1}, depending on data from different authors.

Both SAXS and WAXS results, which yield information, respectively, on bilayer organization and hydrocarbon chain packing, highlight three regions. At temperatures above the broad peak ($T>8$ °C) the lamellar L_α phase is evidenced; the WAXS profiles display a broad band around 1.4 A, like that of liquid paraffin, and the SAXS patterns show a sharp peak arising from regular spacing of bilayers. A second order is also observed. The peak position varies very slightly with temperature (0.102 $Å^{-1} < q < 0.105$ $Å^{-1}$, $q=0.104$ $Å^{-1}$ at $T=15$ °C) and its intensity is almost constant (Fig. 2).

At temperatures below the sharp peak, the SAXS and WAXS data are characteristic of the gel ripple phase $P_{\beta'}$ [1] At wide angles the sharp peak centerd at $q=1.48$ $Å^{-1}$ is indicative of chain crystallization; the 4.25-Å spacing has been associated with a two-dimensional hexagonal lattice of stiff and fully extended chains exhibiting rotational disorder. At low angles small peaks due to the rippled structure appear in addition to the prominent peaks reflecting the lamellar repeat distance ($d=62.6$ A).

These L_α and $P_{\beta'}$ phases have been well characterized previously [1, 2]; more striking is the evolution of the structure between these phases.

Cooling scans (0.1 and 0.2 °C min^{-1})

Figure 3 represents the evolution of lamellar spacing ("long" spacing d) (Fig. 3a) and the "interchain" spacing ("short" spacing) (Fig. 3c) as a function of

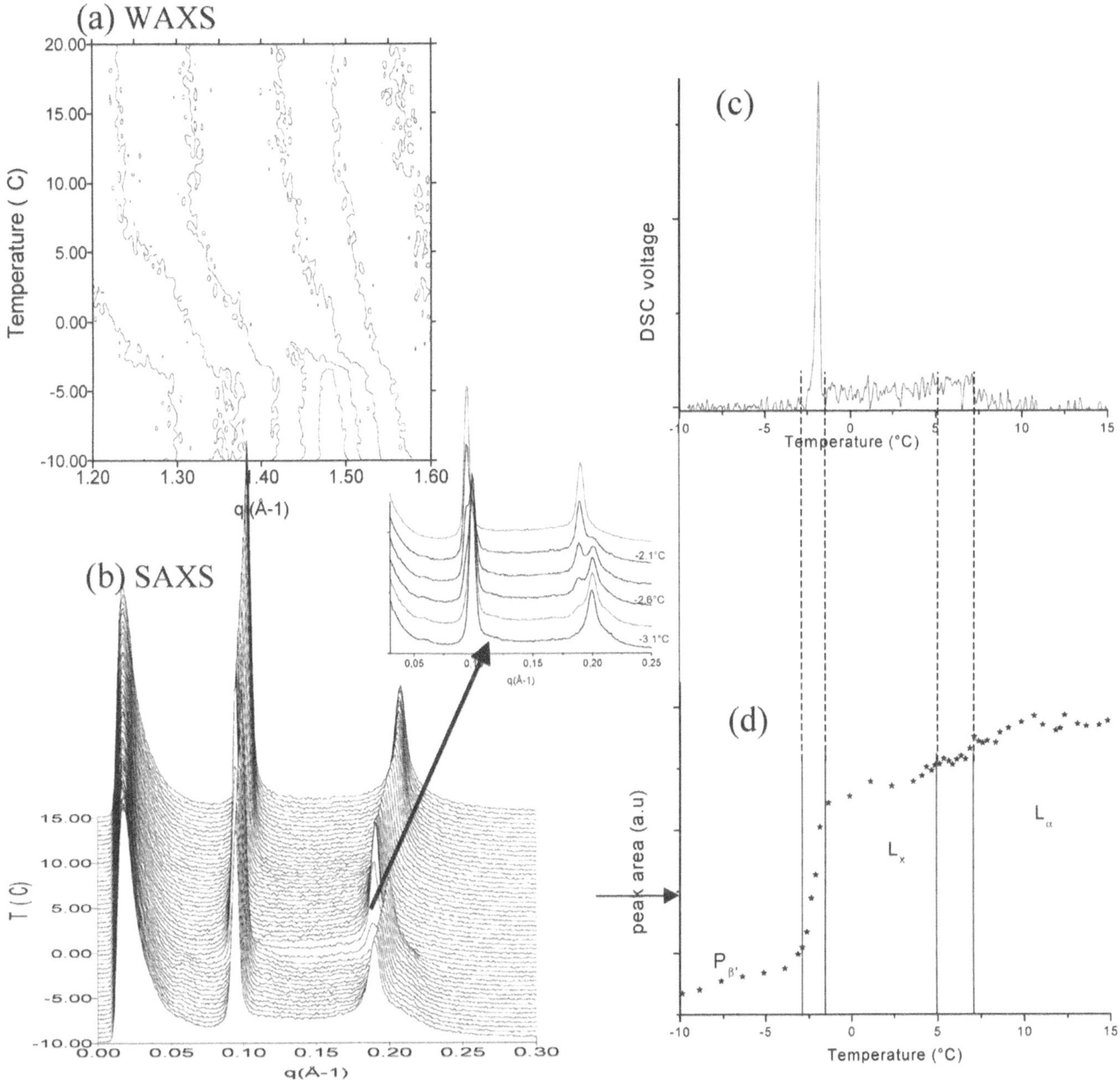

Fig. 3a–d Simultaneous structural and thermal evolutions of DLPC in excess water as recorded by coupled DSC–time-resolved XRD as a function of temperature (*XRDT*) during a heating scan at 0.1 C min^{-1} between −10 and 15 °C. **a** Wide-angle X-ray scattering (*WAXS*) contour plot. **b** 3D plot of the evolution of small-angle X-ray scattering (*SAXS*) patterns, *insert* is a zoom in the (−3.1 °C, −1.9 °C) domain and **c** corresponding DSC scan. **d** Integrated intensity (SAXS) as a function of temperature (onsets and offsets of DSC peaks are delimited by *vertical lines*)

temperature obtained from SAXS and WAXS data, respectively, and simultaneous DSC recording (Fig. 3b and d, respectively) obtained during cooling at 0.1 C min^{-1} (SAXS) and 0.2 °C min^{-1} (WAXS).

Upon cooling the short spacing gradually changes from 4.52 Å at 8 °C to 4.65 at 15 °C. The wide-angle peak remains broad till about −2 °C; it sharpens over the temperature range corresponding to the sharp DSC peak. Thus, during crystallization there is first a decrease of the interchain spacing followed by a change to an ordered hexagonal packing.

The lamellar spacing increases continuously, with an inflexion at about 2 °C, up to the value of 66.3 A (there is a clear breakpoint between two linear regimes at 7.8 °C). This increase is followed by a sharp discontinuity: the lamellar spacing falls from 66.3 Å at −3.6 °C to 62.7 Å at −4.9 °C. The L_i and $P_{\beta'}$

phases coexist over a narrow temperature range, between −3.9 and −4.9 °C, as shown by the coexistence of peaks characteristic of these two phases (insert of Fig. 2b); this range of temperature corresponds to the offset of the sharp peak and the offset of the broad peak located at 7.5 °C. In contrast, the evolution of L_α towards L_i appears continuous, the change in slope observed in the long spacing *d* corresponds exactly to the onset of the broad peak.

The temperature correlation between the SAXS and DSC events becomes obvious when observing brutal changes on cooling in the DSC recording (sharp peak at −2.9 °C) and in the SAXS long-spacing evolution (drastic jump at −3.6 °C), which is interpreted as the transition from L_i to $P_{\beta'}$ (Fig. 3a). Before this dramatic event, an increase in the periodicity of the bilayers (lamellar spacing *d*) during cooling from L_α to L_i is observed. The "one step" increase of the lamellar distance *d* of the transition between L_α and $P_{\beta'}$, which is observed for the other C*n*PCs, is replaced for DLPC by a two-step increase that starts at a clear breakpoint between two linear regimes at 7.8 °C (Fig. 3a). The second linear regime is followed by a quasi-asymptotic increase of *d* to a maximum value that precedes the *d* jump discussed earlier. The bilayer periodicity increases in the temperature range corresponding to the broad DSC transition up to a value of 66.3 Å, which exceeds lamellar periods of both $P_{\beta'}$ and L_α phases (62.6 and 60 Å, respectively). However, it is to be noted that this event superimposes on the increase usually observed for C*n*PC in the $L_\alpha \rightarrow P_{\beta'}$ transition and that the period observed for L_α ($d = 60.5$ Å at 10 °C) exceeds the bilayer spacing $d = 58.5$Å which is expected from a double linear extrapolation as a function of *T* and *n* of the thickness of C*n*PC (from data of the C*n*PC literature $12 \leq n \leq 1.8$ and $25 \leq T \leq 56$; $d = 3.48 + 0.197n - 0.000743T$, $r = 0.925$ [17]). The passage through such a maximum spacing value (when decreasing temperature) was also observed by Hatta et al. [7]. This behavior is not observed for any C*n*PCs and further evidences the specific behavior of DLPC [3].

The broad peak recorded by DSC is accompanied by a continuous increase of the lamellar spacing during cooling, whereas the interchain spacing decreases regularly (from $d = 4.5$ to 4.3 Å), this decrease is indicative of a change in the lateral packing of the chains from a quasi-melted state to that where they are crystallized. Just next to the sharp transition, a small DSC peak is observed (stars in Fig. 1). These DSC peaks, the origin of which are not known, correspond to a maximum in the SAXS intensity peak.

The coupling of DSC and XRDT allowed us to observe that the $P_{\beta'}$ phase appears once the sharp peak transition has occurred on cooling (i.e. once the maximum of the DSC peak has been reached), so the lag between thermal and structural evolution suggested by Hatta et al. [7] has been highlighted thanks to the simultaneous recording of small-angle (or wide-angle) and XRD and high-sensitivity DSC. The structural changes are rather observed on the completion side of the DSC-recorded transition on cooling and start before the onset of the DSC peak on heating. Simultaneous recording of XDRT and DSC warrants against any temperature lag or difference in sample time–temperature history.

As already discussed, at $T < -2.9$ °C, the formation of the gel ripple phase $P_{\beta'}$ is observed; however, a slight bump is also observed close to and preceding this sharp transition at about −1 °C by DSC whatever the scanning rate (stars in Fig. 1, bump in Fig. 2d). This small peak is also reflected by a maximum of intensity of the SAXS peaks in this intermediate regime. This bump could neither correspond to water crystallization nor to melting since the energies recorded are about 2 orders of magnitude smaller than that of the ice crystallization/melting transition. The determination of the origin of such a bump and intensity maximum needs further studies.

Heating scan AT 0.1 °C min^{-1}

Figure 3a and b represents the evolutions of the SAXS and WAXS diffraction patterns as a function of temperature as 3D and contour plots, respectively, during the heating of DLPC in excess water at 0.1 °C min^{-1} together with the simultaneous DSC acquisitions (Fig. 3d). The variation of the SAXS intensity integrated versus temperature during the same heating, which is shown Fig. 3c, highlights the correlation between the phase transition recorded by DSC and lamellar spacing changes. The contour plot for which the three domains are clearly seen underlines the lateral packing change. A continuous linear variation of the interchain spacing is observed in the L_i phase.

Indeed, the DSC recording displays the onset of the "main" transition at −2.5 °C, whereas the sharp intensity inflexion is observed between −2.5 and −1.5 °C (Fig. 3c and insert of b); this latter temperature corresponds to the offset of the sharp "main" transition peak. We observed for the heating scan the same phenomenon as for cooling but reversed; the intermediate phase starts to appear before the sharp DSC peak develops. In the same way, wide-angle and small-angle intensity diffraction profiles do not evolve for temperature higher than 7.5 °C; this latter temperature representing the offset of the broad peak during the heating scan (Fig. 3a, b). We noticed that during the complete heating scan the interchain short spacing decreases with temperature, but it increased for the cooling scan, along the intermediate phase domain, which is unusual (Fig. 2b and contour plot Fig. 3b).

Conclusion

In conclusion, the simultaneous recording of both SAXS and WAXS and DSC allowed us to confirm the existence of an intermediate phase during the $P_{\beta'} \rightarrow L_{\alpha}$ transition. The progressive evolutions of the SAXS and WAXS periods that accompany the existence of a broad peak in the DSC recordings as well as the values of their respective enthalpies demonstrate that the chain melting is not complete at the sharp transition. On heating, persistence of crystalline domains that locally rigidify the chains and maintain them in an extended conformation might explain the increase in the bilayer period observed at about −2 °C. At $T > -2$ °C, the progressive decrease of the bilayer period that is observed during the temperature increase from −2 to 5 °C is counterbalanced by an increase in the lateral chain packing period. The persistence of rigid domains together with that of liquid crystals is likely compensated at the molecular volume scale by water domain increases that progressively diminish on heating from −2 to 5 °C. This point needs to be clarified further.

Acknowledgements The authors thanks CNRS for the BDI grant to L.D. and LURE for hosting the research project.

References

1. Luzzati V, Tardieu A (1974) Annu Rev Phys Chem 25:79
2. Janiak MJ, Small DM, Shipley GG (1976) Biochemistry 15:4575–4580
3. Small DM (1986) Handbook of lipid research, vol 4. The physical chemistry of lipids, from alkanes to phospholipids. Plenum, New York
4. Grabielle-Madelmont C, Perron R (1983) J Colloid Interface Sci 95:471–483
5. Mabrey S, Sturtevant JM (1976) Proc Natl Acad Sci USA 73:3862–3866
6. Finegold L, Shaw WA, Singer MA (1990) Chem Phys Lipids
7. Hatta I, Matuoka S, Singer MA, Finegold L (1993) Chem Phys Lipids 69:129–136
8. Huang CH, Lapides JR, Levin IW (1982), J Am Chem Soc 104:5926–5930
9. Tenchov BG, Yao H, Hatta I (1989) Biophys J 56:757–768
10. Yao H, Matuoka S, Tenchov B, Hatta I (1991) Biophys J 59:252–255
11. Matuoka S, Kato S, Akiyama M, Amemiya Y, Hatta I (1990) Biochim Biophys Acta 1028:103–109
12. Rappolt M, Pabst G, Rapp G, Kriechbaum M, Amenitsh H, Krenn C, Bernstoff S, Laggner P (2000) Eur Biophys J 29:125–133
13. Keller G, Lavigne F, Forte L, Andrieux K, Dahim M, Loisel C, Ollivon M, Bourgaux C, Lesieur P (1998) J Thermal Anal 51:783–791
14. Ollivon M, Perron R (1979) Chem Phys Lipids 25:395
15. Morrow RM, Davis JH (1987) Biochim Biophysic Acta 904:61–70
16. Vervegaert T, De Kruyff B, Verklrij AJ, Tocanne JF, Van Deenen LLM (1975) Chem Phys Lipids 14:97–100
17. Marsh D (1990) Handbook of lipid bilayers, CRC, Boca Raton, pp 186–189

Progr Colloid Polym Sci (2004) 126: 184–189
DOI 10.1007/b93976

Nissim Garti
Imad Zakharia
Aviram Spernath
Anan Yaghmur
Abraham Aserin
Roy E. Hoffman
Lewis Jacobs

Solubilization of water-insoluble nutraceuticals in nonionic microemulsions for water-based use

N. Garti (✉) · I. Zakharia · A. Spernath
A. Yaghmur · A. Aserin
Casali Institute of Applied Chemistry,
Givat Ram Campus,
The Hebrew University of Jerusalem,
91904 Jerusalem, Israel
e-mail: garti@vms.huji.ac.il
Tel.: +972-2-6586574/5
Fax: +972-2-6520262

R.E. Hoffman
Department of Organic Chemistry,
The Hebrew University of Jerusalem,
91904 Jerusalem, Israel

L. Jacobs
Archer-Daniels-Midland Company
(ADM), Decatur, JL, USA

Abstract Winsor IV water-dilutable food-grade microemulsions based on ethoxylated sorbitan esters, water, (*R*)-(+)-limonene, ethanol, and propylene glycol have been prepared. Tocopherols and lycopene, two active lipophilic antioxidants, have been solubilized in the water-in-oil, bicontinuous, and oil-in-water microemulsions several times more than the oil (*R*)-(+)-limonene dissolution capacity. Dilution of the aqueous phase and decrease of the surfactant content strongly affect the solubilization capacity and solubilization efficiency of the microemulsion. Self-diffusion coefficients of the water and the oil in the presence of the solubilizates were correlated to the solubilization capacity, and are indicative of the locus of solubilization of the lycopene in the different microstructures.

Keywords Food microemulsion · Lycopene · Self-diffusion NMR · Solubilization capacity · Vitamin E · Tocopherols

Introduction

The capability of microemulsions to solubilize lipophilic and hydrophilic food additives, to enhance reaction efficacy and to allow selective extraction has attracted the attention of colloid scientists [1, 2].

In our previous studies we examined the phase behavior of food-grade microemulsions based on nonionic surfactants [3, 4, 5]. The effect of cosurfactant and other components on the water solubilization capacity (SC) of water-in-oil (W/O) microemulsions was also evaluated. Recently we also prepared oil-in-water (O/W) food-grade microemulsions [5]. Addition of short-chain alcohols (such as ethanol) and polyols (glycerol or propylene glycol, PG) induced the formation of a large isotropic region of continuous W/O and O/W microemulsions [5]. These unique microemulsions (known also as L phase) are characterized by a single continuous isotropic region starting from a pseudobinary solution (micellar system containing surfactant and oil phase) and proceeding to the water/PG (1/1) corner of the phase diagram [5]. All the structural transformations occurring in the isotropic phase develop continuously with no phase separation.

Food supplements with nutritional and health benefits are often termed "nutraceuticals". Tocopherols and lycopene are important members of this category.

The bioavailability of nutraceuticals is known to be affected by several factors:

1. The amount and type of fat present in the intestine. Fat enhances the formation of micelles that solubilize vitamins and nutraceuticals.
2. Interactions between active molecules that may reduce absorption of either of the compounds [6].
3. Molecular configuration (*cis*/*trans*) of the lycopene molecules. The bioavailability of the cis isomer is higher than that of the trans isomer [7, 8].
4. Particle size [9], *i.e.*, tomatoes converted into tomato paste enhance the bioavailability of lycopene, as the processing includes particle size reduction and heat treatment [6].

There has been much interest in fortifying food products with both vitamin E and lycopene, two very important lipophilic antioxidants in the diet of humans. As revealed by many observations, the average dietary intake of vitamin E and lycopene is insufficient to protect against long-term health risks associated with oxidative stress. Water-insoluble compounds, such as vitamin E and lycopene, have to be solubilized in self-assembled dispersions in order to be added homogeneously to aqueous media.

The present study explores the SC of these new five-component food-grade microemulsions of Winsor IV type. In the first example free *d*-α-tocopherol (TocOH) and *d*-α-tocopherol acetate, (TocAc), which are oil-soluble liquids, are solubilized in O/W microemulsions. These microemulsions are prepared with an oil phase of either (*R*)-(+)-limonene, or a mixture of (*R*)-(+)-limonene and tocopherols, or tocopherols alone as the oil phase. The second example is of a nonsoluble powdered nutraceutical antioxidant, lycopene [with solubility of 700 ppm in (*R*)-(+)-limonene], which is solubilized in the core, and at the interface, of W/O, bicontinuous and O/W microemulsions. The significant differences in SCs are stressed.

An additional objective of the present study is to investigate the influence of solubilized material on the microemulsion microstructure.

Materials and methods

Materials

R(+)-limonene (98%) and Tween 60 [polyoxyethylene(20) sorbitan monostearate] were commercial grade and were purchased from Sigma Chemical Co. (St. Louis, MO). Ethanol (EtOH) was obtained from Frutarom (Haifa, Israel). PG (1,2-propanediol) was purchased from BDH (Poole, UK). Lycopene (oleoresin with 7% lycopene) was obtained from Lycored (Beer-Sheva, Israel). TocAc (E6-100) and TocOH (E5-100) were obtained from ADM Nutraceuticals (Decatur, IL). All components were used without further purification. The water was double distilled.

Phase diagrams

The microemulsion systems were described on pseudo-ternary phase diagrams. They were constructed (at 25 °C) as recently reported [5].

Solubilization measurements

Lycopene and *R*-(+)-limonene were heated to 140 °C for 5 min, in the presence of N_2. The water, PG, EtOH and surfactant were added dropwise to obtain a microemulsion with the desired composition. The samples were cooled and stored at 25 °C. Samples that remained transparent for at least 10 days were considered to be microemulsions. TocAc and TocOH (Vitamin E) were solubilized at room temperature during the preparation of the phase diagrams and were always added as part of the oil phase.

Self-diffusion measurements

NMR measurements were performed at 25 °C using a Bruker DRX-400 spectrometer with a BGU *z*-gradient amplifier unit and a 5-mm broad band inverse probe equipped with a *z*-gradient coil, providing a *z*-gradient strength, *g*, of up to 55 G cm^{-1}. The self-diffusion (SD) coefficients were determined using pulsed-field-gradient stimulated spin echo. For details see Ref. [10].

Results and discussion

Microemulsions are not inert vehicles. Adding components such as lycopene or vitamin E to an isotropic system may affect its phase behavior. It is of vital importance to examine both the influence of the solubilized material on the microstructure and the stability of the microemulsion [11].

The basic system in which the solubilization was explored consists of *R*(+)-limonene, EtOH as the oil phase, water/PG (1/1) as the aqueous phase and Tween 60 as the surfactant (Fig. 1).

d-α-Tocopherol acetate

When TocAc was incorporated into the oil phase (Fig. 2b, c), the two-phase region of the blank microemulsion, made of *R*(+)-limonene and Et-OH alone as the oil in the phase diagram (Fig. 2a), was enhanced. TocAc incorporation seems to decrease the SC of water and oil, modifying the interactions with the surfactant and thus increasing the interfacial tension between water and oil. This can be seen from the decrease in dilution capacity (DC) at both dilution lines T37 and T73 (Fig. 2), as more TocAc replaced (*R*)-(+)-limonene.

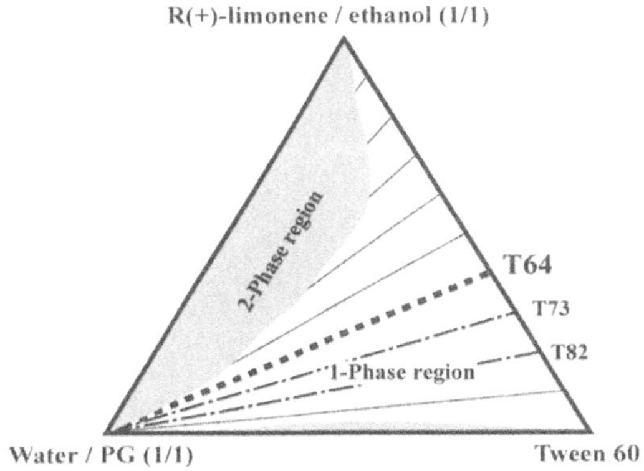

Fig. 1 Pseudo-ternary phase diagram (25 °C) of water/PG/*R*(+)-limonene/ethanol/Tween 60 system with a constant weight ratio of water/PG (1:1) and a constant weight ratio of *R*(+)-limonene/ethanol (1:1). Solubilization of lycopene was studied along dilution line T64

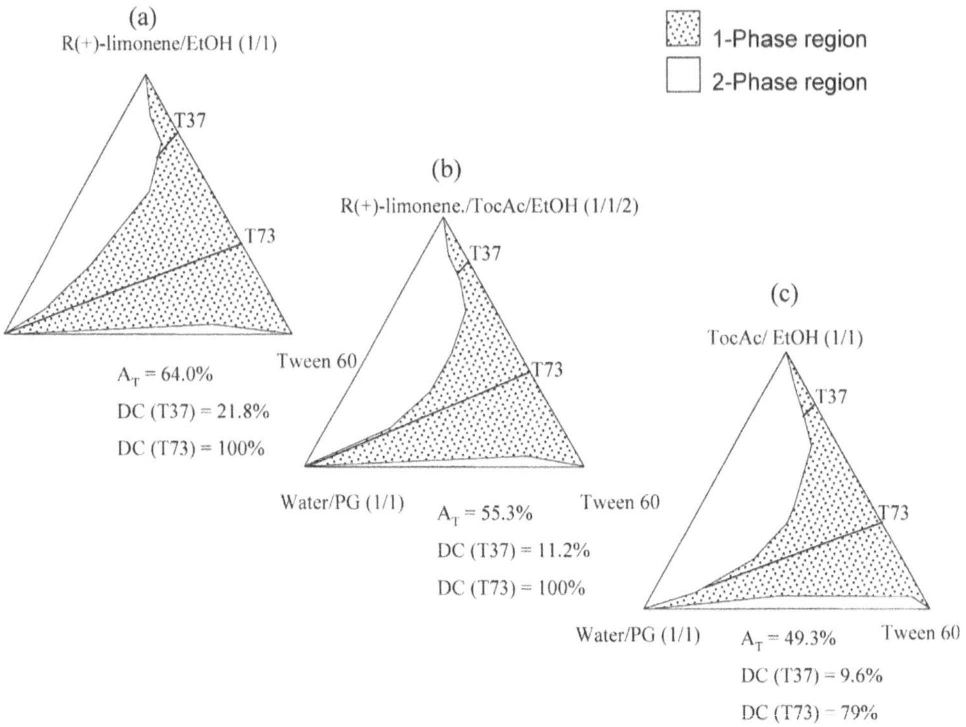

Fig. 2a–c Phase diagrams of the system water/PG/oil phase/Tween 60 at 25 °C with constant weight ratio of water/PG (1:1). The oil phase contain: (**a**) *R*(+)-limonene/ethanol (1:1), (**b**) *R*(+)-limonene/TocAc/ ethanol (1:1:2), (**c**) TocAC/ethanol (1:1)

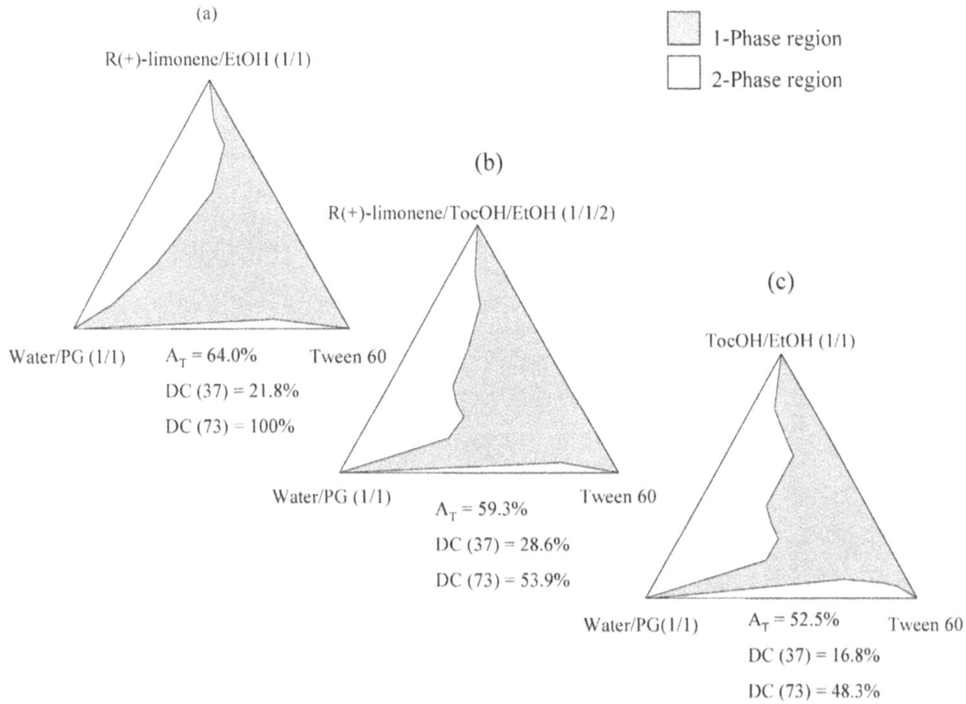

Fig. 3a–c Phase diagrams of the system water/PG/oil phase/Tween 60 at 25 °C with constant weight ratio of water to PG (1:1). The oil phase contains **a** *R*-(+)-limonene/EtOH (1:1), **b** *R*-(+)-limonene/*d*-α-tocopherol (*TocOH*)/EtOH (1:1:2), **c** TocOH/EtOH (1:1)

d-α-Tocopherol

When free TocOH was introduced into the oil phase, replacing part of the (*R*)-(+)-limonene (Fig. 3), it enhanced the SC of water in the oil-rich region (seen as a larger isotropic region in the upper corner of Fig. 3b); however, it reduced the solubilization of oil in water closer to the aqueous corner, even to a greater extent than TocAc. This is seen from a narrow two-phase region near the oil corner and a wider one near the aqueous phase corner. The DC of dilution line T28 in Fig. 3a is 7.8%. With *R*-(+)-limonene/TocOH/EtOH (1/1/2) the DC is 11%, and with TocOH it is 14% (data not shown). Upon increasing the concentration of the surfactant, with TocOH being a part of the oil phase, the DC of the surfactant/oil solution decreases. This behavior may be attributed to the polar nature of the TocOH, which accounts for its penetration ability into the interfacial palisade layer closer to the hydrophilic heads of the surfactant. In the oil-rich region where W/O structures are formed, there is enough space for the TocOH to penetrate into the reverse micelles. The hydrophobic tails of the surfactant move freely. This enables the TocOH to be distributed at the interface and provides some flexibility to the surfactant film. A larger amount of the aqueous phase is thus solubilized inside the swollen reversed micelles.

In the aqueous-rich region, where O/W structures are most likely to be formed, the hydrophobic tails entangle inside the micelles and the hydrophilic heads are in contact with the aqueous phase. The penetration of TocOH into the interface may result in decreased flexibility of the surfactant film caused by the steric effect of the long tail of TocOH. This result in less effective interaction between TocOH and the surfactant molecules, and therefore less stable microemulsions. For surfactant-concentrated compositions near the aqueous corner (Fig. 3c), the formulations tend to be more translucent, indicating larger structures. This may be ascribed to the localization of TocOH within the interface and the possible hydration of the polar hydroxyl group on the chroman ring, at high dilution with water, giving rise to more swollen micelles.

In O/W microemulsions, *R*-(+)-limonene appears to be a better oil than both tocopherols, owing to its smaller molecular volume and the more favorable interactions it undergoes with the hydrophobic portions of the surfactant, leading to a dispersed oil phase with a higher thermodynamic stability.

Note that the solubilization of lipidic nutraceuticals, like TocOH and TocAc, is quite easy since the molecules reside mainly in the oil phase and only partly at the interface. Such a possibility has some benefits since it allows dilution with other oils as per application request. The situation is somewhat more complicated when a nonsoluble nutraceutical, like lycopene, is to be solubilized. As yet, only Winsor IV microemulsions are known for introducing lycopene into water-continuous systems.

Lycopene

The SC and efficiency, α, were measured along dilution line T64 of the same blank system as described for the solubilization of the tocopherols, consisting of *R*(+)-limonene, EtOH, Tween 60, PG and water (Fig. 1). The SC was defined as the amount (part per million) of lycopene solubilized in the microemulsion, and α was defined as the amount (parts per million) of lycopene solubilized in the microemulsion per amount of *R*(+)-limonene in the microemulsion (normalized against oil).

Figure 4 shows the SC of lycopene along water dilution line T64, of the same phase diagram shown in Fig. 1 [at this line the constant weight ratio of *R*(+)-limonene/EtOH/Tween 60 is 1/1/3]. Four different regions can be identified along this dilution line. At 0–20 wt% aqueous phase (region I), the SC of lycopene decreases dramatically from 500 to 190 ppm (a reduction of 62%). This dramatic decrease in the SC can be related to enhanced interactions between the surfactant and water molecules. When water is introduced into the core the micelle swells, and more surfactant and cosurfactant participate at the interface, replacing the lycopene and therefore decreasing its solubilization. We conclude that in region I the reverse micelles swell gradually and turn more hydrophobic, causing less free volume to be available to the solubilized lipophilic lycopene, thereby reducing the SC.

At 20–50 wt% aqueous phase (region II), the SC remains almost unchanged (decreased by only an additional 7%). This fairly small decrease in the SC could be associated with the fact that the system transforms gradually into a bicontinuous phase, and the interfacial area remains almost unchanged when the aqueous phase concentration increases.

Surprisingly, in region III (50–67 wt% aqueous phase), the SC increases from 160 to 450 ppm (an increase of 180%). In region IV the SC decreases to 312 ppm (a decrease of 30%).

To explain the changes in the SC of lycopene, we characterized the microstructure of the microemulsions along dilution line T64, using the SD-NMR technique. Each microemulsion contains the maximum amount of lycopene that it can solubilize.

To evaluate the SD data in terms of microstructure, the calculation of the relative diffusion coefficient, D/D_0,

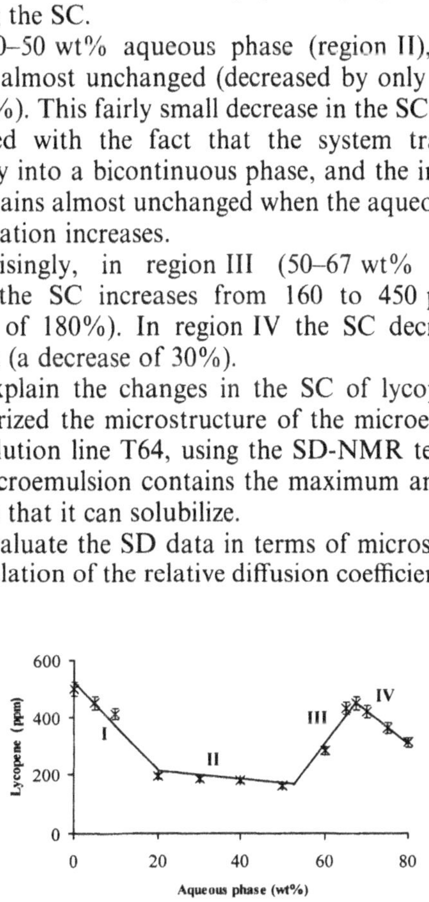

Fig. 4 Solubilization capacity curve of lycopene along dilution line T64 at 25 °C. The four regions along the curve are *I* water-in-oil (*W/O*) microstructure, *II* bicontinuous microstructure, *III* oil-in-water (*O/W*) microstructure, *IV* diluted O/W microstructure. The *line* is only for guidance

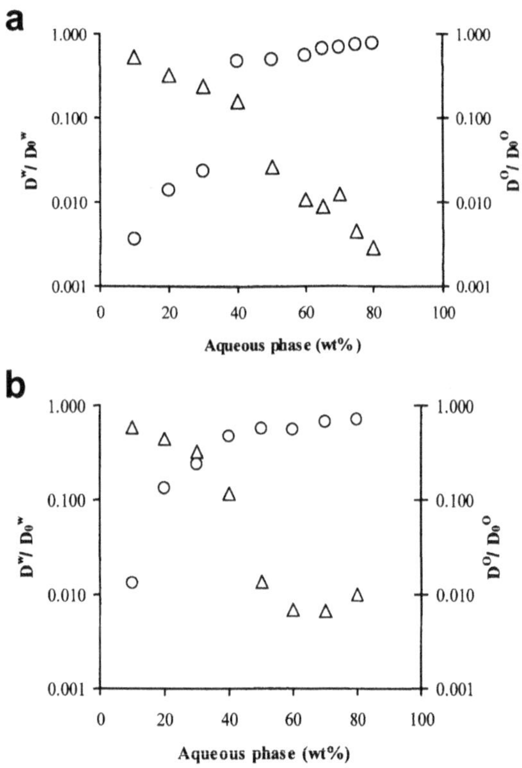

Fig. 5 Relative diffusion coefficient of water (○) and *R*(+)-limonene (△) in microemulsions without (5a) and with (5b) lycopene, as calculated from SD-NMR results at 25 °C. $D^W{}_0$ was measured in a solution containing water/PG (1:1), and determined to be 55.5*10^{-11} m^2s^{-1}. $D^O{}_0$, the pure diffusion coefficient of *R*(+)-limonene was determined to be 38.3*10^{-11} m^2s^{-1}

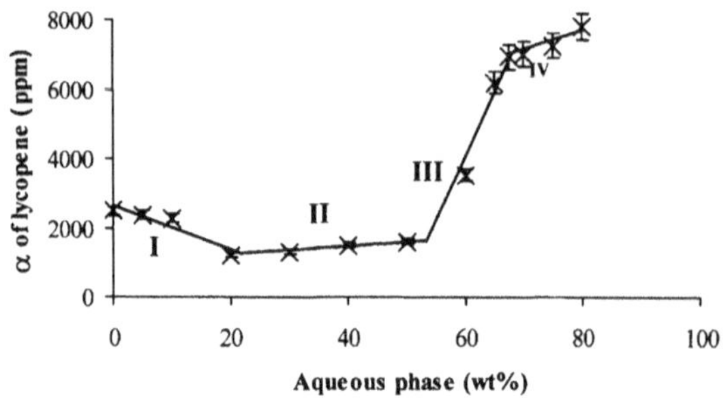

Fig. 6 Solubilization efficiency, α, curve along dilution line T64 at 25 °C. The four regions along the curve are *I* W/O microstructure, *II* bicontinuous microstructure, *III* O/W microstructure, *IV* diluted O/W microstructure. The *line* is only for guidance

of the two solvents is needed. Relative diffusion coefficients were obtained by dividing water, D^W, and oil, D^O, diffusion coefficients in the microemulsion by the diffusion coefficient of water in the pure water phase, $D^W{}_0$, and oil in the net phase, $D^O{}_0$. It is well documented [12] that if D/D_0 values of water and oil differs by more than 1 order of magnitude, discrete particles of the slowly diffusing solvent are expected, whereas D/D_0 values of water and oil of the same order of magnitude suggest a bicontinuous structure.

The relative diffusion coefficients of water and *R*(+)-limonene in empty microemulsions and microemulsions solubilizing lycopene are shown in Fig. 5 as a function of the aqueous phase concentration (weight per weight). One can clearly see that the general diffusion coefficient behavior of microemulsion ingredients [*R*(+)-limonene and water] with or without lycopene is not very different. The total amount of lycopene does not cause dramatic changes in the diffusion patterns of the ingredients.

It can also be seen that in the two extremes of aqueous phase concentrations (up to 20 wt% and above 70–80 wt% aqueous phase) the $D^W/D^W{}_0$ values are easily interpreted, while the in-between regions are somewhat more difficult to explain since gradual changes take place. Regions II and III are difficult to distinguish; however, the structural changes in the presence of lycopene (Fig. 5b) are more pronounced than those in the absence of lycopene (Fig. 5a).

As Fig. 5b indicates, microemulsions containing up to 20 wt% aqueous phase, and solubilizing lycopene, have a discrete W/O microstructure, since the relative diffusion coefficients of water and *R*(+)-limonene differ by more than 1 order of magnitude.

Microemulsions solubilizing lycopene and containing 20–50 wt% aqueous phase have a bicontinuous microstructure, as the diffusion coefficients of water and (*R*)-(+)-limonene are of the same order of magnitude. Increasing the aqueous phase concentration to above 50 wt% induces the formation of a discrete O/W microstructure, as the relative diffusion coefficients of water and (*R*)-(+)-limonene differ by more than 1 order of magnitude.

From the SC and SD-NMR results it is clear that the lycopene SC is structure-dependent.

In order to compare the solubility of lycopene in *R*-(+)-limonene to its solubilization in microemulsions, one must calculate the amount of lycopene per amount of *R*-(+)-limonene in the microemulsion (solubilization efficiency designated as α). The enhanced solubilization of lycopene is remarkable. Yet, swollen micelles have no practical value without the capability of being diluted by water, since many of the final food applications will be made in a water-continuous phase. Moreover, such mixtures cannot be diluted with any type of oil phase [including *R*(+)-limonene], and therefore are not practical for oil-continuous phase applications either. It is, therefore, essential to construct micellar concentrates capable of dilution in both oil and water phases. The microemulsions described in this paper are unique in having this property.

α is shown as a function of the aqueous phase content in Fig. 6. The α of lycopene solubilized in a microemulsion is higher than the solubility of lycopene in $R(+)$-limonene. The α curve (as seen in Fig. 6) can also be divided into four regions, similarly to the SC curve in Fig. 4. The four regions correspond to the microstructure changes (Fig. 5b).

In the first region (0–20 wt% aqueous phase), α decreases by 50% (from 2,500 to 1220 ppm). The decrease in solubilization efficiency can be attributed to the decrease in the interfacial area, as the micelles swell. In the second region (20–50 wt% aqueous phase) α remains almost unchanged (increased by 20%). Systems containing 20–50 wt% aqueous phase seem to have a bicontinuous microstructure (as indicated in Fig. 5). The interfacial area in a bicontinuous microemulsion does not change as a result of the increase in the aqueous phase content. In the third region α increases dramatically by 340%. This increase suggests that lycopene prefers the O/W interface. In the fourth and final region (67–80 wt% aqueous phase) α is almost unchanged (increased by 11%). The unchanged solubilization efficiency indicates that no further microstructure changes occur.

Conclusions

This study has demonstrated the use of food-grade microemulsions as vehicles for solubilizing both tocopherols and lycopene, which are nutraceuticals with health benefits.

TocOH and TocAc can be solubilized in W/O and O/W microemulsions and can serve as part or all of the oil phase. The difference in the solubilization behavior results from the difference in their lipophilicity, which drives them to the core or the interface respectively.

Lycopene, being a very insoluble compound (with only limited lipophilicity), cannot serve as an oil phase, yet it can be solubilized up to 10 times better than its dissolution capacity in R-(+)-limonene or any other edible oil. The lycopene solubilization along any dilution line is microstructure-dependent. Solubilized lycopene influenced both the microstructure and the composition in which a transformation from W/O to bicontinuous and from bicontinuous to O/W microstructure occurred.

References

1. Dungan SR (1997) In: Solans C, Kunieda H (eds) Industrial applications of microemulsions. Surfactant science series, vol 66. Dekker, New York, pp 148–170
2. Holmberg K (1998) In: Shah O (ed) Micelles, microemulsions, and monolayers. Dekker, New York, pp 161–192
3. Garti N, Clement V, Fanun M, Leser ME (2000) J Agric Food Chem 48:3945
4. Garti N, Aserin A, Fanun M (2000) Colloid Surface A 164:27
5. Garti N, Yaghmur A, Leser ME, Clement V, Watzke HJ (2001) J Agric Food Chem 49:2552
6. Van het Hof KH, West CE, Weststrate JA, Hautvast JGAJ (2000) J Nutr 130:503
7. Clinton SK (1998) Nutr Rev 56:35
8. Boileau AC, Merchen NR, Wasson K, Atkinson A, Erdman JW (2000) J Nutr 129:1176
9. Paiva SAR, Russell RM (1999) J Am Coll Nutr 18:426
10. Spernath A, Yaghmur A, Aserin A, Hoffman RE, Garti N (2002) J Agric Food Chem 50: 6917
11. von Corswant C, Thoren EG (1999) Langmuir 15:3710
12. Lindman B, Olsson U, Söderman O (1999) In: Kumar P, Mittal KL (eds) Handbook of microemulsion science and technology. Dekker, New York, pp 309–356

Progr Colloid Polym Sci (2004) 126: 190–193
DOI 10.1007/b93965

Lionel Massi
Frédéric Guittard
Serge Géribaldi

Quaternary bisammonium fluorosurfactants for antimicrobial devices

L. Massi · F. Guittard (✉) · S. Géribaldi
Chimie des matériaux organiques et métalliques, Faculté des sciences, Université de Nice Sophia-Antipolis, Parc valrose, 06108 Nice Cedex 2, France
e-mail: guittard@unice.fr
Tel.: +33-4-92076159
Fax: +33-4-92076156

Abstract Cationic surfactants became important when their strong antimicrobial properties were discovered, including the highlight of the importance of surfactants properties for antimicrobial activity. The aim of this work is to know the impact of surfactant properties of such molecules on antimicrobial activity. We describe the synthesis, surface activity properties and antimicrobial characteristics of highly fluorinated Gemini surfactants. We have studied a series of amide types through the variation of the spacer bonding the two ammonium heads. We have found this parameter to have a modulating effect on antimicrobial activity. The fluorinated chains are the main contributors to the hydrophobicity of such surfactants and can be put in correlation with bacteriostatic activities.

Keywords Fluorosurfactants · Antimicrobial agents · Quaternary bisammonium surfactants · Critical micelle concentration · Gemini surfactants

Introduction

During the second part of the twentieth century, many industrial uses were discovered for cationic surfactants, raising them from the position of speciality chemicals to bulk industrial chemicals. The fields of applications for such chemicals include their use as antistatic agents, fabric softeners, corrosion inhibitors, foam depressants, flotation chemicals, and additives for asphalt and petroleum [1]. But cationic surfactants first became important when their strong antimicrobial properties were discovered by Dogmak [2]. From this discovery, hundreds of products were developed and became of commercial importance [3]. Today, antimicrobial cationic surfactants continue to play an important role as sanitizing and antiseptic agents [4], as components in cosmetic formulations, and as germicides and fungicides [3]. Since the first work of Dogmak on long-chainquaternary ammonium surfactants, a great deal of research has led to numerous structural modifications of such surfactants in order to increase their surface activity [5, 6, 7, 8, 9, 10, 11]. These researches led to performance augmentation in various fields of application, including antimicrobial agents for surface treatment [12, 13, 14]. As reported previously, the synthesis and studies of micellization properties of a particular generation of bifunctionnal cationic surfactants, called Gemini surfactants [15, 16], showed unusual physicochemical properties [17, 18, 19] of these molecules, i.e., low critical micelle concentration (cmc), high solubilizing capacity, etc. There is a well-established relationship between cytolytic action and surface tension [20]; hence, the introduction of a perfluorinated tail instead of a hydrocarbon one drastically modifies the physical properties and confers to the resulting material peculiar properties: low values of free surface energy compared to those obtained within the hydrocarbon series, providing high surface activity to these surfactants, low cmc, etc. [21, 22]. The molecular design allows the careful choice of the type of moiety which will be the most suitable for the amplification of the desired qualities of the perfluorinated tail [23]. From these facts, we present here the synthesis and evaluation

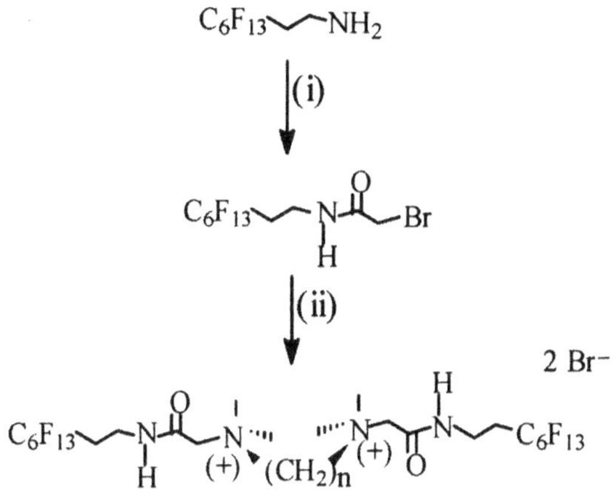

$\mathbf{T_n}$: n=2,3,4,6,8,10,12

Fig. 1 Synthetic pathways to highly fluorinated Gemini surfactants T_n: *i* $BrCH_2CO_2H$, DCC, CH_2Cl_2, room temperature; *ii* *N,N,N′,N′*-tetramethyl-α,ω-alkyldiamine, Et_2O, reflux

of a series of quaternary bisammonium Gemini fluorosurfactants, called T_n, including antimicrobial and physicochemical studies. The general structure of the surfactants synthesized is shown in Fig. 1. Compounds T_n differ in the spacer length (2–12 methylene units).

Experimental

Synthesis

N,N,N′,N′-Tetramethyl-α,ω-alkyldiamine (greater than 97%) and *N,N′*-dicyclohexylcarbodiimide (DCC, 99%) were purchased from Aldrich and were used as received. The preparations of 2-*F*-hexylethylamine [24] and 2-*F*-hexylethylbromoacetamide [25] are described elsewhere. Unless specified the solvents were of unpurified reagent grade.

Synthesis of T_n. The precursor (2-*F*-hexylethylbromoacetamide), 10 mmol, was reacted with 0.5 Eq *N,N,N′,N′*-tetramethyl-α,ω-alkyldiamine in diethyl ether (30ml). The mixture was stirred at reflux for 12 h, then filtrated and the residue was washed with diethyl ether to give white solids (T_2–T_8) or oils (T_{10}–T_{12}). The yields ranged from 80 to 90%.

All the Gemini surfactants synthesized were analysed by NMR spectroscopy and mass spectrometry (MS). The NMR spectra were recorded with a Brucker Advance 200 MHz NMR spectrometer. The mass spectra were recorded using a Finnigan Mat LCQ Classic Electrospray source API1 ESI-MS mass spectrometer in positive mode. In this case, only the cationic part of the molecule was detected. As these cations are bischarged, the observed sole *m*/2*z* peak (where *m* is the mass of the cationic part of the molecule) is in agreement with the structure. Characterization of T_2, T_3, T_4 and T_6 is given elsewhere [26].

T_8: ^{1}H NMR (methanol-d_4, δ): 1.4 (8H, m, [N–CH_2CH_2*CH_2CH_2*]$_2$), 1.95 (4H, m, [N–CH_2*CH_2*]$_2$), 2.55 (4H, tt, [C_6F_{13}–*CH_2*]$_2$), 3.3 (12H, s, [*CH_3*–N^+–*CH_3*]$_2$), 3.6 (8H, m, [*CH_2*–NH–C(O)]$_2$ + [$^+$N–*CH_2*CH_2]$_2$), 4.15 (4H, s, [C(O)–*CH_2*–N^+]$_2$). MS *m*/2*z* (%): 504.3 (100).

T_{10}: ^{1}H NMR (methanol-d_4, δ): 1.4 (12H, m, [N–CH_2CH_2*$CH_2CH_2CH_2$*]$_2$), 1.95 (4H, m, [N–CH_2*CH_2*]$_2$), 2.55 (4H, tt, [C_6F_{13}–*CH_2*]$_2$), 3.3 (12H, s, [*CH_3*–N^+–*CH_3*]$_2$), 3.6 (8H, m, [*CH_2*–NH–C(O)]$_2$ + [$^+$N–*CH_2*CH_2]$_2$), 4.15 (4H, s, [C(O)–*CH_2*–N^+]$_2$). MS *m*/2*z* (%): 518.2 (100).

T_{12}: ^{1}H NMR (methanol-d_4, δ): 1.4 (16H, m, [N–CH_2CH_2*$CH_2CH_2CH_2CH_2$*]$_2$), 1.95 (4H, m, [N–CH_2*CH_2*]$_2$), 2.55 (4H, tt, [C_6F_{13}–*CH_2*]$_2$), 3.3 (12H, s, [*CH_3*–N^+–*CH_3*]$_2$), 3.6 (8H, m, [*CH_2*–NH–C(O)]$_2$ + [$^+$N–*CH_2*CH_2]$_2$), 4.15 (4H, s, [C(O)–*CH_2*–N^+]$_2$). MS *m*/2*z* (%): 532.3 (100).

Antimicrobial evaluation

Antibacterial and antifungal activities of quaternary ammoniums compounds were evaluated using measurements of the minimal inhibitory concentrations (MIC) [27] expressed in micromoles per litre. These tests were realized on four microorganisms, i.e., *Pseudomonas aeruginosa* (CIP A22) class II type Gram(–), *Candida albicans* (ATCC 2091) class II Yeast, *Staphylococcus aureus* (ATCC 9144) class II type Gram(+) and *Aspergillus niger* (ATCC 6275) class I Mould. Bacteria were incubated on trypcase soy agar slants for 18 h at 30 °C. MIC tests were run with the third generation of bacteria; samples were taken during the exponential phase of bacterial growth. This third generation of bacterial strains obtained was incubated in an M9 minimal phosphate medium suspension for *P. Aeruginosa* and trypcase soy broth for *S. Aureus*. In the case of the fungal strains, the incubating protocol was identical, only the incubation time was changed (7 days in the case of the fungal strains). The third generation samples were incubated exclusively in an M9 suspension. Benzalkonium chloride (Empigen Bac 50 purchased from Albright & Wilson) was used as a commercial reference. Initial solutions of biocides were prepared at a 500 ppm concentration of active ingredient in distilled water. Bacterial inocula were prepared at 5×10^6–5×10^7 colony forming units/ml. Fungal inocula were prepared at 5×10^6–5×10^7spores/ml. Adjustment was made by optical density measurement for bacteria (absorbance 0.05 at a wavelength of 660 nm) and by counting under a microscope using a Neubauer cell for fungi and yeasts. The Biomek 1000 (Beckman) apparatus used for our experiments realized automatically dilutions of the biocide solutions tested in an inoculated 96-well microtitration plate. After 24-h incubation for bacteria and 72-h incubation for fungi, growth (or its lack) was determined visually. The lowest concentration at which no growth was observed (turbidity) was taken as the MIC.

Surface activity

The CMCs were measured using a conductimetry technique with a Tacussel CDM 210 conductimeter. Samples were thermostated at 25 °C. Conductimetry cells were calibrated using a 0.1 M potassium chloride solution at 25 °C. All the samples were measured at a constant temperature of 25 °C. The CMCs were determined from the breakpoint of each conductimetry (function of the concentration) curve obtained.

Results and discussion

Physicochemical properties

For quaternary ammonium antimicrobial surfactants, it is assumed that the mechanism of action is based on a surfactant-like interaction which has a perturbation effect on the permeability properties of biological membranes; hence, tensioactive properties have to be considered in the design of such antimicrobial agents. As a consequence, the amphiphilic nature and ability to form micelles are closely related to the cmc, which is an important parameter to take into account for the efficacy of such antimicrobial surfactants. It is well known that the cmc of surfactants decreases exponentially with increasing hydrophobic chain length according to the empirical relation [21]

$$\log(\mathrm{cmc}) = A - B(n) \tag{1}$$

where the concentrations are expressed in moles per litre, n is the number of carbon atoms in the spacer, and A and B are constants for a particular and homologous series of molecules at a given temperature. When studying the impact of spacer length, bonding the two cationic heads, on the cmc values for Gemini surfactants this relationship (Eq. 1) is not valid [28, 29]. We observe here a curve with a maximum value of the cmc for four methylene units in the spacer (T_4). The cmc range for the T_n series is not large, showing that the contribution of the spacer length for the modification of the cmc is minor. The results obtained in the present work confirm that the main contribution to the hydrophobicity of the molecule is mainly from the perfluorinated chains. From conductivity measurements, it is possible to obtain the apparent degree of dissociation of the micelles, α_{ap}. This parameter is given by the ratio of the slopes from the plot of conductivity versus concentration above and below the cmc. The degree of binding of the micelles, β_{ap}, is strictly obtained from the variation of the cmc of an ionic surfactant with an added electrolyte containing a common counterion. From the literature, it is assumed that if α_{ap} determined from electrical conductivity measurements (to be the degree of dissociation of the micelles) the value $1-\alpha_{ap}$ is usually nearly close to the experimental estimate of β_{ap}. The cmc measurements also allow us to calculate the thermodynamic parameters of micellization, such as the Gibbs free energy of micellization, ΔG^0_M, using the relation from Ref. [30]:

$$\Delta G^0_M = RT\left(\frac{3}{2} - \alpha_{ap}\right)\ln(\mathrm{cmc}) \tag{2}$$

This relation (Eq. 2) takes into account the peculiar structure of Gemini surfactants, i.e., two chains, two heads and the apparent degree of ionization of the micelles, α_{ap}. In this relation the cmc is expressed in moles of fluorinated chains per litre. All the parameters obtained are given in Table 1. It appears that the variation of spacer length does not strongly affect cmc values (same decade) but modifies the micellization parameters.

Table 1 Thermodynamic parameters of micellization for T_n

Molecules	Critical micelle concentration (mol/l)	ΔG^0_m (kJ/mol)	α_{ap}
T_2	5.91×10^{-4}	−21.56	0.26
T_3	5.46×10^{-4}	−21.26	0.21
T_4	7.58×10^{-4}	−19.97	0.24
T_6	5.88×10^{-4}	−18,90	0.37
T_8	3.61×10^{-4}	−15.76	0.62
T_{10}	2.96×10^{-4}	−14.94	0.69
T_{12}	2.69×10^{-4}	−14.74	0.71

Antimicrobial characteristics

All the MIC results are shown in Table 2. The efficacy of these compounds was evaluated against bacterial and

Table 2 Minimal inhibitory concentration (*MIC*) measurements for T_n

Molecules	Gemini type	*n*	MIC (μmol/l)			
			Staphylococcus aureus ATCC 9144	*Candida albicans* ATCC 2091	*Aspergillus niger* ATCC 6275	*Pseudomonas aeruginosa* CIP A22
T2[a]	Amide	2	6.73	6.73	3.99	5.80
T3[a]	Amide	3	4.55	6.65	4.55	5.19
T4[a]	Amide	4	3.15	7.64	4.95	5.94
T6[a]	Amide	6	2.96	4.39	3.79	5.52
T8	Amide	8	3.69	4.28	5.38	2.89
T10	Amide	10	3.61	5.26	17.54	2.83
T12	Amide	12	17.14	4.08	17.14	2.76
References						
Cetyl pyridinium chloride			1.98	2.32	2.96	17.60
Empigen BAC 50[b]			3.42	3.59	3.59	47.72

[a]Ref. [31]
[b]Benzalkonium chloride

fungal strains. The T_n series is of amide type. We studied the variation of the spacer length (*n* value). In view of the results, it appears that, in the range of spacer length studied, this factor weakly influences the MIC value; however, it has a significant effect on activity. Increasing the spacer length decreases the MIC, indicating a slight enhanced activity. For the microbial strains, *S. aureus* and *A. niger*, we found a minimum of activity corresponding to a spacer length of ten carbons. For *P. aeruginosa* and *C. albicans* the MIC still decreases until a spacer length of 12 carbons, and the MIC values are 2.76 and 4.08 μmol/l, respectively. A comparison to quaternary ammonium commercial references was performed and shows a particularly strong activity against Gram negative bacteria. Whereas the references Empigen BAC 50 and cetyl pyridinium chloride (purchased from ACIMA) have MICs against *P. aeruginosa* of 47.72 and 17.60 μmol/l, respectively, the active Gemini surfactants synthesized have MICs from 2.76 to 7.31 μmol/l, showing enhanced activity against Gram- bacteria.

Conclusion

Elaboration of biocide surfactant structures leads us to consider the impact of the spacer binding the two ammonium heads of Gemini fluorosurfactants. All members of the series exhibit strong bacteriostatic activity compared to available commercial references or a hydrocarbon homologue [31]. The most important parameter was the introduction into our molecules of hydrophobic perfluorinated chains. The study shows the spacer length has a weak influence on MIC values (same decade). Gram negative bacteria of the genus *Pseudomonas* can metabolize quaternary ammonium compounds using them as a carbon, nitrogen, and energy source. For example, it is well known that cetrimide agar is used as a selective medium for cultivation of *Pseudomonas*. According to the activity of the fluorinated bisammoniums, it seems that *Pseudomonas* cannot metabolize such products, leading to interesting activities.

Acknowledgements The authors thank Atofina for 2-*F*-hexylethyliodide and Rohm & Haas European laboratories for biological evaluations. L.M. thanks MEN for a grant.

References

1. Jungermann E (1970) Cationic surfactants. Surfactants science series, vol 4. Dekker, New York
2. Dogmak G (1935) Dtsh Med Wochenschr 61:829
3. Merianos JJ (1992) In: Block SS (ed) Disinfection, sterilization, preservation, 4th edn. Lea & Febiger, Philadelphia, pp 225–255
4. Ascenzi JM (1996) Handbook of disinfectants and antiseptics. Dekker, New York
5. Pernak J, Rogoza J, Mirska I (2001) Eur J Med Chem 36:313
6. Pernak J, Kalewska J, Ksycinska H, Cybulski J (2001) Eur J Med Chem 36:899
7. Skrzypczak A, Brycki B, Mirska I, Pernak J (1997) Eur J Med Chem 32:661
8. Brycki B, Skrzypczak A, Mirska I, Pernak J (1996) Arch Pharm Pharm Med Chem 329:279
9. Pernak J, Mirska I, Kmiecik R (1999) Eur J Med Chem 34:765
10. Pernak J, Skrzypczak A (1996) Eur J Med Chem 31:901
11. Pernak J, Arndt A, Brycki B (1997) Arch Pharm Pharm Med Chem 330:253
12. Isquith AJ, Abbott EA, Walters PA (1972) Appl Microbiol 24:859
13. Speier JL, Malek JR (1982) J Colloid Interface Sci 89:68
14. Tiller JC, Liao CJ, Lewis K, Klibanov AM (2001) Proc Natl Acad Sci USA 98:5981
15. Menger FM, Keiper JS (2000) Angew Chem Int Ed Engl 39:1906
16. Menger FM (2002) Proc Natl Acad Sci USA 99:4818
17. Diz M, Manresa A, Pinazo A, Erra P, Infante MR (1994) J Chem Soc Perkin Trans 21871
18. Pavlikova-Moricka M, Lacko I, Devinsky F, Masarova L, Mlynarcik D (1994) Folia Micobiol 39:176
19. Pavlikova M, Lacko I, Devinsky F, Mlynarcik D (1995) Collect Czech Chem Commun 60:1213
20. Kopecky F (1996) Pharmazie 3:135
21. Kissa E (2001) Fluorinated surfactants and repellents. Surfactant science series, vol 97. Dekker, New York, and references therein
22. Banks RE, Smart BE, Tatlow JC (1994) Organofluorine chemistry. Principles and commercial applications. Plenum, New York
23. Guittard F, Geribaldi S (2001) In: Glaser R, Kaszynski (eds) Anisotropic organic materials – approaches to polar order. ACS symposium series, vol 798. American Chemical Society, Washington, DC, pp 180–194
24. Szönyi F, Guennouni F, Cambon A (1991) J Fluorine Chem 55:85
25. Sismondi A, Abenin PS, Joncheray L, Cambon A (1992) J Fluorine Chem 59:127
26. Guittard F (1994) PhD thesis. University of Nice
27. Bruch MK (1992) In: Block SS (ed) Disinfection, sterilization, preservation, 4th edn. Lea & Febiger, Philadelphia, pp 1028–1046
28. Zana R (2002) J Colloid Interface Sci 248:203
29. Zana R, Benrraou M, Rueff R (1991) Langmuir 7:1072
30. Zana R (1996) Langmuir 12:1208
31. Massi L, Guittard F, Levy R, Duccini Y, Géribaldi S (2003) Eur J Med Chem 38:519

Progr Colloid Polym Sci (2004) 126: 194–196
DOI 10.1007/b94016

E. López Cabarcos
J. Rubio Retama
B. López-Ruiz

Immobilization of glucose oxidase in cross-linked poly(acrylamide/acrylic acid) microgels

E. López Cabarcos (✉) · J. Rubio Retama
Departamento Físico Química
Farmacéutica, Facultad de Farmacia,
UCM, 28040 Madrid, Spain
e-mail: cabarcos@farm.ucm.es

B. Lopez-Ruiz
Departamento Química Analítica,
Facultad de Farmacia, UCM,
28040 Madrid, Spain

Abstract Glucose oxidase has been immobilized in cross-linked poly(acrylamide/acrylic acid) microgels using the concentrated emulsion polymerization method. We have observed that the immobilization of the enzyme is effective when the number of cross-links is equal to or greater than 0.021 mol/g. Microgels with entrapped glucose oxidase have been used as the biological component of a biosensor which presents optimal performance for glucose detection.

Keywords Poly(acrylamide/acrylic acid) microgels · Immobilization of Glucose oxidase · Glucose biosensors

Introduction

Immobilized enzymes have been used in chemical analysis and more recently they have found application as biological components of biosensors [1, 2, 3]. Effective enzyme immobilization can be achieved using several techniques [4, 5, 6], one of which is entrapment within polymer microgels prepared by polymerization from a water-in-oil (W/O) concentrated emulsion. This method presents the advantage of using small quantities of surfactants and an organic phase. The use of polyacrylamide (PAA) gels for entrapping enzymes goes back to the 1960s, however entrapment in PAA microgels and nanogels produced by polymerization in microemulsions, concentrated emulsions, nanoemulsions [7, 8] and micelles [9] has gained increased attention over the last decade because of the potential applications. The incorporation of acrylic acid (AA) into the microgels contributes to negatively charge the microparticles, facilitating the discrimination of molecules when the microgels are employed as biosensors. We report in this paper the preparation of cross-linked PAA and poly(acrylamide/AA) microgels (PAA/AA) with entrapped glucose oxidase (GOx) and its use in a glucose biosensor.

Experimental

Materials

Acrylamide (Panreac), AA (Fluka), ammonium persulfate (Fluka), *N,N′*-methylene bisacrylamide (Aldrich), *N,N,N′,N′*-tetramethylethylendiamine (TEMED), Span 80 (Fluka) and GOx from *Aspergillus niger* (Sigma) were used as received.

Methods

The microgel particles were examined by using scanning electron microscopy, JEOL JSM-6400. Replicas were produced by shadowing with gold, deposited with a Balzers sputter coater (SCD-004). Amperometric measurements at constant potential were carried out using a Metrohm Polarecord E-506 potentiostat. Electrochemical measurements were performed in 0.1 M phosphate buffer in a three-electrode cell with a platinum electrode as the working electrode, an Ag/AgCl reference electrode and a platinum counter electrode.

Preparation of microgels

PAA microgels were prepared using the concentrated emulsion pathway [10, 11, 12]. The concentrated W/O emulsion were produced by dropwise addition, with a syringe, of the dispersed aqueous phase (aqueous solution of monomer, initiator and cross-linker) to the continuous oil phase (dodecane plus Span 80). The emulsion was homogenized under magnetic stirring and purged with nitrogen to remove residual oxygen. The encapsulation of GOx was carried out by solubilization of the enzyme in phosphate buffer of pH 7.2, which prevents enzyme denaturation and constitutes the aqueous dispersed phase where monomer, initiator and cross-linker are dissolved. The addition of new components must be performed drop by drop under stirring because the equilibrium hydrophilic/lipophilic balance can be broken, preventing the emulsion from forming. Redox polymerization of the concentrated W/O gel-like emulsion was started by adding TEMED and after 1 h of reaction, microgel particles were precipitated with freezing methanol and were isolated by centrifugation (10,000 rpm for 10 min). Two types of polymeric particles were prepared; the first one contained acrylamide as a monomer and bisacrylamide as a cross-linker and in the second case, 50% of the acrylamide was substituted by AA with the aim to negatively charge the microparticles. The cross-linking, η (given as the molar concentration of cross-links per gram of polymer), was varied between 0.005 and 0.039 mol/g. During polymerization, the temperature was controlled and kept below 35 °C to preserve enzyme activity. After 1 h of reaction, the microgel particles were isolated, washed with phosphate buffer of pH 7.2 and the supernatant was analyzed. The supernatant showed enzymatic activity for $\eta = 0.005$ and 0.011 mol/g but not for the other cross-linking fractions. On this basis, we selected $\eta = 0.021$ mol/g to prepare the microgel particles. The microgels were washed with acetone, freeze-dried and stored at 4 °C. PAA microgels with GOx are named PMP1 and PAA/AA microgels with GOx are named PMP2.

Results and discussion

As shown in Fig. 1, the polymerization of acrylamide with GOx produces microgel particles with near bimodal size distribution (diameters between 0.9 and 12 μm) with peaks at 5 and 8 μm which approximate the size of the precursor emulsion polyhedral cells [12].

The particles with smaller diameter correspond to the polymerization of the small droplets filling the holes between the emulsion polyhedral cells. Microgel particles prepared with 50% of AA, shown in Fig. 2, are larger than acrylamide microbeads and their diameters lie between 1.5 and 23 μm.

The enzymatic activity detected in the supernatant for microgels with cross-linking equal to or smaller than 0.011 mol/g is attributed to the large pore size of these samples which would allow GOx to escape the beads, whereas for higher cross-linking the enzyme is entrapped in the microgels and no enzymatic activity was detected in the supernatant.

Determination of glucose

The amount of encapsulated enzyme was determined by using the microgels with GOx as the biological component of an amperometric biosensor. Immobilized GOx activity was evaluated amperometrically using a Metrohm E-506 voltammetric analyzer. For the experiments a 10-ml electrochemical three-electrode cell was employed: The working electrode (3-mm diameter, platinum), the reference electrode (Ag/AgCl) and the platinum wire auxiliary electrode were inserted into the cell and in the experiments the working electrode was potentiostated at +0.6 V (versus Ag/AgCl). For determination of immobilized enzyme activity, an electrode was prepared by depositing an exact weight (3 mg) of PMP2 microgel around the platinum electrode, which was covered with a dialysis membrane. The resulting electrode was washed thoroughly with 0.1 M phosphate buffer and overoxidized at +0.6 V. The principle of the determination of the current response is based on the formation of hydrogen peroxide during the enzyme-catalyzed reaction. The hydrogen peroxide was detected by amperometry.

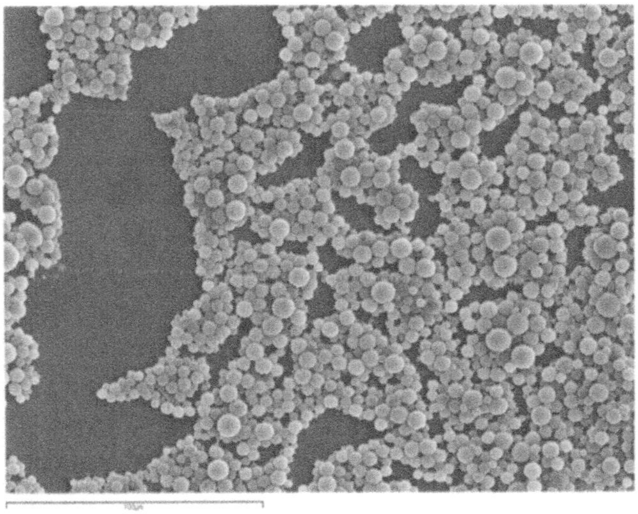

Fig. 1 Scanning electron microscope (*SEM*) micrograph of polyacrylamide microgel particles with glucose oxidase (*GOx*) immobilized (cross-linking $\eta = 0.021$ mol/g)

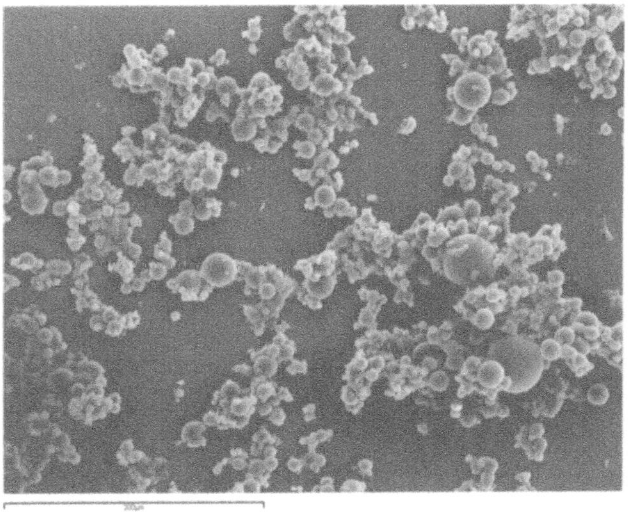

Fig. 2 SEM micrograph of poly(acrylamide/acrylic acid) microgel particles with GOx immobilized ($\eta = 0.021$ mol/g)

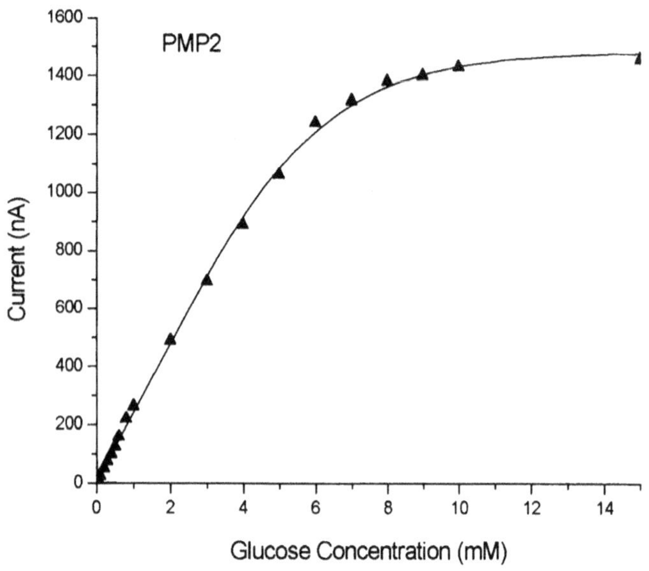

Fig. 3 Response of the biosensor, prepared with poly(acrylamide/acrylic acid) microgels with GOx studied by a calibration plot for glucose in stirred 0.1 M phosphate buffer and potential +0.6 V versus Ag/AgCl

As can be seen in Fig. 3, the kinetics of the reaction catalyzed by GOx has a sigmoidal dependence on the concentration of the substrate (glucose) and can be described by the Hill equation:

$$V = \frac{V_{max}[S]^n}{K_M + [S]^n},$$

where $[S]$ is the substrate concentration, K_M the Michaelis–Menten constant, $n = 1.5$ is the Hill cooperativity coefficient and $V_{max} = k_{catalytic}[GOx]$. As can be seen in Fig. 3 the response of the biosensors prepared with PAA/AA microgels follows the Hill equation and is linear up to 6 mM glucose concentration. From the fitting of the curve we obtain $V_{max} \approx 1,450$ nA which is clear verification that GOx has been entrapped in the microgel. Once the biosensor is prepared its response can be reproduced exactly for several months. The enzyme electrode operates at a potential of 0.6 V and when the PMP1 electrode is used in real samples, the glucose detection is disturbed by the presence of interfering species, such as ascorbic and uric acid. However, because of the negative charge of these substances at biological pH, the incorporation of AA into the polymeric matrix of the PMP2 electrode led to a considerable improvement of the sensor, which only responds to glucose and is unaffected by ascorbic and uric acid. In a separate paper we have reported a detailed study of the preparation of glucose biosensors based on GOx entrapment in polyacrylic microgels [16].

Conclusions

Acrylic microgels were synthesized using the concentrated emulsion polymerization method. The entrapment of GOx inside the microgel was performed by adding the enzyme to the aqueous dispersed phase of the concentrated emulsion before redox polymerization. The size of the microgel particles slightly increases by the entrapment of the enzyme. Microgels of PAA and AA provide an excellent matrix for GOx immobilization when this biomaterial is part of a biosensor. With this immobilization method the enzymatic activity of the microgels remains constant for at least 8 months.

Acknowledgements The authors acknowledge financial support from DGI of the Spanish Science and Technology Ministry (grants BMF2000-0620, MAT2002-10233-E and MAT2003-03051-C03-03).

References

1. Wang J, Leech D, Ozsoz M, Martinez Smith R (1991) Anal Chim Acta 245:139
2. Bu H, Mikkelsen SR, English AM (1998) Anal Chem 70:4320
3. Kros A, van Hövel SW, Sommerdijk JM, Nolte R J (2001) Adv Mater 13:1555
4. Gonzalez-Saiz JM, Pizarro C (2001) Eur Polym J 37:435
5. Bera P, Saha SK (1998) Polymer 39:1461
6. Bu H, Mikkelsen SR, English AM (1995) Anal Chem 67:4071
7. Hong-Ru L (2001) Eur Polym J 37:1507
8. Silva MESR, Dutra ER, Mano V, Machado JC (2000) Polym Degrad Stab 67:491
9. Jingcheng H, Liqiang Z, Ganzuo L, Hanqing W, Zhengwei D (1996) Polymer 37:3117
10. Bhakta A, Ruckenstein E (1995) Langmuir 11:1486
11. Kunieda H, Evans DF, Solans C, Yoshida M (1990) Colloids Surf 47:35
12. Terreros A, Galera Gomez PA, López-Cabarcos E (2000) Prog Colloid Polym Sci 115:50
13. Princen HM, Aronson MD, Moser JC (1980) J Colloid Interface Sci 75:246
14. Ruckenstein E, Ebert G, Platz G (1990) J Colloid Interface Sci 133:432
15. Chen HH, Ruckenstein E (1980) J Colloid Interface Sci 75:246
16. Rubio Retama J, López Cabarcos E, Lopez Ruiz B (2004) Analyst (in press)

AUTHOR TITLE INDEX

Progr Colloid Polym Sci (2004) 126: 199

KEY WORD INDEX

GPSR Compliance

The European Union's (EU) General Product Safety Regulation (GPSR) is a set of rules that requires consumer products to be safe and our obligations to ensure this.

If you have any concerns about our products, you can contact us on ProductSafety@springernature.com

In case Publisher is established outside the EU, the EU authorized representative is:

Springer Nature Customer Service Center GmbH
Europaplatz 3
69115 Heidelberg, Germany

Batch number: 10755910

Printed by Printforce, the Netherlands